QUANTUM ELECTRONICS

Volume 2: Maser Amplifiers and Oscillators

QUANTUM ELECTRONICS

Volume 2: Maser Amplifiers and Oscillators

V. M. FAIN and YA. I. KHANIN

TRANSLATED BY

H. S. H. MASSEY

EDITED BY

J. H. SANDERS

THE M.I.T. PRESS

MASSACHUSETTS INSTITUTE OF TECHNOLOGY

CAMBRIDGE, MASSACHUSETTS

First English edition 1969

This book is Volume 14 of the International Series of Monographs in Natural Philosophy published by Pergamon Press Ltd., Oxford, England.

Distributed in the United States and Canada
by M. I. T. Press, Cambridge, Massachusetts

This is a translation of Part II of the original Russian Квантовая Радиофизика by V. M. Fain and Ya. I. Khanin published in 1965 by Sovetskoye Radio, Moscow, and includes corrections and revisions supplied by the authors.

Library of Congress Catalog Card No. 67–22832

PRINTED IN GERMANY

08 012238 8

Contents of Volume 2

Contents of Volume 2

Contents of Volume 1

VOLUME 1. BASIC THEORY

Foreword

THE BIRTH of a new independent field of physics, now known by the name of quantum electronics, was heralded about ten years ago by the creation of the molecular oscillator. This field at once attracted the attention of a large number of research workers, and rapid progress took place. Extensive experimental and theoretical material has now been accumulated. The present book attempts to give a résumé of this material and, to a certain extent, to generalize it.

We have tried to arrange the material so that, as far as is possible, the reader need not continually refer elsewhere. The references to literature of a theoretical nature make no pretence of completeness, but when citing experimental work we have to give as full a list as possible since readers may be interested in details for which there is no space in the book.

The theoretical sections of the books are by no means a survey of present work. We have tried to highlight the basic principles and their results. It is natural that slightly more attention has been paid to fields in which the authors themselves have been involved. The experimental material is given in the form of a survey, with only a brief description of the technical details of devices.

The book as a whole is designed for the reader with a knowledge of theoretical physics (quantum mechanics in particular) at university level. It should be pointed out that the material in the various sections of the book is of differing degrees of complexity. Readers will need less preparation for Volume 2. The most difficult paragraphs are marked with an asterisk.

In conclusion we should mention that §§ 1–20, 22–40 and Appendix I were written by V. M. Fain; §§ 41–49 and 51–58 by Ya. I. Khanin. At the author's request § 21 was written by V. N. Genkin, § 50 by E. G. Yashchin, Appendix II by V. I. Talanov and Appendix III by Ye. L. Rozenberg.

We are grateful to Professor A. V. Gaponov and Professor V. L. Ginzburg for reading the book in manuscript and making a number of useful suggestions. We are also grateful to A. P. Aleksandrov, V. N. Genkin, G. M. Genkin, N. G. Golubeva, G. L. Gurevich, G. K. Ivanova, M. I. Kheifets, Yu. G. Khronopulo, Ye. E. Yakubovich and E. G. Yashchin for their great help in reading the proofs.

Radiophysics Scientific Research Institute,
Gor'kii

V. M. FAIN
YA. I. KHANIN

Preface to the English Edition

We were very pleased to learn that our book was to be translated into English and would thus become available to a wide circle of English-speaking readers.

Research in the field of quantum electronics has continued since our manuscript was handed over to the Soviet publishers, but there has been no essential change in the basic theory or understanding of the physical processes in quantum devices. Among the most interesting questions on which work has been done of late, mention should be made of the development of coherence theory and holography, which is based on it. Non-linear optics and its application are developing rapidly. Although the book treats the fundamentals of non-linear optics, one of the important non-linear optical effects—the phenomenon of self-trapping—is not discussed in the book.

Among the other important problems recently worked on and not reflected in the book are the electrodynamics of gas lasers, the theory of the natural width of lasers (gas lasers in particular) and the important work on semiconductor lasers.

Despite the great importance of these questions we did not include them in the present edition, not merely for lack of time but also because one is here dealing with subjects under development, not all of whose facets are completely clear and whose discussion would be previous.

We are preparing material on the majority of these questions, and also a detailed treatment of the theory of the non-linear properties of materials (mainly solids) which is of great importance for non-linear optics and quantum electronics; this will be included in the second Soviet edition, which is planned for 1969.

In the present edition we have confined ourselves to correcting any errors that have been found and making some slight additions.

February 1967

V. M. Fain — Institute of Solid-State Physics, Moscow, Academy of Sciences U.S.S.R.
Ya. I. Khanin — Radiophysics Scientific Research Institute, Gor'kii.

Introduction

QUANTUM electronics as an independent field of physics came into prominence in the middle fifties when the first quantum oscillators and amplifiers were made. The immediate precursor of quantum electronics was radiofrequency spectroscopy, which is now one of its branches. An enormous quantity of experimental material concerned with the resonant properties of substances had been accumulated by radiofrequency spectroscopy. Such research had made it possible to establish the structure of levels, the frequencies and intensities of transitions, and the relaxation characteristics of different substances. Investigations of paramagnetic resonance spectra in solids and the inversion spectrum of ammonia have been of particular importance to quantum electronics.

During radiofrequency investigations the state of a substance is not, as a rule, subject to significant changes and remains close to thermodynamic equilibrium. But besides the investigation of substances under undisturbed conditions, other methods began to appear which were connected with the action of strong resonant fields on a substance. These methods, which we can call active, were first applied in nuclear magnetic resonance. They include nuclear magnetic induction spin echo and the Overhauser effect. The main outcome of these methods was the possibility of producing strongly non-equilibrium states in quantum systems which could emit coherently. Therefore the actual material was accumulated through radiofrequency spectroscopy, and resulted in the birth of experimental ideas which were then used as the basis of quantum oscillators and amplifiers.

The concept of stimulated emission, which is important for quantum electronics, was first formulated by Einstein as early as 1917. Ginzburg (1947) pointed out the importance of this phenomenon in radiofrequency spectroscopy.

The idea of amplifying electromagnetic waves by non-equilibrium quantum systems was first mooted by Fabrikant, Vudynskii and Butaeva. The patent (Fabrikant, Vudynskii and Butaeva, 1951) obtained by this team in 1951 contains a description of the principle of molecular amplification. Slightly later, in 1953, Weber made a suggestion about a quantum amplifier. Basov and Prokhorov (1954) discussed an actual design for a molecular oscillator and amplifier operating with a beam of active molecules and developed their theory. Gordon, Zeiger and Townes independently had the same idea and, in the same year, 1954, published a report on the construction of an oscillator that operated with a beam of ammonia molecules. Gordon, Zeiger and Townes introduced the now well-known term "maser".†

The successful operation of a beam molecular oscillator stimulated the search for new methods and results were not long in coming. Basov and Prokhorov (1954) suggested the principle of a three-level gas-beam oscillator. In 1956 Bloembergen discussed the possibility of making a quantum amplifier with a solid paramagnetic working medium. The estimates he made confirmed that the idea was feasible and in 1957 such an instrument was made by Scovil, Feher and Seidel. After this, reports appeared on the production of a whole series of similar instruments based on different paramagnetic crystals.

Instruments based on quantum principles have a number of exceptional properties when compared with ordinary amplifiers and oscillators. The molecular beam maser oscillator is not particularly powerful but its stability is far better than the stability of the best quartz oscillators. This has brought about the use of the maser as a frequency standard. The paramagnetic maser amplifier has an extremely low noise level and satisfactory gain and bandwidth characteristics.

The next stage in the development of quantum electronics was the extension of its methods into the optical range. In 1958 Schawlow and Townes discussed the question theoretically and came to the conclusion that it was perfectly possible to make an optical maser oscillator. They suggested gases and metal vapour as the working substances. The question of possible working substances and the methods of producing the necessary non-equilibrium states in them was also discussed in a survey by Basov, Krokhin and Popov (1960). These authors discussed paramagnetic crystals and semiconductors as well as gases.

In 1960 Maiman made the first pulsed ruby quantum optical generator which is called a "laser".‡ For the first time science and technology had available a coherent source of light waves. The future prospects of devices of this kind were obvious and in a very short time a large number of teams had come onto the scene of laser research. The list of crystals suitable for use in lasers quickly grew. Then certain luminescent glasses and liquids were used for the same purpose. In 1961 Javan, Bennett and Herriott made the first continuous laser operating with a mixture of the inert gases neon and helium.

† Maser is an acronym formed from *M*icrowave *A*mplification by *S*timulated *E*mission of *R*adiation.

‡ The term laser is an acronym from *L*ight *A*mplification by *S*timulated *E*mission of *R*adiation. It must be pointed out that there is not yet any firmly established terminology in quantum electronics. Besides "laser" the name "optical maser" is frequently used.

Introduction

Quantum electronics is very young; its basic trends are still far from clear. A whole series of problems is still unsolved. Under these conditions the writing of a monograph discussing the theoretical and experimental basis of quantum electronics is a rather complex affair. It must be understood that the book reflects to only a limited extent the position as it is today.

Quantum electronics as a theoretical science possesses a number of characteristic features which separate it both from quantum physics and from electronics. Unlike ordinary "classical" electronics, quantum electronics is characterized by the extensive application of the methods of quantum theory. However, the application of quantum field theory to quantum electronics has a specific feature which distinguishes it from ordinary quantum electrodynamics (see, e.g., Heitler, 1954*; Akhiezer and Berestetskii, 1959*).

Quantum electronics makes wide use of the resonant properties of matter both for the study of matter itself (radiofrequency spectroscopy, paramagnetic resonance), and for its use in quantum amplifiers and oscillators. It is obvious that resonances with a high Q-factor in a substance are essential for both purposes. To obtain sharp resonances discrete energy levels must exist in the substance. The presence of discrete electron levels means that these electrons cannot be free but must be in bound states in the atoms, molecules or solid. We notice that the characteristic feature of ordinary "classical" electrodynamics is the interaction of the radiation field with free electrons. It is true that quasiclassical systems (harmonic oscillator, electron in a magnetic field, etc.) may also have a discrete spectrum but the essential feature of classical and quasi-classical systems is that the energy levels of such systems are quasi-equidistant. For example, the harmonic oscillator has equidistant levels with no upper limit. The energy spectrum of quantum systems is much more diverse than the spectrum of quasi-classical systems. In particular the energy levels may be so arranged that there are two levels whose spacing is not the same as the spacing of any other levels in the same system. Under certain conditions no attention need be paid (during an interaction with radiation of the corresponding frequency) to any other levels of the system and we can use the idealization of a two-level system. The idealizations of a three-level system, etc., are introduced likewise.

As we have already pointed out, wide use is made of the resonant properties of matter in quantum electronics. As may be easily understood, during the resonant interaction of matter with a field it is particularly important to allow for different kinds of dissipative relaxation processes. Unlike ordinary quantum electrodynamics in which, as a rule, we are not interested in relaxation processes in matter, in quantum electronics the concept, and thus the description, of the different relaxation processes plays a major part.

The concept of stimulated emission plays an important and even predominant part in quantum electronics. All the active quantum-electronic instruments—maser amplifiers and oscillators—use the phenomenon of stimulated emission. The phenomenon of stimulated emission is closely linked (as will become clear from the appropriate sections of the book) with the non-linear properties of quantum systems used in quantum electronics. The non-linear properties, in their turn, are caused by the non-equidistant nature of the energy levels.

When describing the processes of the interaction of matter with radiation we must, strictly speaking, use quantum theory, i.e. quantum theory is used to treat the matter and the field. For many problems, however, the classical description of an electromagnetic field is a fully justified approximation. This is because the fields discussed in quantum electronics are large, and because the mean quantum values of the electric and magnetic fields are accurately described by the classical Maxwell equations. It is essential to allow for the quantum properties of the field when investigating the quantum fluctuations of the field, in particular when studying the noise properties of amplifiers and oscillators.

The arrangement and selection of the material in the present book have been made with these features of quantum electronics in mind. The book is composed of two parts: Volume 1 "Basic Theory" and Volume 2 "Maser Amplifiers and Oscillators". A large amount of material has been kept for the Appendixes.

In Volume 1 an attempt is made to give the basic theory of quantum electronics. In this part we have tried to show how the concepts and equations used in quantum electronics follow from the basic principles of theoretical physics. When doing this we make frequent use of very simple models so as not to complicate the treatment. Such models are necessary for the understanding of a particular process, but the models can often not be used for direct comparison with experiment.

The first chapter of the book deals with general questions of the interaction of radiation with matter. The basic concepts of quantum theory are briefly treated in this chapter. The reader's attention is particularly drawn to the density matrix description of the quantum state. This is because in its various applications quantum electronics deals with mixed states and not with pure states. Quantum theory allows us, by the use of the density matrix, to give a unified description of both pure and mixed states. In the first chapter we discuss in sequence the quantum theory of fields in resonators, in waveguides and in free space and also the concept of phase in quantum field theory, of the indeterminacy relation between the phase and the number of particles, the question of the transition to classical physics, etc. Section 4 discusses in more detail than usual the question of the different forms of interaction energy between a field and charged particles.

The second chapter deals with the general question of relaxation. When there are relaxation processes present the behaviour of quantum systems is governed by the cause of the dissipation—a dissipative system which possesses a continuous spectrum and an infinite number of degrees of freedom. In this case we must derive approximate equations which will take into account the relaxation processes (the transport equations). Therefore Chapter II deals essentially with the applicability of the different equations used in quantum electronics. In particular, by proceeding from basic principles, we derive the conditions for applicability of the frequently used population balance equations. The same chapter discusses the questions of the irreversibility of real systems and the principle of the increase of entropy. We also show how the transport equations can be used to describe fluctuations. Some long calculations are given in this chapter but they may be omitted on the first reading without making it difficult to understand the other parts of the book. The results which are necessary for reading subsequent chapters are given in the introduction to this chapter.

In Chapter III we have gathered together the possible quantum effects in ordinary electronics which may appear at very high frequencies and at low temperatures. These effects, as a rule, are small. In Chapter III an account is also given of the quantum theory of real resonators with finite Q.

In Chapters IV and V we discuss the behaviour of quantum systems in fields, which are here described classically. Particular attention is paid to the response of a system to such fields. This response, for example in the form of the mean magnetization of the system, is described in terms of the susceptibility. A number of general susceptibility properties are discussed, particularly the dispersion relations, the fluctuation-dissipation theorem, and the symmetry properties. In these same chapters we treat the idealizations of two- and three-level systems and find the equations of motion for these systems. In § 20 of Chapter IV we show how it is possible to give a rigorous description of systems which are not subject to the equations derived in Chapter II. The method of moments is used; the rigorous basis of this method is given in § 20. In § 21 it is used to examine cross-relaxation processes.

In Chapters VI, VII and VIII we deal with a number of questions concerning the theory of spontaneous and stimulated emission. In particular we discuss the connection with classical theory, the part played by non-linearity, the phase relations, etc. Chapter VII treats the theory of coherent spontaneous emission in free space and the theory of the natural line width. In Chapter VIII we deal with the physical nature of the processes of spontaneous and stimulated emission in a resonator.

Recently a new branch of quantum electronics—non-linear optics—has appeared. The development of non-linear optics, connected with success in the field of optical quantum light generators (lasers), is only just beginning. However, in our view a number of the essential features of the interaction of matter with optical waves can already be stated. The ninth and last chapter of the first part of the book is devoted to relating these features to the general scheme of quantum electronics.

In Chapter X and XI of Volume 2 we discuss the elements of the theory of quantum oscillators and amplifiers working in the microwave region and review the practical achievements in this field. A relatively large amount of attention is paid to two-level paramagnetic masers although they have not been put to practical use. This is done because two-level systems are simpler, and their theoretical analysis can be carried out in detail; in addition, this material is not contained in other books on the physics of quantum electronics (Singer, 1959; Troup, 1959; Vuylsteke, 1960).

The quantum paramagnetic amplifier theory discussed in Chapter X is of a general nature and its results are also fully applicable to the case of multi-level amplifiers. In the section dealing with quantum oscillators most attention is paid to the dependence of the form of the emitted signal on the different parameters. Unfortunately there is at present no satisfactory theory of transient modes in quantum oscillators. Material is therefore presented which, although it reflects the present level of the theory, is more illustrative in nature. Questions connected with methods of exciting the working substance are discussed in a fair amount of detail in Chapters X and XI. In all cases when approximate methods of calculation are used we have tried to explain the limits of their applicability, since this is generally not discussed in published works.

The maser oscillator operating with a beam of active molecules is described somewhat briefly in Chapter XI. We considered that we could limit ourselves to a short description by assuming that this material is known to the reader from the books of Singer (1959), Troup (1959) and Vuylsteke (1960).

Chapter XII is devoted to optical masers. Most of the space here is occupied by a survey of experimental achievements and a description of the features of laser operation. Theoretical questions are touched upon in so far as the present state of the theory permits. It should be pointed out that the stream of original papers about lasers is so thick and fast at present that the material in Chapter XII will probably be out of date by the time this book sees the light of day.

The book contains three Appendixes. To them is relegated material which lies a little outside the general plan of the book. Nevertheless the importance of this material is quite clear.

In Appendix II we give the elements of the theory of optical resonators. It should be pointed out that interest in problems concerning the electrodynamics of the optical waveband has been aroused only quite recently because of the development of laser technology. As far as we know the attempt made in Appendix II to give a systematic treatment of the material is one of the first.

Appendix III discusses the spectra of the paramagnetic crystals used in maser amplifiers and oscillators. This chapter is by way of a short review and assumes the reader's acquaintance with the problems discussed. In this chapter there is a detailed bibliography and list of sources which must be used as an introduction to the subject.

In conclusion we would remark that the reading of the book requires the reader's acquaintance with quantum theory at the level of a university course in the theoretical physics.

We make frequent reference to the excellent course of theoretical physics by Landau and Lifshitz. This does not mean, of course, that the reader has to know the whole of Landau and Lifshitz. It is sufficient merely to understand those parts of the book mentioned here. Among other books we can recommend *The Quantum Theory of Radiation* by Heitler, to which we make frequent reference.

It must be pointed out that the present book does not discuss a number of problems in quantum electronics such as the radio-spectroscopy of gases, nuclear magnetic resonance and paramagnetic resonance. These are dealt with in a number of monographs (Townes and Schawlow, 1959*; Andrew, 1955; Ingram, 1955*; Al'tshuler and Kozyrev, 1961; Gordy, Smith and Trambarulo, 1953) to which we refer the interested reader.

In conclusion we should like to say that a number of chapters in the book can to a certain extent be read independently of the others. For example, the reader chiefly interested in masers and lasers can concentrate his attention on the second volume of the book. From the first volume he may need § 24 and an acquaintance with Chapters I and VIII.

Note: These references marked with an asterisk will be found in the reference list of Volume 1.

VOLUME 2

Maser Amplifiers and Oscillators

CHAPTER X

Paramagnetic Maser Amplifiers

THE OPERATION of a whole class of electronic devices which have been given the name of quantum amplifiers and oscillators, or masers,† is based on the use of the phenomenon of stimulated emission. The idea of a maser amplifier is extremely simple. Let us imagine a section of a waveguide filled with a certain medium which in future we shall call the working substance. As a result of the interaction between the working substance and an electromagnetic wave in the waveguide, the intensity of the wave varies as

$$I = I_0 \, e^{-2\alpha(\omega) z}.$$

The variation of the coefficient 2α with frequency depends upon the dispersion properties of the working substance. In the natural state, i.e. near thermal equilibrium, all substances absorb energy from an electromagnetic wave ($\alpha > 0$). The substance must be brought into a non-equilibrium state in order to obtain amplification. This principle is not new: it happens in any electronic vacuum tube. The electron beam, which is the working substance in the devices of classical electronics, is not itself an equilibrium system. If we take the concrete example of a klystron, there is spatial bunching of the electrons in the beam. In a magnetron the electrons emitted by the cathode travel along certain curvilinear trajectories.

The principal difference between quantum and classical electronics is that the former deals with interacting charges. For the sake of simplicity let us assume that the working substance consists of identical molecules‡ which have only two energy levels with energies $E_1 < E_2$. An electromagnetic field at the resonance frequency stimulates transitions between these levels. Here the transition $2 \rightarrow 1$ is accompanied by an increase in the energy of the field and the reverse transition $1 \rightarrow 2$ by a decrease. The net effect is determined by the populations of the levels in question. When $N_2 > N_1$ the

† The first term is widely used in the Soviet literature and the second in the non-Soviet literature.

‡ By a molecule we understand any quantum object: an atom, an ion or any polyatomic molecule.

number of transitions from the upper to the lower level exceeds the number of reverse transitions, since the direct and reverse transitions are equally probable. Therefore the case of $N_2 > N_1$ corresponds to amplification when electromagnetic waves with the resonance frequency $\omega_{21} = (E_2 - E_1)/\hbar$ are propagated in the working substance. When the quantum system is in thermal equilibrium the populations of its levels obey the Boltzmann law

$$N_2/N_1 = e^{-\hbar\omega_{21}/kT},$$

which gives $N_2 < N_1$. Therefore a quantum system close to equilibrium always absorbs the energy of an electromagnetic field incident upon it, converting it into thermal energy. Obtaining non-equilibrium states in the working substance is a major problem in quantum electronics.

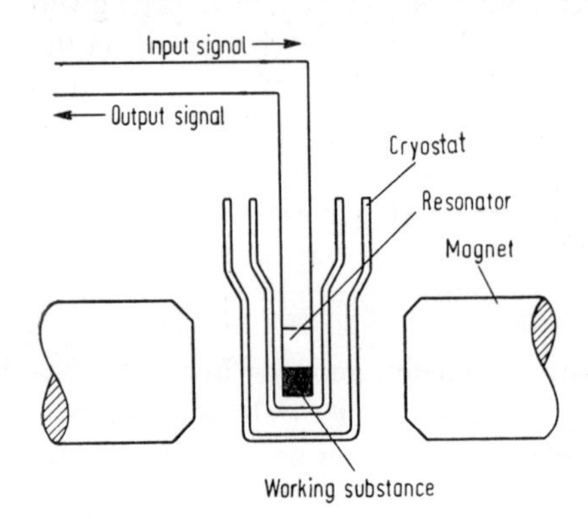

FIG. X.1. Simplified diagram of a paramagnetic maser amplifier.

In all cases of practical importance it is a matter of obtaining states with a population inversion or, as it is often called, with a negative temperature. A medium in which inversion is achieved we shall call active. In view of their practical significance paramagnetic crystals are an important class of known working substances.

A waveguide cell filled with an active medium is a very simple (but far from the best) model of a maser amplifier. For more effective use of the working substance the velocity of the wave should be reduced, or it should be made to pass many times through the sample (in fact a standing wave should be created in it). Devices in which the principle of slowing down the wave is used are called travelling wave masers. Masers using resonators, to which special attention will be paid below, form another class. A simplified diagram of a resonator paramagnetic maser is shown in Fig. X.1. A resonator containing a paramagnetic crystal is placed in the gap between the poles of a magnet. Liquid helium (more rarely, nitrogen) cooling of the sample is used to increase the equilibrium population difference and the spin–lattice relaxation time. A feature of this circuit is that a single waveguide acts as both the input and the output of the amplifier. Amplifiers of this kind are called reflex amplifiers or reflection cavity amplifiers. There are also amplifiers with transmission cavities, which we shall refer to below.

It should be pointed out that a number of surveys have been made of paramagnetic maser amplifiers (Wittke, 1957; Weber, 1959; Zverev *et al.*, 1962; Karlov and Manenkov, 1964).

41. Equations of Motion of a Paramagnetic placed in a High-frequency Field

The operation of paramagnetic maser amplifiers is based on the use of the phenomenon of paramagnetic resonance in a solid. It is impossible, of course, to understand the operation of the amplifier without some knowledge of this field of radiofrequency spectroscopy. Some information on radiofrequency spectroscopy is given in other sections of this book: Chapter II—general theory of relaxation, Appendix III—elements of the theory of the paramagnetic spectra of crystals. The problem of the behaviour of quantum systems in high-frequency fields has been discussed in a general way (Chapters IV–V, Vol. 1). We shall now return anew to this problem, but in the more limited form applied to a two-level paramagnetic substance.

The present section is on the one hand a concrete illustration of the general theory, and on the other it serves as an introduction to the theory of a two-level maser, which we shall treat later.

The behaviour of a two-level quantum system interacting with a field obeys the equation for the density matrix (Chapter II, Vol. 1). In the case of paramagnetics this equation is equivalent to the Bloch equation (19.36):

$$\dot{\boldsymbol{M}} = -\gamma[\boldsymbol{M} \wedge \boldsymbol{H}] - \boldsymbol{i}\frac{M_x}{T_2} - \boldsymbol{j}\frac{M_y}{T_2} - \boldsymbol{k}\frac{M_z - M_0}{T_1}. \tag{41.1}$$

Here $\boldsymbol{M}$ is the magnetization of the paramagnetic, γ is the modulus of the gyromagnetic ratio of an electron, T_1 is the longitudinal relaxation time, and T_2 is the transverse relaxation time. The magnetic field $\boldsymbol{H}$ is the sum of the static field $\boldsymbol{H}_0$ parallel to Oz and the high frequency field $\boldsymbol{H}_1$ perpendicular to Oz.

Let us examine the case, typical in radiofrequency spectroscopy, of a specimen placed in a resonator. The oscillations in the resonator are excited by a source called a pump oscillator. Generally speaking, the high-frequency field in the resonator is unknown since its amplitude and phase are determined not only by the pump oscillator but also by the state of the paramagnetic. Let us take this fact into consideration by dividing $\boldsymbol{H}_1$ into two parts:

$$\boldsymbol{H}_1 = \boldsymbol{H}_c + \boldsymbol{H}_s, \tag{41.2}$$

where $\boldsymbol{H}_c$ is in the field in the resonator neglecting the reaction of the specimen; $\boldsymbol{H}_s$ is the field radiated by the paramagnetic.

Since $\boldsymbol{H}_s$ is contained in equation (41.1) as an unknown quantity the number of unknowns in (41.1) exceeds the number of equations, and for the system to become closed it must be supplemented by the field equations. Using the method of expanding the field in the eigenmodes of the resonator (§ 19, Vol. 1) we can write these equations in the form

$$\ddot{q}_\nu + \frac{\omega_\nu}{Q_{L_\nu}} \dot{q}_\nu + \omega_\nu^2 q_\nu = -\omega_\nu \int_V (\boldsymbol{H}_\nu \cdot \boldsymbol{M})\, dV + F(t). \tag{41.3}$$

We recall that $\boldsymbol{H}_1(\boldsymbol{r}, t) = -\sum_\nu \omega_\nu \boldsymbol{H}_\nu(\boldsymbol{r})\, q_\nu(t)$, and that Q_L is the loaded Q-factor of the circuit. The term $F(t)$ in the right-hand side of (41.3) describes the action of a subsidiary e.m.f., e.g. of the pump oscillator or the signal to be amplified.

In the majority of problems relating to the microwave range only one of the modes of the resonator† is significant. The frequencies of other modes are sufficiently different that they are not excited. Considering only one mode and changing from a vector equation (41.1) to a scalar equation we obtain

$$\begin{aligned}
\dot{M}_x &= -\gamma H_0 M_y - \gamma\omega_c H_y(\boldsymbol{r})\, q_1 M_z - M_x/T_2, \\
\dot{M}_y &= \gamma H_0 M_x + \gamma\omega_c H_x(\boldsymbol{r})\, q_1 M_z - M_y/T_2, \\
\dot{M}_z &= \gamma\omega_c q_1 (H_y(\boldsymbol{r})\, M_x - H_x(\boldsymbol{r})\, M_y) - (M_z - M_0)/T_1, \\
\ddot{q}_s &+ \frac{\omega_c}{Q_L} \dot{q}_s + \omega_c^2 q_s = -\omega_c \int_V [H_x(\boldsymbol{r})\, M_x + H_y(\boldsymbol{r})\, M_y]\, dV.
\end{aligned} \tag{41.4}$$

Here H_x and H_y are the projections of the vector function $\boldsymbol{H}_\nu(\boldsymbol{r})$ onto the coordinate axes, ω_c is the eigenfrequency of the resonator, $q_1 = q_c + q_s$.

Equation (41.3) separates into two parts: one of them describes the unperturbed field q_c and the other the reaction field q_s produced by the paramagnetic. Considering q_c to be given, we have left only the equation for q_s in (41.4).

Equations (41.4) are the starting points of the solution of many problems and in particular those connected with the theory of a two-level maser. In many of these we can limit ourselves to a paramagnetic specimen whose dimensions are small compared with the wavelength. Consequently, the high-frequency field within the paramagnetic may be considered to be uniform; we shall take the polarization to be linear, let us say $H_y = 0$.‡ These limi-

† If the dielectric constant of the paramagnetic specimen is large, the modes may differ significantly from the modes of an empty cavity.

‡ In the problems discussed below the nature of the polarization plays no significant part. Only the orientation of the plane of polarization is significant.

tations are of no significance and slightly simplify the calculations. Equations (41.4) now reduce to

$$\dot{M}_x = -\omega_0 M_y - M_x/T_2,$$

$$\dot{M}_y = \omega_0 M_x - \gamma H_1 M_z - M_y/T_2,$$

$$\dot{M}_z = \gamma H_1 M_y - (M_z - M_0)/T_1, \tag{41.5}$$

$$\ddot{H}_s + \frac{\omega_c}{Q_L}\dot{H}_s + \omega_c^2 H_s = \omega_c^2 a M_x V_s,$$

where V_s is the volume of the specimen, $a = H_x^2(0)$ is the square of the resonator eigenfunction for the field at the antinode where the paramagnetic is located, and $\omega_0 = \gamma H_0$ is the free precession frequency.

The characteristic form of the motion of the magnetic moment is precession around the direction of the constant magnetic field. Assuming that the external source creates a field in the resonator

$$H_c = h_c \cos \omega t, \tag{41.6}$$

it is natural to look for a solution of the set (41.5) in the form of oscillations with slowly changing amplitudes and phases:

$$H_s = h_s(t) \cos [\omega t + \psi(t)],$$

$$M_x = m(t) \cos [\omega t + \varphi(t)] = u(t) \cos \omega t - v(t) \sin \omega t, \tag{41.7}$$

$$M_y = m(t) \sin [\omega t + \varphi(t)] = u(t) \sin \omega t + v(t) \cos \omega t.$$

Substituting (41.7) in the equation (41.5) we obtain, after averaging over the period of the oscillations, the set of equations for the amplitudes and phases ($\omega = \omega_c$):

$$\dot{v} - (\omega_0 - \omega)\, u + v/T_2 = -\tfrac{1}{2}\gamma M_z(h_c + h_s \cos \psi),$$

$$\dot{u} + (\omega_0 - \omega)\, v + u/T_2 = \tfrac{1}{2}\gamma h_s M_z \sin \psi,$$

$$\dot{M}_z - \frac{M_0 - M_z}{T_1} = \frac{1}{2}\gamma h_c v + \frac{1}{2}\gamma h_s(v \cos \psi - u \sin \psi), \tag{41.8}$$

$$\dot{h}_s + (\omega/Q_L)\, \dot{h}_s = \omega^2(u \cos \psi + v \sin \psi)\, V_s a,$$

$$\dot{h}_s + (\omega/2Q_L)\, h_s = \omega(v \cos \psi - u \sin \psi)\, V_s a/2.$$

The coefficient a here is determined by the condition for normalization of the field in the resonator. Very often paramagnetic resonance in a specimen under investigation is observed by placing it in a rectangular resonator in which TE_{01n} type oscillations are excited. Using normalization condition

(3.9) and the expressions for the eigen-fields (see, for example, Gurevich, 1952) we find for TE_{01n} modes

$$a = \frac{16\pi}{V_c}\left(\frac{\lambda}{\lambda_g}\right)^2, \tag{41.9}$$

where λ_g is the wavelength in the waveguide and λ the wavelength in free space.

The change of variables (41.7) has a clear geometrical meaning: transformation from a fixed system of coordinates to one rotating with a frequency ω around the z-axis. The variables u and v are the transverse components for the magnetic moment in a rectangular system of coordinates, whilst m and φ describe the same transverse component in a cylindrical system.

42. Susceptibility. The Shape of the Paramagnetic Resonance Line

42.1. The Susceptibility of a Homogeneous Paramagnetic

We now find the stationary states of a physical system described by equation (41.8). To do this, having made all the derivatives in (41.8) equal to zero and eliminated h_s and ψ we obtain the following relations between the components of the vector $\boldsymbol{M}$:†

$$\begin{aligned} u &= \frac{1}{2}\gamma M_z \frac{(\omega_0 - \omega)\, h_c}{(\omega_0 - \omega)^2 + (T_2^{-1} + xM_z)^2}, \\ v &= -\frac{1}{2}\gamma M_z \frac{(T_2^{-1} + xM_z)\, h_c}{(\omega_0 - \omega)^2 + (T_2^{-1} + xM_z)^2}, \end{aligned} \tag{42.1a}$$

$$\gamma^2 h_c^2 M_z T_1/4T_2 - (M_0 - M_z)\left[(\omega_0 - \omega)^2 + (T_2^{-1} + xM_z)^2\right] = 0. \tag{42.1b}$$

Here we have used the abbreviated notation

$$x = \gamma Q_L a V_s/2. \tag{42.2}$$

Since we wished to allow for the reaction of the specimen we had to supplement the system of Bloch equations with the circuit equation. The consequence of this is the appearance of the term xM_z in (42.1). The other terms in the relations (42.1.) do not depend on the circuit parameters. This fact permits us to formulate the criterion for being able to neglect the reaction of the specimen in the form of the inequality $|xM_z| \ll T_2^{-1}$, or in the expanded form:

$$\gamma Q_L a V_s |M_z|/2 \ll T_2^{-1}. \tag{42.3}$$

† Bloom (1957) obtained the relations in this form.

Usually, the electrodynamic properties of a paramagnetic are characterized by its magnetic susceptibility. The susceptibility is a tensor connecting the magnetic moment of the substance with the applied magnetic field (the general definition of susceptibility is given in § 14 *et seq.*):

$$\boldsymbol{M} = \overleftrightarrow{\chi}\boldsymbol{H}. \tag{42.4}$$

Since a paramagnetic, as can be seen from (42.1), is a dispersive medium its susceptibility depends upon the frequency. The definition of $\overleftrightarrow{\chi}$ is not unambiguous until we have agreed upon the appropriate value of the magnetic field. For $\boldsymbol{H}$ in (42.4) we can take the unperturbed field $\boldsymbol{H}_c$ or the active field $\boldsymbol{H}_1$. Both cases are found in published papers. In order to distinguish them we shall write $\overleftrightarrow{\chi}^{(c)}$ in the first case and $\overleftrightarrow{\chi}^{(1)}$ in the second.

Let us determine the components of the tensor $\overleftrightarrow{\chi}^{(c)}$ starting with the diagonal component $\chi_{xx}^{(c)}$. The phase of the oscillations of both M_y and M_x is different from the phase of the forcing field H_c. Therefore the most convenient form of notation is the complex one, in which $H_c = h_c\, \mathrm{e}^{-i\omega t}$ and $M_x = m\, \mathrm{e}^{-i(\omega t + \varphi)}$. The complex susceptibility in relation to the unperturbed field is

$$\chi_{xx}^{(c)} = \chi' + i\chi'' = \frac{M_x}{H_c} = \frac{m}{h_c}\, \mathrm{e}^{-i\varphi}, \tag{42.5}$$

and it is easy to check, by using (41.7), that

$$\chi' = u/h_c, \quad \chi'' = -v/h_c. \tag{42.6}$$

The rest of the components are found in an identical way, and on completing the calculations we find the tensor

$$\overleftrightarrow{\chi}^{(c)} = \begin{vmatrix} \chi & i\chi & 0 \\ -i\chi & \chi & 0 \\ 0 & 0 & 0 \end{vmatrix}. \tag{42.7}$$

There is no need for any additional work to find the tensor $\chi^{(1)}$. In this case we consider the field $\boldsymbol{H}_1$ to be known, without breaking it down into parts. Then the Bloch equations themselves form a closed system, and this means that it is sufficient to put $x = 0$ in $\overleftrightarrow{\chi}^{(c)}$ to obtain $\overleftrightarrow{\chi}^{(1)}$. It is clear that when the reaction of the specimen can be neglected the difference between the two definitions of the susceptibility disappears.

It is sufficient to know the values of u and v in order to be able to write an explicit expression for the components of the tensor (42.7). Expressions (42.1a) are not yet in their final form since they define u and v in terms of an unknown longitudinal component M_z. Finding M_z in the general case involves the solution of the cubic equation (42.1 b), but there is no particular interest in discussing the general case. Let us examine two special cases:

(a) $M_z \approx M_0$.

The state of the paramagnetic differs little from Boltzmann equilibrium and this case can, with justification, be called the case of a weak field. Expressions (42.6) for the susceptibility components are as follows:

$$\chi' = \frac{1}{2}\gamma M_0 \frac{\omega_0 - \omega}{(\omega_0 - \omega)^2 + (T_2^{-1} + xM_0)^2},$$
$$\chi'' = \frac{1}{2}\gamma M_0 \frac{T_2^{-1} + xM_0}{(\omega_0 - \omega)^2 + (T_2^{-1} + xM_0)^2}. \tag{42.8}$$

(b) $xM_z \ll T_2^{-1}$.

The presence of a paramagnetic in a resonator has little effect on the field in it. Equations (42.1 b) become linear, which leads to

$$M_z = M_0 \frac{1 + (\omega_0 - \omega)^2 T_2^2}{1 + (\omega_0 - \omega)^2 T_2^2 + \gamma^2 h_c^2 T_1 T_2/4}, \tag{42.9a}$$

$$\chi' = \frac{1}{2}\gamma M_0 T_2 \frac{(\omega_0 - \omega) T_2}{1 + (\omega_0 - \omega)^2 T_2^2 + \gamma^2 h_c^2 T_1 T_2/4},$$
$$\chi'' = \frac{1}{2}\gamma M_0 T_2 \frac{1}{1 + (\omega_0 - \omega)^2 T_2^2 + \gamma^2 h_c^2 T_1 T_2/4}. \tag{42.9b}$$

The shape of the curves of $\chi'(\omega)$ and $\chi''(\omega)$ is shown in Fig. X.2. As the field h_c increases both the longitudinal component of the magnetic moment

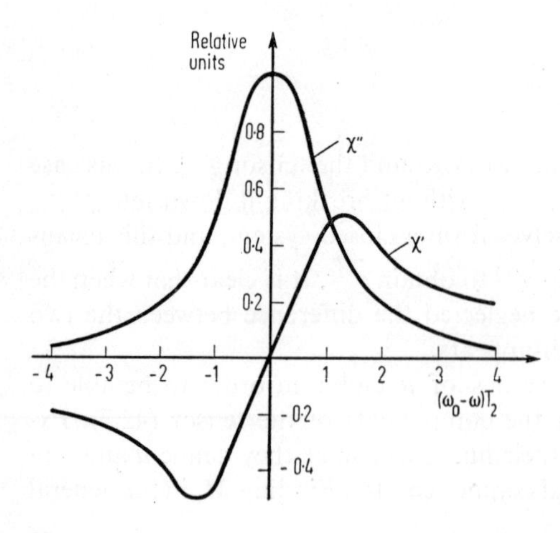

FIG. X.2. Curves of absorption and dispersion near the paramagnetic resonance frequency.

(and thus the population difference of the levels in question) and the susceptibility decrease, approaching zero at the limit. This is the so-called saturation effect. The parameter determining the degree of saturation is $\gamma^2 h_c^2 T_1 T_2/4$. When there is no saturation

$$\gamma^2 h_c^2 T_1 T_2/4 \ll 1 \tag{42.10}$$

and the susceptibility does not depend on the high-frequency field:

$$M_z = M_0, \tag{42.11a}$$

$$\chi' = \frac{1}{2}\gamma M_0 T_2 \frac{(\omega_0 - \omega) T_2}{1 + (\omega_0 - \omega)^2 T_2^2},$$

$$\chi'' = \frac{1}{2}\gamma M_0 T_2 \frac{1}{1 + (\omega_0 - \omega)^2 T_2^2}. \tag{42.11b}$$

A real paramagnetic is an absorbing medium. Its internal energy, on the one hand, increases at the expense of the energy of the electromagnetic field and, on the other hand, decreases due to relaxation processes which convert it into the energy of the thermal motion of the lattice atoms. When they balance each other these two processes give the state of equilibrium of the paramagnetic in the presence of a field described by (42.9).

Let us calculate the power absorbed by a unit volume of a paramagnetic substance. It can be determined by time-averaging the quantity $(4\pi)^{-1}[\boldsymbol{H}\cdot(d\boldsymbol{B}/dt)]$ (see, for example, Landau and Lifshitz, 1957). Remembering that $\boldsymbol{B} = \boldsymbol{H} + 4\pi\boldsymbol{M}$ we can also write

$$P = \frac{\omega}{2\pi}\int_0^{2\pi/\omega}\left[\frac{1}{4\pi}\left(\boldsymbol{H}\cdot\frac{d\boldsymbol{H}}{dt}\right) + \left(\boldsymbol{H}\cdot\frac{d\boldsymbol{M}}{dt}\right)\right] dt. \tag{42.12}$$

The linearly polarized field $H_x = h_c \cos\omega t$, in accordance with (41.7) and (42.6), corresponds to

$$M_x = h_c\chi' \cos\omega t + h_c\chi'' \sin\omega t. \tag{42.13}$$

It is obvious that a finite contribution to the absorbed power is made only by the term proportional to χ'', and as a result

$$P = \tfrac{1}{2}\omega\chi'' h_c^2, \tag{42.14}$$

which is in agreement with the general expression (17.9) for the energy lost by a field in a substance. Since χ'' is a quantity that is essentially positive the equilibrium state of a two-level paramagnetic is unsuitable for amplification purposes.

42.2. The Width of Resonance Line of a Paramagnetic

As can be seen from (42.9b), the function $\chi''(\omega)$ describes an absorption line that has a Lorentzian form (Fig. X.2). Maximum absorption occurs when $\omega = \omega_0$. The line width $\Delta\omega_0$ can be determined from the equation $\chi''(\omega) = \frac{1}{2}\chi''(\omega_0)$. A long way from saturation, as can be seen from (42.11),

$$\Delta\omega_0 = 2/T_2. \tag{42.15}$$

Taking into account the effect of the specimen leads to a broadening of the observed line and, in accordance with (42.8),

$$\Delta\omega_0 = 2(T_2^{-1} + xM_0). \tag{42.16}$$

Line broadening also occurs because of saturation. Provided that $\gamma^2 h_c^2 T_1 T_2/4 \gg 1$, (42.9b) gives

$$\Delta\omega_0 = \gamma h_c (T_1/T_2)^{1/2}. \tag{42.17}$$

The line width in a strong field defined by (42.17), as Bloembergen has pointed out, agrees with the uncertainty relation between the energy and the time only in the case $T_1 = T_2$. In practice the life of a paramagnetic ion at any of the levels in question in the presence of a strong field is chiefly determined by the magnitude of this field and is (as will be clear from what follows) $\Delta t = (\gamma h_c)^{-1}$. The uncertainty relation $\Delta E\, \Delta t \geqslant \hbar$ leads to

$$\Delta\omega_0 = \gamma h_c. \tag{42.18}$$

The invalidity of equation (41.1) in the strong field range has been experimentally confirmed by Redfield (1955). He has put forward a hypothesis which shows a way out of the difficulty. It is that T_2 is not constant, but depends on the amplitude of the high-frequency field, and $T_2 \to T_1$ as $h_c \to \infty$. In order to simplify the mathematical formulation of this hypothesis we first write (41.1) in the rotating system of coordinates $x'y'z$. The change to this system is made in accordance with the well-known relation (see, for example, Landau and Lifshitz, 1958)

$$\frac{d\boldsymbol{M}}{dt} = \left(\frac{d\boldsymbol{M}}{dt}\right)_{\text{rot}} + [\boldsymbol{\omega} \wedge \boldsymbol{M}], \tag{42.19}$$

where $\boldsymbol{\omega}$ is the gyration vector.

Replacing $d\boldsymbol{M}/dt$ by the above expression we arrive, in accordance with (41.1), at

$$\left(\frac{d\boldsymbol{M}}{dt}\right)_{\text{rot}} = -\gamma\left[\boldsymbol{M} \wedge \left(\boldsymbol{H} - \frac{\boldsymbol{\omega}}{\gamma}\right)\right] - \boldsymbol{i}_{\text{rot}}\frac{u}{T_2} - \boldsymbol{j}_{\text{rot}}\frac{v}{T_2} - \boldsymbol{k}\frac{M_z - M_0}{T_1}. \tag{42.20}$$

The vector

$$H_{\text{eff}} = H - \frac{\omega}{\gamma}, \tag{42.21}$$

which is called the effective field, lies in the $x'z$ plane and has $H_x = \frac{1}{2}h_c$ and $H_z = H_0 - \omega/\gamma$ (Fig. X.3) as its components along the axes.

The component $H_{x'}$ is caused by the presence of a high-frequency linearly polarized field $h_c \cos \omega t$ which is equivalent to the sum of two rotating ones. One of them turns in the direction of the precession and is therefore fixed in

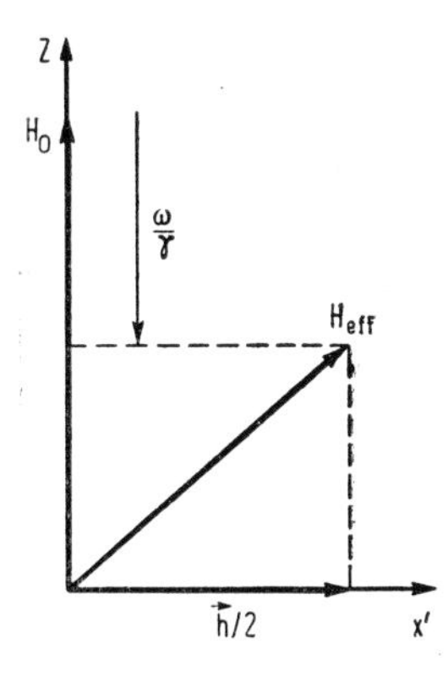

FIG. X.3. The effective magnetic field in a rotating system of coordinates.

the rotating system of coordinates. The other causes only a slight high-frequency perturbation of the motion M. When averaged over the period of the precession the effect of this field is zero. Consequently, an oscillating field of amplitude h is equivalent in its action to a rotating field of amplitude $\frac{1}{2}h$.

Let us return to Redfield's hypothesis. If

$$h_c \gg \Delta H, \tag{42.22}$$

then the external field has a far greater effect on the spin than the field of the adjacent spins. This undoubtedly has an effect on the relaxation of the component of M and is in the same direction as $H_{x'}$.

Bloch's equation modified in accordance with this view becomes

$$\left(\frac{dM}{dt}\right)_{\text{rot}} = -\gamma[M \wedge H_{\text{eff}}] - i_{\text{rot}}\frac{u}{T_{2e}} - j_{\text{rot}}\frac{v}{T_2} - k\frac{M_z - M_0}{T_1}, \tag{42.23}$$

where T_{2e} is the relaxation time of the component u which is parallel to the rotating magnetic field.

The real and imaginary parts of the susceptibility, allowing for Redfield's effect, can be written as

$$\chi' = \frac{1}{2}\gamma M_0 T_2 \frac{(\omega_0 - \omega) T_{2e}}{1 + (\omega_0 - \omega)^2 T_2 T_{2e} + \gamma^2 h_c^2 T_1 T_2/4},$$
$$\chi'' = \frac{1}{2}\gamma M_0 T_2 \frac{1}{1 + (\omega_0 - \omega)^2 T_2 T_{2e} + \gamma^2 h_c^2 T_1 T_2/4}. \tag{42.24}$$

These give the required expression (42.18) for the width of the saturated line if we take $T_{2e} = T_1$ in a strong field. In a weak field $T_{2e} = T_2$.

Corrections to equation (41.1) such as those made by Redfield do not contradict the spirit of Bloch's theory. It follows from the relaxation theory discussed in Chapter IIof Vol. 1, t hat when the fields are strong enough the relaxation coefficients de pend on the magnitude of the field. The effects observed in experiment, however, are not connected with spin–lattice relaxation (for which Bloch's equations are derived) but with spin–spin relaxation.

42.3. *A Paramagnetic Specimen in a Non-uniform Field*

Up till now we have considered only the so-called homogeneous broadening of spectral lines. Amongst the mechanisms causing homogeneous broadening according to Portis (1953) are:

(a) dipole interaction between spins;
(b) spin–spin relaxation;
(c) interaction of spins with the radiation field;
(d) diffusion of excitation through the specimen;
(e) local field fluctuations.

It is characteristic of all the mechanisms listed that their action on all spins without exception is exactly the same. The broadening of the resonance line of a sample is the consequence of the broadening of the i ndividual responses of each of the spins and is therefore called homogeneous.

Inhomogeneous broadening covers:

(a) hyperfine interaction between electron and nuclear spins;
(b) anisotropic broadening;
(c) the presence of a non-uniform magnetic field $H_0(\boldsymbol{r})$.

Hyperfine interaction leads to a multiplet spectrum and as a result to broadening of the spectral line. Anisotropic broadening is connected with the dependence of the resonance frequency ω_0 on the orientation of the crystal axis relative to the external magnetic field. The scatter of the directions of the optic axis due to the non-ideal nature of the crystal structure or due to the polycrystalline structure of the specimen leads to local shifts in ω_0, and this appears as a broadening of the resonance curve of the specimen as a whole. A non-uniform magnetic field (Abragam, 1961), of which we shall speak later, influences the shape of the curve in just the same way.

The spins contained in a specimen placed in a non-uniform field can be divided into groups with slightly different resonance frequencies (Fig. X.4). The relative weight of each spin group which has its eigenfrequencies in the range between $\omega_0 + x$ and $\omega_0 + x + dx$ is given by the function $g(x)$ and

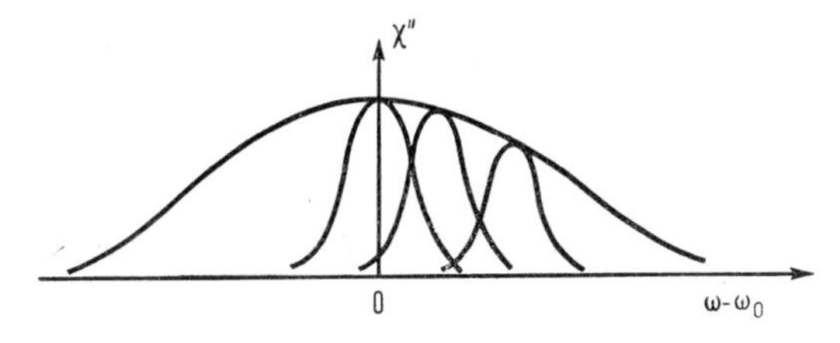

FIG. X.4. The shape of a line with inhomogeneous broadening (envelope). Each elementary volume of the sample has a Lorentzian line shape.

is equal to $g(x)\,dx$. The action of all the spin groups must be summed to obtain the susceptibility of the whole specimen, i.e. the susceptibility of an individual spin group $\chi''(\omega_0 - \omega + x)$ is multiplied by the weight function $g(x)$, and the resulting function is integrated over all values of x:

$$\chi'' = \int_{-\infty}^{\infty} \chi_j''(\omega_0 - \omega + x)\, g(x)\, dx. \tag{42.25}$$

In the cases when non-uniformity of the field is the main factor determining the line width the weight function $g(x)$ varies slowly compared with χ'' and

$$\chi'' = g(\omega - \omega_0) \int_{-\infty}^{\infty} \chi_j''(\omega_0 - \omega + x)\, dx. \tag{42.26}$$

If we substitute the value of χ_j'' from (42.9b) in (42.26), then

$$\chi'' = \frac{\pi \gamma M_0}{2(1 + \gamma^2 h_c^2 T_1 T_2/4)^{1/2}}\, g(\omega - \omega_0). \tag{42.27}$$

We shall consider a field non-uniformity given by the Lorentzian function $g(x)$:

$$g(x) = \frac{T_2^*}{\pi} \frac{1}{1 + (xT_2^*)^2}. \tag{42.28}$$

When writing (42.28) we used the obvious normalization condition $\int_{-\infty}^{\infty} g(x)\, dx = 1$. If the amplitude h_c is negligibly small, then

$$\chi'' = \frac{1}{2} \gamma M_0 T_2^* \frac{1}{1 + [(\omega_0 - \omega)\, T_2^*]^2}. \tag{42.29}$$

Expression (42.29) described a resonance curve of width $\Delta\omega_0^* = 2/T_2^*$, which becomes (42.11 b) on replacing T_2^* by T_2.

The saturation effect is manifested in a particular way in the case of an inhomogeneously broadened line. Until the field becomes anomalously strong and we can use (42.26) and the subsequent expressions, saturation, which reduces the magnitude of the absorption, has no effect on the line width:

$$\chi'' = \frac{M_0 T_2^*}{h_c(T_1 T_2)^{1/2}} \frac{1}{1 + [(\omega_0 - \omega) T_2^*]^2}. \tag{42.30}$$

43. Methods of Inversion in Two-level Paramagnetic Substances

43.1. Possible Ways of Exciting Two-level Systems

The necessary condition for the working of a maser is the creation at a selected pair of levels of the working substance of a state of inversion, or some other emitting state. A number of ways of doing this are known. Two-level masers acquired their name from the fact that no level apart from the two working ones takes part in the excitation process. In this method the

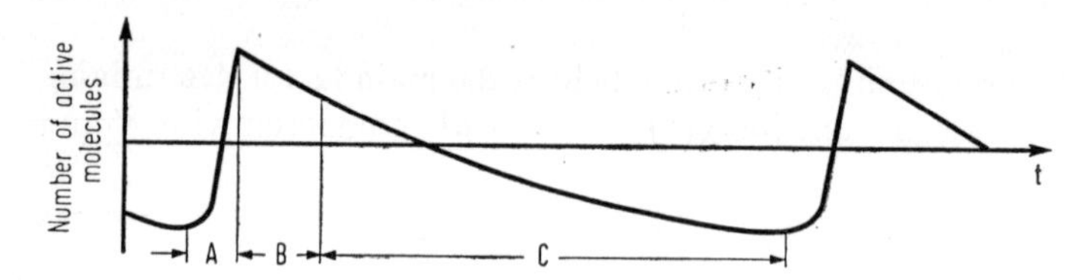

FIG. X.5. Sequence of phases in the operation of a two-level paramagnetic amplifier: *A*—excitation phase of the working substance, *B*—amplification phase, *C*—time taken to return to thermal equilibrium.

processes of excitation and emission do not occur at the same time. The molecule must first be excited and then brought into conditions favourable for emission.

If a molecular system is "transportable", i.e. can be moved in an excited state from one region of space to another, the above processes can, in addition, be separated spatially. In this case the two-level maser can operate continuously since active molecules are continuously entering the resonator and ones that are no longer active are leaving. This principle is used in a molecular beam oscillator. In the case of a solid working substance the practical realization of the principle of spatial separation has considerable difficulties, although the possibility cannot be excluded in principle (Bolef and Chester, 1958).

In existing two-level solid-state masers the working substance is immobile, so only pulsed operation is possible. Any of the processes below may create excited states in paramagnetic systems:

1. pulsed excitation;
2. adiabatic rapid passage;
3. non-adiabatic rapid passage;
4. sudden rotation of the magnetic field.

The first two methods are used in practice. Non-adiabatic passage is unsuitable when compared with them, and sudden rotation of the field is very difficult to achieve in experiments with electron spins.

The whole cycle of maser operation can be divided into three basic stages: excitation, whose duration is short when compared with the longitudinal relaxation time T_1; emission (amplification or oscillation), also limited to a fraction of T_1; and restoration of equilibrium, for which a time of the order of T_1 is necessary. This sequence is shown graphically in Fig. X.5.

43.2. Pulsed Excitation

The theory of pulsed excitation has been treated in fair detail by Vuylsteke (1960). We shall limit ourselves to discussing processes developing in a time interval $\Delta t \ll T_1; T_2$. This means that we can put $T_1 = T_2 = \infty$ in the Bloch equations. Let an alternating field which is switched on at a certain time remain constant in amplitude and phase. This way of stating the problem covers the necessity of neglecting the reaction field of the specimen and equations (41.8) are reduced to

$$\begin{aligned} &\dot{u} + (\omega_0 - \omega)\, v = 0, \\ &\dot{v} - (\omega_0 - \omega)\, u + \gamma h_c M_z/2 = 0, \\ &\dot{M}_z - \gamma h_c v/2 = 0. \end{aligned} \tag{43.1}$$

Since the process in question is not stationary the initial conditions must be stated at the time the field is switched on ($t = 0$). These conditions correspond to the paramagnetic being in a state of thermal equilibrium when there is no field, i.e.

$$u(0) = v(0) = 0, \quad M_z(0) = M_0. \tag{43.2}$$

We shall first find the solution for the function v, eliminating the variables u and M_z from (43.1) to do this. This operation leads to the oscillator equation (see also § 37.1, Vol. 1)

$$\ddot{v} + \Omega^2 v = 0, \tag{43.3}$$

where we use the notation

$$\Omega^2 = (\omega_0 - \omega)^2 + (\gamma h_c/2)^2. \tag{43.4}$$

The solution of (43.3) is the function

$$v = C \sin(\Omega t + \varphi), \tag{43.5}$$

which depends on the constant parameters C and φ which have to be found from the initial conditions (43.2). After this it is not difficult to find the rest of the unknowns. We finally obtain the following solutions:

$$\begin{aligned} v &= -\frac{\gamma h_c M_0}{2\Omega} \sin \Omega t, \\ u &= \frac{\gamma h_c M_0(\omega_0 - \omega)}{2\Omega^2} (1 - \cos \Omega t), \\ M_z &= \frac{(\gamma h_c/2)^2}{\Omega^2} M_0 \cos \Omega t + \frac{(\omega_0 - \omega)^2}{\Omega^2} M_0. \end{aligned} \tag{43.6}$$

The form of the notation can be altered very conveniently by introducing into the discussion the parameter

$$\delta = \frac{\omega_0 - \omega}{\gamma h_c/2}, \tag{43.7}$$

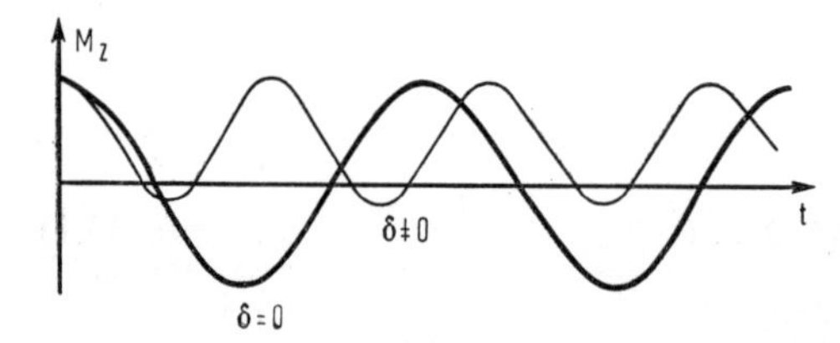

FIG. X.6. The change in the longitudinal magnetization of a paramagnetic due to a high-frequency field, in the absence of relaxation.

which is generally called the detuning. After this substitution we have

$$\begin{aligned} \Omega^2 &= (\gamma h_c/2)^2 (1 + \delta^2), \\ v &= -\frac{M_0}{(1 + \delta^2)^{1/2}} \sin \Omega t, \\ u &= \frac{\delta M_0}{1 + \delta^2} (1 - \cos \Omega t), \\ M_z &= \frac{M_0}{1 + \delta^2} (\cos \Omega t + \delta^2). \end{aligned} \tag{43.8}$$

Figure X.6 shows how the longitudinal component M_z varies with time. It is easiest to see the nature of the motions described by (43.8) when the detuning is zero; then

$$
\begin{aligned}
u &= 0, \\
v &= -M_0 \sin \Omega_0 t, \\
M_z &= M_0 \cos \Omega_0 t, \\
\Omega_0 &= \gamma h_c/2.
\end{aligned}
\tag{43.9}
$$

The motion described by (43.9) is the precession of the vector $\boldsymbol{M}$ with an angular frequency Ω_0 about the x'-axis, with which the vector $\boldsymbol{M}$ makes a right angle. In the general case of an arbitrary disturbance the nature of the motion cannot be seen so clearly from the equations (43.8) and it is more convenient to use the vector form of the Bloch equations (42.23).

In a rotating coordinate system the vector $\boldsymbol{M}$ precesses around the field $\boldsymbol{H}_{\text{eff}}$ with a frequency

$$\Omega = \gamma \, |H_{\text{eff}}| = \gamma \left[\left(H_0 - \frac{\omega}{\gamma} \right)^2 + \frac{1}{4} \, h_c^2 \right]^{1/2}, \tag{43.10}$$

defined by (43.4). The effective field makes an angle θ_0 with H_0 equal to

$$\theta_0 = \arctan (1/\delta). \tag{43.11}$$

If we return to a fixed system of coordinates, it turns out that the vector $\boldsymbol{M}$ takes part in two forms of motion: it rotates around $\boldsymbol{H}_{\text{eff}}$ and, together with the latter, around $\boldsymbol{H}_0$. The second motion, which is far more rapid, takes place at a frequency ω, and it is this that we have in mind when we speak of precession. The frequency of the first, Ω, is generally much less than ω. This motion has been given the name of nutation. From the quantum point of view (see § 37.1, Vol. 1) Ω is the frequency of transitions between levels acted upon by a field.

The maximum angle through which the magnetic moment is deflected in the process of motion is $2\theta_0$. When the resonance conditions are satisfied $H_{\text{eff}} = \frac{1}{2} h_c$ and $\theta_0 = \pi/2$, and only in this case is a state of complete inversion achieved. It occurs at the times

$$\Omega_0 t_n = (2n + 1) \pi. \tag{43.12}$$

If the high-frequency field is switched off at one of these points in time, the magnetic moment vector remains at an angle of 180° to the initial position. Therefore a resonance radiofrequency pulse that satisfies condition (43.12) can be called an inverting pulse. Generally, however, a pulse that rotates the vector $\boldsymbol{M}$ through an angle θ from the initial position is called a θ-pulse.

The shortest of the inverting pulses is thus a 180° pulse. Its duration can be calculated from (43.12) for $n = 0$ and is

$$t_0 = \frac{2\pi}{\gamma h_c}. \tag{43.13}$$

Unless relaxation times are taken into consideration the duration of the inverting pulse is not limited in any way. In practice the above remains valid only within short periods of time $\Delta t \ll T_2$. Pulsed excitations are essentially based on the use of transients.

We have thus given all the conditions under which pulsed inversion occurs. Let us collect them together:

$$\omega = \omega_0,$$
$$t_0 = (2\pi/\gamma h_c) \ll T_2. \tag{43.14}$$

We do not give the inequality $t_0 \ll T_1$ here since $T_1 \geqslant T_2$ and satisfaction of (43.14) guarantees it automatically. If we transform inequality (43.14), expressing T_2 in terms of the line width ΔH, it becomes

$$h_c \gg \pi \, \Delta H, \tag{43.15}$$

which is close to condition (42.22).

What is the order of the duration and magnitude of the 180° pulse in an actual experiment? Let us make an estimate for a substance with a line width of $\Delta H = 1$ oersted. From (43.15) $h_c \gg \pi$ and $t_0 \gg 2/\gamma \approx 10^{-7}$ sec ($t_0 h_c = 2\pi/\gamma$ = const.). These figures show one of the basic difficulties in making practical use of a 180° pulse for exciting a maser. A short pulse excludes the use of a resonator with a high Q-factor since the duration of the transients (the "ringing time" of the resonator) increases as Q rises. In a resonator with a low Q-factor, on the other hand, a field of the necessary amplitude can be produced only by a very powerful oscillator. In addition, a high Q-factor is required to obtain the amplification and oscillation effect in the actual maser. Singer's book (1959) discusses some ways of reconciling these contradictory requirements. Amongst them is Q-spoiling for the duration of the inversion, or using different modes of the resonator for the purposes of inversion and amplification.

The power P of the subsidiary radiation which must be introduced into the resonator to maintain the field h_c can be estimated roughly by using the definition of the Q-factor $Q = \omega W/P$. The energy of the field in the resonator is $W = h_c^2 V_c/8\pi$ and finally for a matched resonator

$$P \simeq \omega h_c^2 V_c/8\pi Q.$$

For $Q = 10^2$, $\omega/2\pi = 10^{10}$ Hz, $V_c = 10$ cm^{-3} and $h_c = \pi$ Oe, the formula gives us $P \sim 250$ W.

43.3. Adiabatic Rapid Passage (ARP)

The determination of the shape of a spectral line is one of the basic measurements in radiofrequency spectroscopy. In general outline the procedure of taking an absorption curve is as follows: the oscillator frequency is taken through a certain sequence of values and the magnitude of the absorbed power is simultaneously measured. In order to eliminate distortion due to transients the rate of change of the frequency should be small during an interval comparable with the characteristic times associated with the various motions of the system, i.e. precession, nutation and relaxation. This type of passage through resonance is called slow and the state of the system at any time is described by (42.1). When each measurement is made the system can be considered to be in a state of equilibrium. If the rate of passage exceeds any of the values indicated, the passage becomes rapid. For example, in the case of infinite relaxation times any passage which occurs at a finite rate can be considered to be rapid.†

The rapid category also includes the adiabatic passage discussed below. It is rapid since it takes place in a time short compared with T_1 and T_2. At the same time the adiabatic condition (see Landau and Lifshitz, 1957) is satisfied:

$$T\left|\frac{d\lambda}{dt}\right| \ll \lambda. \tag{43.16}$$

The latter means that the parameter λ which characterizes the system in question, or the external field, should vary little during the period of motion of the system T.

To explain the physics of the processes that occur we shall turn to the case $T_1 = T_2 = \infty$ and find the stationary states of the system in a given field. It follows from the equations (43.1) that under the conditions under discussion

$$v = 0. \tag{43.17a}$$

For the set of equations to be closed it must be supplemented by the law of the conservation of magnetic moment which is valid when there is no relaxation:

$$u^2 + M_z^2 = M_0^2. \tag{43.17b}$$

† In practice it is more convenient to alter the frequency $\omega_0 = \gamma H_0$ and not the oscillator frequency ω since in the second case we must simultaneously retune the resonator and allow for the variations in the response of the high-frequency section of the apparatus.

On solving (43.17) we obtain

$$M_z = \frac{\delta M_0}{\pm(1+\delta^2)^{1/2}} = M_0 \cos\theta_0, \tag{43.18}$$

$$u = \frac{M_0}{\pm(1+\delta^2)^{1/2}} = M_0 \sin\theta_0.$$

The connection between the quantities θ_0 and δ is given by the equation we already know

$$\tan\theta_0 = \delta^{-1}. \tag{43.19}$$

In just the same way [see (43.11)] the effective magnetic field is oriented in a rotating coordinate system.

Generally speaking, in a system without loss transients last for an infinitely long time and it is not possible to find a steady state with arbitrary initial conditions. We can satisfy (43.18) only with completely defined initial conditions, although there is ambiguity in their selection. For a given field disturbance and amplitude two stationary angles of precession differing by π are possible. For each of these angles there is a definite sign in front of the square roots in (43.18).

Adiabatic rapid passage is a practical means of obtaining any of the stationary states contained in (43.18) for a paramagnetic. Let the magnetic moment at the start be parallel to the static field and the detuning be very great, let us say $\delta \ll 0$. The initial conditions $M_z(0) > 0$ and $\delta(0) < 0$ indicate the choice of the negative value of the roots in (43.18). In the case of $\delta(0) > 0$ we take the positive value of the roots. We shall then start to reduce the disturbance. If we carry out this process slowly enough, then at an arbitrary time t the state of the system is given by (43.18) with $\delta = \delta(t)$. The motion as a whole is the successive passage of a whole combination of stationary states. In other words, the magnetic moment follows the changes in direction of the effective field. The adiabatic condition must be satisfied in order to avoid distortion due to the rate of passage. In our case the part of the parameter λ is played by the stationary precession angle θ_0 and condition (43.16) can be rewritten in the form

$$\left|\frac{d\theta_0}{dt}\right| \ll \theta_0 \Omega, \tag{43.20}$$

where Ω is the frequency of rotation of $\boldsymbol{M}$ around $\boldsymbol{H}_{\text{eff}}$ given by the formula (43.10).

Using (43.10) and (43.11) we find from (43.20) that

$$\left|\frac{d(\omega_0 - \omega)}{dt}\right| \ll \Omega_0^2\,(1+\delta^2)^{3/2} \arctan\left(\frac{1}{\delta}\right). \tag{43.21}$$

The form of the adiabatic criterion can be considerably simplified in two special cases.

(a) $\theta_0 \ll 1$, which is true for $\delta \gg 1$. In this case condition (43.21) becomes

$$\left|\frac{d(\omega_0 - \omega)}{dt}\right| \ll \Omega_0^2 \delta^2; \tag{43.22}$$

(b) $\theta_0 \sim 1$. This case also includes zero detuning, when the right-hand side is a minimum:

$$\left|\frac{d(\omega_0 - \omega)}{dt}\right| \ll \Omega_0^2. \tag{43.23}$$

Inequality (43.23) is more restrictive than (43.22). Therefore (43.23) should be taken as the adiabatic condition for the whole process of the passage.

When the rate of the passage is greater than permitted by the adiabatic condition the nature of the motion of the magnetic moment changes since it does not have time to follow the rapidly changing disturbance exactly. This means, in the first place, that $\theta(t) < \theta_0(t)$ and, in the second place, that nutation may appear.

Any orientation of the magnetic moment in space can be achieved in the adiabatic process. At resonance $\delta = 0$ and $\theta_0 = \pi/2$, and when $\delta(t) = -\delta(0)$ the moment is inverted ($\theta_0 = \pi$). This method of using ARP to obtain a state with a negative temperature is also used in masers.

Up to now we have been dealing with an ideal paramagnetic with infinite relaxation times. For adiabatic inversion to occur in a real paramagnetic the process should be carried out in a time short compared with the relaxation times of the specimen. In this case the ideas developed above remain valid. A quantitative criterion of rapidity can be obtained from the following. It can be seen from (43.18) that the system is subjected to the greatest changes when the disturbance changes in the range $|\delta| \lesssim 1$ and, to make an estimate, we can put $|\Delta(\omega_0 - \omega)| \approx \Omega_0$. The time of the passage is $\Delta t \ll T_2$, so the criterion of rapidity will be

$$\left|\frac{d(\omega_0 - \omega)}{dt}\right| \gg \frac{\Omega_0}{T_2}. \tag{43.24}$$

The rapidity criterion obtained for the passage is compatible with the adiabatic criterion only in the cases when $T_2^{-1} \ll \Omega_0$ or, expressing T_2 in terms of the resonance line width ΔH,

$$h_c \gg \Delta H. \tag{43.25}$$

We have arrived at the Redfield condition (42.22) which we already know.

Here the transverse relaxation time is comparable with the longitudinal. The conditions for adiabatic rapid passage can be stated as

$$\frac{\Omega_0}{T_1} \ll \left|\frac{d(\omega_0 - \omega)}{dt}\right| \ll \Omega_0^2, \quad h_c \gg \Delta H. \tag{43.26}$$

If conditions (43.26) are satisfied, the picture drawn above of the process of adiabatic rapid passage is a good enough reflection of the actual state of affairs. There is a more general formulation and discussion of the problem of passage through resonance in the paper by Gvozdoder and Magazanik (1950). There are mathematical difficulties in its solution, with the exception of the above-mentioned cases of slow and adiabatic rapid passage and also of rapid non-adiabatic passage (i.e. in all the intermediate cases). Weger (1960) gives a detailed analysis of the different variants of rapid passage.

43.4. Other Methods

A two-level system can be brought into an excited state by non-adiabatic passage (Weger, 1960; Jacobsohn and Wangsness, 1948). When the velocity of the passage is greater than permitted by the adiabatic condition (43.23) the vector $\boldsymbol{M}$ deviates from the initial position but does not have time to follow the effective field. The magnetic moment makes a complex motion which can be looked upon as the superposition of a forced oscillation with a frequency ω and a free oscillation with a frequency ω_0. In other words, the motion is composed of precession and nutation. Since the mean precession angle lags behind θ_0 complete inversion cannot be achieved by non-adiabatic passage. It is therefore reasonable to use non-adiabatic excitation only when the adiabatic and rapidity conditions (43.26) are difficult to meet simultaneously because of the smallness, for example, of T_1, and a small deviation of $\boldsymbol{M}$ must be obtained in a short time. This requirement arises in certain experiments with ferrites (Aleksandrov *et al.*, 1960). This method is not used in practice in paramagnetic amplifiers.

The last of the methods of excitation upon which we shall touch consists of fairly rapid alteration of the direction of the magnetic field $\boldsymbol{H}_0$ (Singer, 1959; Fain, 1958a and 1958b). It is attractive in that it requires no high-frequency pumping source. Of the two possible ways of altering the field direction a physical rotation can at once be rejected since the turning time should not exceed the precession period (Yashchin, 1960). The other method is more practical; it consists of altering the field strength from H_0 to $-H_0$ without rotation. Since this process is not accompanied by the appearance of transverse components of $\boldsymbol{M}$ its rate is limited at the lower end only by the time T_1. The basic difficulties appear because there are always some fields $\boldsymbol{H}_i$ with a different orientation which take part in the experiment besides the field $\boldsymbol{H}_0$. These may be stray fields or simply the field of the earth. The presence of these small fields has a considerable effect at times when $|\boldsymbol{H}_0|$ runs through a range of small values during the change (Bunkin, 1959; Rodak, 1959). Therefore the weak field range $|\boldsymbol{H}_0| \lesssim |\boldsymbol{H}_i|$ should be passed through in a time $\Delta t \sim 1/\gamma\, |\boldsymbol{H}_i|$ which is generally far shorter than T_1.

In practice the idea of inversion with field rotation has been achieved in an experiment with the nuclear paramagnetic LiF with $T_1 = 15$ sec (Purcell and Pound, 1951). In this connection it is interesting to note that before the appearance of masers all the above methods had been used in nuclear magnetic resonance experiments for the investigation of transients. Andrew's monograph (1955) surveys this problem in detail.

44. The Theory of the Resonator-type Two-level Amplifier

44.1. The Amplifier Equations

From the point of view of methods the problem of a two-level maser is very similar to that discussed above (§ 42) of the stationary state of a paramagnetic in a resonator. We have to determine the transmission coefficient

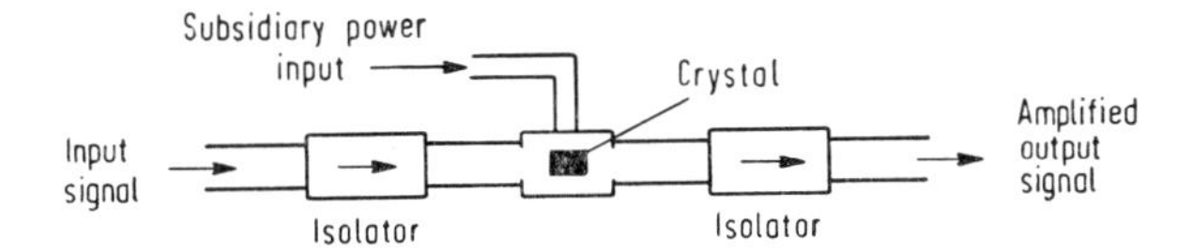

FIG. X.7a. Connection of a straight-through amplifier to a receiver.

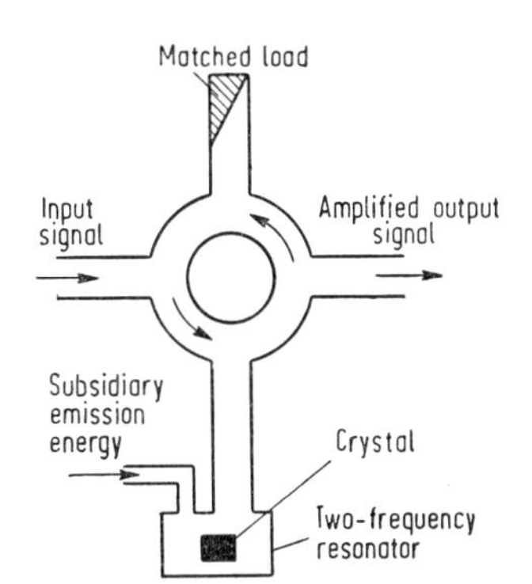

FIG. X.7b. Connection of a reflex amplifier to a receiver.

and the bandwidth of a resonator containing an active medium. Let us start with the straight-through type of amplifier (Fig. X.7a) since the method of calculation in this case is more easily followed than in the case of a reflex circuit (Fig. X.7b). The fundamental equations are (41.1) and (41.3). The term $F(t)$ in the right-hand side of (41.3) can be written in the form

$$F(t) = \left(\frac{8P_1\omega_c}{Q_{e1}}\right)^{1/2} \cos \omega t, \tag{44.1}$$

where ω is the frequency of the input signal; P_1 is its power; Q_{el} is the Q-factor component due to the coupling between the resonator and the input waveguide.

We can arrive at a similar way of writing the input signal if we consider resonator theory (Slater, 1950). In the calculations that follow it is more convenient to use the complex form of the equations obtained by changing to the variables $M_{\pm} = M_x + iM_y$:

$$\dot{M}_+ = i\omega_0 M_+ + \alpha M_z q - M_+/T_2, \tag{44.2a}$$

$$\dot{M}_- = -i\omega_0 M_- + \alpha^* M_z q - M_-/T_2, \tag{44.2b}$$

$$\ddot{q} + \frac{\omega_c}{Q_L}\dot{q} + \omega_c^2 q = -\frac{i}{2\gamma}\int_{V_c} (M_+\alpha^* - M_-\alpha)\, dV + \left(\frac{8P_1\omega_c}{Q_{e1}}\right)^{1/2} \cos \omega t. \tag{44.2c}$$

We use α to denote the quantity

$$\alpha = i\omega_c\gamma[H_x(\boldsymbol{r}) + iH_y(\boldsymbol{r})]. \tag{44.3}$$

Let us examine the quasi-stationary mode of an amplifier with a small input signal. The latter limitation allows us to consider M_z to be a given quantity which is independent of the field amplitude and is a function only of the spatial coordinates. The slow change in M_z due to relaxation processes can be taken into consideration in the final result.

Up to now we have considered only a point paramagnetic specimen. In this case the integration of the right-hand side of (44.2c) is not a problem. But the dimensions of an amplifying paramagnetic crystal are comparable with λ_g and we must remember that $\boldsymbol{M} = \boldsymbol{M}(\boldsymbol{r}, t)$. A convenient method of calculation in this case is the expansion of the transverse components $M_+(\boldsymbol{r}, t)$ in a set of functions that is complete and orthogonal in the volume of the resonator. In a rectangular resonator these requirements are, obviously, satisfied by the functions $\alpha_\lambda(\boldsymbol{r})$, $\alpha_\lambda^*(\boldsymbol{r})$, defined by (44.3). Therefore, having carried out the expansion

$$\begin{aligned} M_+(\boldsymbol{r}, t) &= \sum_\lambda \alpha_\lambda(\boldsymbol{r})\, M_\lambda^+(t), \\ M_-(\boldsymbol{r}, t) &= \sum_\lambda \alpha_\lambda^*(\boldsymbol{r})\, M_\lambda^-(t), \end{aligned} \tag{44.4}$$

we substitute in (44.2). Multiplying both parts of (44.2a) by α^* and (44.2b) by α, we integrate them over the volume of the resonator (α is that function of the α_λ which relates to the one mode of the resonator which is excited). By virtue of the orthogonality condition

$$\int_{Vc} \alpha_\lambda^*(\boldsymbol{r})\, \alpha_\mu(\boldsymbol{r})\, dV = I\delta_{\lambda\mu}, \tag{44.5}$$

where because of the normalization condition (3.9)

$$I = \gamma^2\omega_c^2 \int_{Vc} [\boldsymbol{H}(\boldsymbol{r})]^2\, dV = 4\pi\gamma^2\omega_c^2, \tag{44.6}$$

we obtain

$$\dot{M}^+(t) = i\omega_0 M^+(t) + \xi\tilde{M}_z q - M^+(t)/T_2,$$

$$\dot{M}^-(t) = -i\omega_0 M^-(t) + \xi\tilde{M}_z q - M^-(t)/T_2,$$

$$\ddot{q} + \frac{\omega_c}{Q_L}\dot{q} + \omega_c^2 q$$

$$= -\frac{iI}{2\gamma}[M^+(t) - M^-(t)] + (8P_1\omega_c/Q_{e1})^{1/2}\cos\omega t. \tag{44.7}$$

The suffix λ is omitted here since we are dealing with a single mode. We use ξ to denote the quantity

$$\xi = \int_{Vs} \alpha^*(\boldsymbol{r})\, \alpha(\boldsymbol{r})\, M_z(\boldsymbol{r})\, dV/I\tilde{M}_z, \tag{44.8}$$

which is called the filling factor of the resonator. $\tilde{M}_z$ is the maximum value of the longitudinal component of the magnetization. If M_z does not depend on the coordinates within the limits of integration, the expression can be simplified and

$$\xi = \int_{Vs} \alpha^*(\boldsymbol{r})\, \alpha(\boldsymbol{r})\, dV/I. \tag{44.9}$$

The filling factor is the ratio of the field energy contained in the volume of the working substance to the total energy of the field in the resonator. In particular for a resonator completely filled with a substance $\xi = 1$; for a point specimen at a field node $\xi = 0$; and for a point specimen at a field antinode (in the case of the TE_{01n} mode of a rectangular resonator)

$$\xi = 4(V_s/V_c)(\lambda/\lambda_g)^2 = \frac{aV_s}{4\pi}. \tag{44.10}$$

44.2. The Amplification Factor and Bandwidth of a Straight-through Amplifier

The stationary solutions of (44.7) are the functions

$$M^{\pm} = m_{\pm}\, e^{\pm i\omega t}, \qquad q = \bar{q}\cos(\omega t + \varphi), \tag{44.11}$$

which upon substitution in (44.7) lead to a system of algebraic equations for the amplitudes:

$$m_+ = \frac{1}{2}\,\xi M_z\bar{q}\,\frac{T_2^{-1} + i(\omega_0 - \omega)}{(\omega_0 - \omega)^2 + T_2^{-2}}\, e^{i\varphi},$$

$$m_- = \frac{1}{2}\,\xi M_z\bar{q}\,\frac{T_2^{-1} - i(\omega_0 - \omega)}{(\omega_0 - \omega)^2 + T_2^{-2}}\, e^{-i\varphi},$$

$$\left\{\left[(\omega_c^2 - \omega^2) + \frac{1}{2\gamma}\,\frac{\xi M_z I(\omega_0 - \omega)}{(\omega_0 - \omega)^2 + T_2^{-2}}\right]^2 + \left[\frac{\omega_c\omega}{Q_L} + \frac{1}{2\gamma}\,\frac{\xi M_z I T_2^{-1}}{(\omega_0 - \omega)^2 + T_2^{-2}}\right]^2\right\}\bar{q}^2 = \frac{8P_1\omega_c}{Q_{e1}}. \tag{44.12}$$

The energy of the field in the resonator $W_c = \omega_c^2\bar{q}^2/2$ is related to the power fed into the output waveguide by the relation

$$P_2 = \frac{\omega W_c}{Q_{e2}}, \tag{44.13}$$

where Q_{e2} is the component of the Q-factor due to the coupling to the output waveguide.

Substituting in (44.13) the value of $\bar{q}$ found from (44.12) we arrive at the expression for the transmission coefficient†

$$G = \frac{P_2}{P_1} = \frac{4/Q_{e1}Q_{e2}}{\left(\dfrac{1}{Q_L} + \dfrac{1}{Q_M}\right)^2 + y^2\left(1 - \dfrac{\omega_c T_2}{2Q_M}\right)^2}. \tag{44.14}$$

The quantity $y = 2[(\omega_c - \omega)/\omega_c]$ is the relative detuning. It is quite obvious that the closer the resonant frequencies of the substance and the resonator the greater is the effect of the working substance on the transmission

† The frequencies ω_c, ω_0 and ω are of necessity so close to each other that the difference of their ratios from 1 can be neglected almost anywhere.

coefficient. In practice we always try to make these frequencies the same and in future we shall put $\omega_0 = \omega_c$, as in (44.14). The quantity

$$Q_M = (4\pi\xi\chi'')^{-1} = \left[\frac{2\pi\gamma\xi T_2 M_z}{1 + (\omega_0 - \omega)^2 T_2^2}\right]^{-1} \tag{44.15}$$

in (44.14) is called the magnetic Q-factor or the Q-factor of the working substance. The basis for this terminology is not only the symmetry of Q_M and Q_L in (44.14) but also the fact that the physical meaning of Q_M is the ratio of the energy stored in the resonator to the energy absorbed by the paramagnetic specimen in one oscillation period $(1/\omega)$. The modulus of the magnetic Q-factor reaches its lowest value at zero detuning, when

$$Q_{M0} = (2\pi\gamma\xi T_2 M_z)^{-1}. \tag{44.16}$$

The signs of Q_M and M_z are the same, so the magnetic Q-factor of an inverted paramagnetic is negative. Physically this means that when there is a population inversion of the levels of the paramagnetic it gives up energy to the resonator field (negative absorption).

The transmission coefficient of the resonator has its largest value at the resonance frequency $\omega = \omega_c$:

$$G_0 = \frac{4/Q_{e1}Q_{e2}}{(Q_L^{-1} + Q_{M0}^{-1})^2}. \tag{44.17}$$

The device described can be called an amplifier if its transmission coefficient (in this case the amplification factor) is between the limits

$$1 < G_0 < \infty. \tag{44.18}$$

In future we shall call the quantity Q_Σ defined by the equation

$$Q_\Sigma^{-1} = Q_L^{-1} + Q_M^{-1} \tag{44.19}$$

the total Q-factor of the resonator.† It is not difficult to find from (44.17) that with a fixed total Q-factor the maximum amplification factor occurs when $Q_{e1} = Q_{e2} = Q_e$, i.e.

$$G_0 = \frac{4Q_e^{-2}}{Q_\Sigma^{-2}}. \tag{44.20}$$

† The power amplification factor of a maser (44.17) is connected by a simple relation with the amplification factor of a field in a resonator (37.18) $G/\overline{G} = k^2$, where $\overline{G}$ is the transmission coefficient of an empty resonator. The parameter η in (37.18): can be expressed in terms of the Q-factors: $\eta = -(Q_L/Q_M)$. It is sometimes called the self-excitation parameter.

It should be noted that the unloaded Q-factor Q_a is usually much greater than the external Q-factor Q_e, and then we may take $Q_L^{-1} = 2Q_e^{-1} + Q_a^{-1} = 2Q_e^{-1}$.

If the amplification factor in (44.18) is replaced by the value given by (44.20), the inequality can be rewritten in the form

$$Q_a^{-1} < -Q_M^{-1} < Q_L^{-1}. \tag{44.21}$$

The interpretation of (44.21) is very simple. For the output signal to be greater than the input signal the emission from the paramagnetic should compensate for the losses in the resonator walls. On the other hand, if the paramagnetic can equal all the sources of losses the maser starts to oscillate even when there is zero signal at the input. The inequality $-Q_M^{-1} > Q_L^{-1}$, which is the inverse of (44.21), is the condition for self-excitation of a maser:

$$M_z < 0, \quad 2\pi\gamma\xi T_2 Q_L |M_z| \geqq 1. \tag{44.22}$$

We shall make a detailed analysis of the oscillation processes in a two-level maser in the next chapter.

By the bandwidth of the amplifier we generally mean the frequency range at the limits of which the power amplification is half the maximum. The bandwidth is found from the relation

$$G = G_0/2. \tag{44.23}$$

The amplifier bandwidth $\Delta\omega$ cannot exceed the width of the paramagnetic resonance line $\Delta\omega_0$ or the resonator pass band $\Delta\omega_c$. If $\Delta\omega \ll \Delta\omega_0$, the parameters of the working substance can be assumed to be constant over the frequency range which is of interest. Then $Q_M = Q_{M0}$ and the bandwidth of a maser amplifier can be found without difficulty:

$$\Delta\omega = \frac{\omega_c/Q_\Sigma}{1 - (\omega_c T_2/2Q_{M0})} = \frac{\omega_c/Q_\Sigma}{1 + (Q_0/|Q_{M0}|)}, \tag{44.24}$$

where $Q_0 = \omega_0 T_2/2 = \omega_0/\Delta\omega_0$ is the Q-factor of the resonance line of the working substance.

We now return to (44.17) and notice that in the amplifier $|Q_M| > Q_L$. Therefore $\omega_c T_2/2|Q_M| < \Delta\omega_c/\Delta\omega_0$ and in the case when $\Delta\omega_0 \gg \Delta\omega_c$ the denominator in (44.24) differs only slightly from unity, i.e. when the inequality $\Delta\omega_c \ll \Delta\omega_0$ is satisfied the amplifier bandwidth is

$$\Delta\omega = \omega_c/Q_\Sigma. \tag{44.25}$$

The resonator-type maser belongs to the category of amplifiers with positive feedback or, as they are often called, regenerative amplifiers. Feedback is provided by the resonator itself and this, rather than the selectivity, is its main function. A general property of regenerative amplifiers is narrowing of the band as the amplification increases. Therefore neither the bandwidth nor the amplification factor characterize the amplifier when taken separately. Instead the capabilities of a maser can be judged from the parameter $G^{1/2}\,\Delta f$, which is called the figure of merit. It is easy to find from (44.17) and (44.25) that when the input and output coupling Q-factors are equal and when $\Delta\omega_0 \gg \Delta\omega_c$

$$(G_0^{1/2} - 1)\,\Delta f = f_c(|Q_{M0}|^{-1} - Q_a^{-1}). \tag{44.26}$$

If $G_0^{1/2} \gg 1$, i.e. $|Q_M| \sim Q_L$, and at the same time $Q_L \ll Q_a$ this expression can be simplified to

$$G_0^{1/2}\Delta f = f_c/|Q_{M0}| \simeq \Delta f_c. \tag{44.27}$$

The figure of merit is a constant for a given amplifier.

To conclude the discussion of a straight-through maser amplifier we note that (44.24) is valid for any relation between $\Delta\omega_0$ and $\Delta\omega_c$ provided that $G_0 \gg 1$. In the extreme case $\Delta\omega_0 \ll \Delta\omega_c$, which is the inverse of what has just been discussed,

$$\Delta\omega \approx \Delta\omega_0 \frac{Q_L}{Q_\Sigma}. \tag{44.28}$$

In this case the figure of merit of the amplifier

$$G_0^{1/2}\Delta f \simeq \Delta f_0 \frac{Q_L}{|Q_{M0}|} = \Delta f_0 \tag{44.29}$$

is determined by the width of the paramagnetic absorption line.

44.3. The Amplification Factor and Bandwidth of a Reflex Amplifier

The power amplification factor of a reflex maser is the ratio of the power reflected from the resonator to the power of the signal incident on the resonator:

$$G = P_2/P_1 = (P_1 - P_a)/P_1. \tag{44.30}$$

Here P_a is the power dissipated within the resonator.

For the calculation of P_a/P_1 we use the well-known expression derived by Slater (1950) for a cavity resonator:

$$\frac{P_a}{P_1} = \frac{4Q_e^{-1}(Q_L^{-1} - Q_e^{-1})}{Q_L^{-2} + y^2}. \tag{44.31}$$

If there is a paramagnetic specimen in the resonator, (44.31) must be modified. From the form of (44.14) we can guess that this consists of substituting

$$Q_a^{-1} \to Q_a^{-1} + Q_M^{-1}, \quad y \to y[1 - (\omega_c T_2/2Q_M)]. \tag{44.32}$$

After these substitutions (44.31) becomes

$$\frac{P_a}{P_1} = \frac{4Q_e^{-1}(Q_\Sigma^{-1} - Q_e^{-1})}{Q_\Sigma^{-2} + y^2[1 - (\omega_c T_2/2Q_M)]^2}. \tag{44.33}$$

Upon substitution in (44.30), (44.33) leads to the following expression for the amplification factor:

$$G = \frac{(Q_e^{-1} + |Q_M|^{-1} - Q_a^{-1})^2 + y^2\left(1 - \dfrac{\omega_c T_2}{2Q_M}\right)^2}{(Q_e^{-1} - |Q_M|^{-1} + Q_a^{-1})^2 + y^2\left(1 - \dfrac{\omega_c T_2}{2Q_M}\right)^2}. \tag{44.34}$$

The amplification factor at resonance is

$$G_0 = \frac{(Q_e^{-1} + |Q_{MO}|^{-1} - Q_a^{-1})^2}{(Q_e^{-1} - |Q_{MO}|^{-1} + Q_a^{-1})^2}. \tag{44.35}$$

The bandwidth of the amplifier is found once more from the relation $G = G_0/2$. The bandwidth of a reflex amplifier with $G_0 \gg 1$ is given by the same expression (44.24) as is valid for a straight-through amplifier. If in addition $\Delta\omega_0 \gg \Delta\omega_c$, then

$$(G_0^{1/2} - 1)\,\Delta f = 2f_c(|Q_{MO}|^{-1} - Q_a^{-1}). \tag{44.36}$$

On comparing (44.36) with (44.26) we come to the conclusion that a reflex amplifier is better than a straight-through amplifier since

$$\frac{[(G_0^{1/2} - 1)\,\Delta f]_{\text{refl}}}{[(G_0^{1/2} - 1)\,\Delta f]_{\text{s.t.}}} = 2.$$

Physically this is connected with the fact that in a straight-through amplifier the power of the amplified signal is shared between the output and input waveguides and part of it is consequently lost.

Although the expressions for the amplification factor and bandwidth of a maser have been obtained from quite definite ideas about the working substance the final results are far more general in nature. Expressions (44.17), (44.25), (44.34), (44.36), which contain only the Q-factors of the working substance, describe a maser with essentially an arbitrary working substance. We must, however, remember that we started with equations with relaxation terms that define a spectral line of Lorentzian shape. This is the only limitation on the class of working substance. This limitation is not too serious since in problems of the type under discussion we can almost always neglect the departure of the shape of the line from Lorentzian. When we discuss amplifiers of other types in future we shall consider the problem to be solved when we have obtained the expression for Q_M.

To conclude the analysis of a two-level paramagnetic maser amplifier we recall that, because of relaxation processes and the presence of the actual signal being amplified, the parameters of an amplifier vary continuously with time and, generally speaking, we should write $Q_M = Q_M(t)$.

45. The Theory of the Resonator-type Three-level Amplifier

Two-level paramagnetic amplifiers have one main disadvantage—they cannot operate continuously. This feature can be overcome by using a different method of obtaining population inversion. Greater scope is offered by a principle of excitation suggested by Basov and Prokhorov (1955a). Their idea consists of saturating the secondary transition between a pair of levels

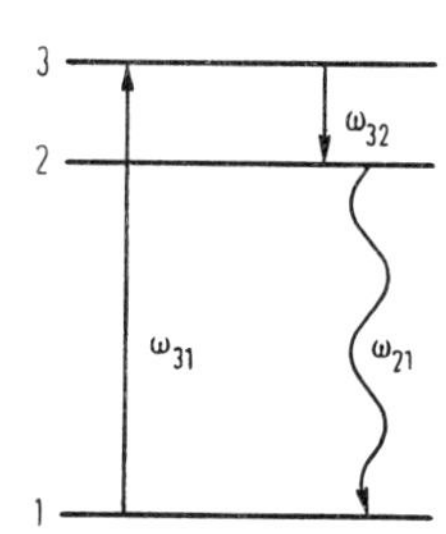

FIG. X.8. Three-level quantum system. The straight arrows indicate the pumping and signal transitions and the wavy arrow the idler transition.

different from the working pair. The fundamental principle can easily be seen by considering the example of a system of molecules with three energy levels (Fig. X.8). The secondary field has a frequency of $\omega_{31} \approx (E_3 - E_1)/\hbar$ and saturates the 3–1 transition. As a result the sign of the susceptibility of the paramagnetic changes at the frequency ω_{21} or ω_{32}, and at this frequency the substance displays amplifying properties. Section 24 provided a theoretical treatment of this effect.

Bloembergen pointed out the class of substances which can be used for multilevel masers and made a quantitative estimate that confirmed the technical feasibility of the method (Bloembergen, 1956). Bloembergen's work stimulated experimental research and soon afterwards a three-level paramagnetic maser was developed by Scovil *et al.* (1958).

Three- and four-level masers have a low noise level and can operate continuously. The promising features of continuous-operation masers in a wide range of applications such as radio astronomy, radiofrequency spectroscopy and radar caused an extensive search to be made for new working substances and circuit designs. At present the list of operating masers is quite extensive and covers the range from decimetric to millimetric waves.

i45.1. *The Conditions for achieving an Inverted Population*

In order to develop his theory of a three-level maser Bloembergen (1956) made use of the population balance equation

$$\dot{\sigma}_{nn} = \sum_k (w_{kn}\sigma_{kk} - w_{nk}\sigma_{nn}) + \sum_k W_{nk}(\sigma_{kk} - \sigma_{nn}). \tag{45.1}$$

Here w_{nk} are the transition probabilites, σ_{nn} are the level populations, and W_{nk} are the probabilities of stimulated transitions caused by the pump and the signal.

The necessary condition for achieving amplification at the 3→2 transition is inversion of the populations of these levels. It can be achieved if the working substance meets certain requirements. Let us assume that there is no signal and the pumping transition is saturated. Then in the steady state (45.1) becomes

$$\begin{aligned} &\sigma_{11} = \sigma_{33}, \\ &w_{12}\sigma_{11} + w_{32}\sigma_{33} - w_{21}\sigma_{22} - w_{23}\sigma_{22} = 0 \end{aligned} \tag{45.2}$$

and gives the ratio of the populations of interest to us:

$$\frac{\sigma_{33}}{\sigma_{22}} = \frac{\dfrac{1}{T_{21}} e^{\hbar\omega_{021}/kT} + \dfrac{1}{T_{32}} e^{-\hbar\omega_{032}/kT}}{1/T_{21} + 1/T_{32}}. \tag{45.3}$$

The ratio σ_{33}/σ_{22} should be as large as possible and should in any case be greater than unity. In accordance with (45.3) this condition in the high-temperature approximation ($h\omega_{0mn} \ll kT$), with which we shall be dealing in future, becomes

$$\frac{T_{32}}{T_{21}} \frac{\omega_{021}}{\omega_{032}} > 1. \tag{45.4}$$

In a more general form this inequality can be written as

$$\frac{T_s}{T_i} \frac{\omega_i}{\omega_s} > 1, \tag{45.5}$$

where the suffixes s and i denote quantities relating to the signal and idler transitions respectively. In future when we have pumping transition parameters we shall denote them by the suffix p.

The meaning of condition (45.5) is quite clear: the signal transition should have the lowest frequency or the longest relaxation time. If we turn to actual working substances (more about these later), two cases can be distinguished: the case of a large frequency ratio

$$\omega_i/\omega_s \gg 1,$$

$$T_s/T_i \sim 1, \tag{45.6}$$

and the case of a large relaxation ratio

$$\omega_i/\omega_s \sim 1,$$

$$T_s/T_i \gg 1. \tag{45.7}$$

We can use the balance equations (45.1) to analyse the operation of a solid-state three-level maser. However, these equations do not reflect all the wealth of physical possibilities contained in three-level systems and are useful only in special cases. We shall therefore adopt the more general approach developed in § 24, Vol. 1, and use the equation for the density matrix. This will help us in particular in answering the question of the validity of the balance scheme.

45.2. The Equations for Steady State Oscillations in a Three-level Maser. The Case of a Two-frequency Resonator

We already know from § 44 a considerable amount about the topics we shall now consider. We shall be dealing with the amplification factor, the amplification bandwidth and the oscillation criterion of a maser. From now on, however, we treat the case of a more complex quantum system containing a large number of levels and subject to the action of several electromagnetic fields with different frequencies. If we consider the system to be non-linear new effects are to be expected. Amongst these is, for example, frequency conversion (Fain *et al.*, 1961 and 1962).

In § 24, Vol. 1, we discussed the behaviour of a three-level system acted upon by known fields, and found the resultant effects. In contradistinction to this the analysis of the operation of a maser consists of solving the self-consistent problem since the high-frequency fields cannot now in principle be considered to be known. We place the three-level quantum system in a resonator having three eigenfrequencies ω_{cmn} close to the resonance frequencies (ω_{0mn}) of the quantum system. The oscillatory state of each of the modes of the resonator is determined by the signal supplied from outside and the state of the working substance. As we already know [see (41.3)], the field in the resonator corresponding to a certain type of oscillation obeys the equation

$$\ddot{q}_{mn} + \frac{\omega_{cmn}}{Q_{Lmn}}\dot{q}_{mn} + \omega_{cmn}^2 q_{mn} = -\omega_{cmn}\int_{Vc}(\boldsymbol{H}_{mn}(\boldsymbol{r})\cdot \boldsymbol{M})\,dV + F(t). \quad (45.8)$$

The magnetization $\boldsymbol{M}$ can be expressed in terms of the density matrix whose components in their turn depend on the magnitude of the applied fields:

$$\boldsymbol{M} = N\,\mathrm{Tr}\,(\hat{\sigma}\hat{\boldsymbol{\mu}}) = N\sum_{mn}\sigma_{mn}\boldsymbol{\mu}_{nm}. \quad (45.9)$$

Here N is the density of the paramagnetic particles.

In the most general statement of the problem we must find the self-consistent state at all three frequencies, but a specific feature of a maser allows us to simplify the matter slightly. The pump field that saturates the $1\to3$ transition can naturally be considered as known thus reducing the number of equations in (45.8) to two. Since the eigenfrequencies of the substance and the resonator are close to each other, one of the spectral components of the magnetization is contained in each of equations (45.8) (see § 24, Vol. 1):

$$\begin{aligned}
\boldsymbol{M}_{21} &= N\lambda_{21}\boldsymbol{\mu}_{12}\,\mathrm{e}^{-i\omega_{21}t} + \text{c.c.}\\
&= N(\lambda_{21}^{(1)}A_{21} + \lambda_{21}^{(2)}A_{23}A_{31})\,\boldsymbol{\mu}_{12}\,\mathrm{e}^{-i\omega_{21}t} + \text{c.c.}\\
\boldsymbol{M}_{32} &= N\lambda_{32}\boldsymbol{\mu}_{32}\,\mathrm{e}^{-i\omega_{32}t} + \text{c.c.}\\
&= N(\lambda_{32}^{(1)}A_{32} + \lambda_{32}^{(2)}A_{12}A_{31})\,\boldsymbol{\mu}_{32}\,\mathrm{e}^{-i\omega_{32}t} + \text{c.c.}
\end{aligned} \quad (45.10)$$

In future, in order not to encumber the calculations with insignificant details, we shall consider the working substance to have a sufficiently small volume V_s that the field in it is uniform. We shall take the position of the specimen to be the origin of the coordinates ($\boldsymbol{r} = 0$). Remembering this and writing $F(t)$ in the form (44.1) we obtain the equations

$$\ddot{q}_{21} + \frac{\omega_{c21}}{Q_{L21}}\dot{q}_{21} + \omega_{c21}^2 q_{21} = -\omega_{c21}(\boldsymbol{H}_{21}(0)\cdot\boldsymbol{M}_{21})V_s + \left(\frac{8P_1^{(21)}\omega_{c21}}{Q_{e1}^{(21)}}\right)^{1/2}\cos\omega_{21}t, \quad (45.11\,\text{a})$$

$$\ddot{q}_{32} + \frac{\omega_{c32}}{Q_{L32}}\dot{q}_{32} + \omega_{c32}^2 q_{32} = -\omega_{c32}(\boldsymbol{H}_{32}(0)\cdot\boldsymbol{M}_{32})V_s + \left(\frac{8P_1^{(32)}\omega_{c32}}{Q_{e1}^{(32)}}\right)^{1/2}\cos\omega_{32}t. \quad (45.11\,\text{b})$$

The steady state of the field is of the form

$$q_{mn} = \bar{q}_{mn} \cos(\omega_{mn} t - \varphi_{mn}), \tag{45.12}$$

and the magnetization can also be expressed in terms of the amplitudes $\bar{q}_{mn}$. From (24.9), (24.29) and (3.10) we have

$$A_{mn} = -\frac{1}{2\hbar} \zeta_{mn} \bar{q}_{mn} e^{i\varphi_{mn}}, \tag{45.13}$$

in which we have introduced the notation

$$\zeta_{mn} = (\boldsymbol{\mu}_{mn} \boldsymbol{H}_{mn} \cdot (0))\, \omega_{mn}. \tag{45.14}$$

Substitution of (45.10) and (45.12) in (45.11) with subsequent omission of the resonance terms gives a system of algebraic equations for determining the amplitudes and phases of the stationary solution:

$$\bar{q}_{21}\left[\left(\omega_{c21}^2 - \omega_{21}^2 - \frac{N\zeta_{21}^2 V_s \operatorname{Re}\lambda_{21}^{(1)}}{\hbar}\right)\cos\varphi_{21} + \frac{\omega_{c21}\omega_{21}}{Q_{\Sigma 21}}\sin\varphi_{21}\right]$$

$$= \bar{q}_{32}\frac{N\zeta_{21}\zeta_{32}V_s}{\hbar}\left[\operatorname{Re}(\lambda_{21}^{(2)}A_{31})\cos\varphi_{32} + \operatorname{Im}(\lambda_{21}^{(2)}A_{31})\sin\varphi_{32}\right]$$

$$+\left(\frac{8P_1^{(21)}\omega_{c21}}{Q_{e1}^{(21)}}\right)^{1/2}, \tag{45.15a}$$

$$\bar{q}_{21}\left[\left(\omega_{c21}^2 - \omega_{21}^2 - \frac{N\zeta_{21}^2 V_s \operatorname{Re}\lambda_{21}^{(1)}}{\hbar}\right)\sin\varphi_{21} - \frac{\omega_{c21}\omega_{21}}{Q_{\Sigma 21}}\cos\varphi_{21}\right]$$

$$= -\bar{q}_{32}\frac{N\zeta_{21}\zeta_{32}V_s}{\hbar}\left[\operatorname{Re}(\lambda_{21}^{(2)}A_{31})\sin\varphi_{32} - \operatorname{Im}(\lambda_{21}^{(2)}A_{31})\cos\varphi_{32}\right], \tag{45.15b}$$

$$\bar{q}_{32}\left[\left(\omega_{c32}^2 - \omega_{32}^2 - \frac{N\zeta_{32}^2 V_s \operatorname{Re}\lambda_{32}^{(1)}}{\hbar}\right)\cos\varphi_{32} + \frac{\omega_{c32}\omega_{32}}{Q_{\Sigma 32}}\sin\varphi_{32}\right]$$

$$= \bar{q}_{21}\frac{N\zeta_{21}\zeta_{32}V_s}{\hbar}\left[\operatorname{Re}(\lambda_{32}^{(2)}A_{31})\cos\varphi_{21} + \operatorname{Im}(\lambda_{32}^{(2)}A_{31})\sin\varphi_{21}\right]$$

$$+\left(\frac{8P_1^{(32)}\omega_{c32}}{Q_{e1}^{(32)}}\right)^{1/2}, \tag{45.15c}$$

$$\bar{q}_{32}\left[\left(\omega_{c32}^2 - \omega_{32}^2 - \frac{N\zeta_{32}^2 V_s \operatorname{Re}\lambda_{32}^{(1)}}{\hbar}\right)\sin\varphi_{32} - \frac{\omega_{c32}\omega_{32}}{Q_{\Sigma 32}}\cos\varphi_{32}\right]$$

$$= -\bar{q}_{21}\frac{N\zeta_{21}\zeta_{32}V_s}{\hbar}\left[\operatorname{Re}(\lambda_{21}^{(2)}A_{31})\sin\varphi_{21} - \operatorname{Im}(\lambda_{32}^{(2)}A_{31})\cos\varphi_{21}\right]. \tag{45.15d}$$

We can now make an obvious simplification which follows from a comparison of the width $\Delta\omega_0$ of the resonance lines of real paramagnetics with the bandwidth $\Delta\omega_c$ of the resonator: usually $\Delta\omega_0 \gg \Delta\omega_c$. It is natural to limit the discussion to the spectral range $|\omega_{cmn} - \omega_{mn}| = \Delta\omega_c$ in which the parameters of the working substance are practically independent of frequency. Let us take the optimum case for a maser of $\omega_{cmn} = \omega_{0mn}$ when, as can be seen from (24.33)–(24.36):

$$\operatorname{Re} \lambda_{mn}^{(1)} = 0, \qquad \operatorname{Im} \lambda_{mn}^{(2)} = 0. \tag{45.16}$$

Remembering (45.16) and the fact that $\omega_{mn}/\omega_{0mn} \approx 1$ we can reduce the set (45.15) to the form:

$$\bar{q}_{21}\left[(\omega_{c21}^2 - \omega_{21}^2)\cos\varphi_{21} + \frac{\omega_{21}^2}{Q_{\Sigma 21}}\sin\varphi_{21}\right] = \bar{q}_{32}\frac{N\zeta_{21}\zeta_{32}V_s\lambda_{21}^{(2)}|A_{31}|}{\hbar}\cos(\varphi_{32} - \varphi_{31}) + \left(\frac{8P_1^{(21)}\omega_{21}}{Q_{e1}^{(21)}}\right)^{1/2}, \tag{45.17a}$$

$$\bar{q}_{21}\left[(\omega_{c21}^2 - \omega_{21}^2)\sin\varphi_{21} - \frac{\omega_{21}^2}{Q_{\Sigma 21}}\cos\varphi_{21}\right] = -\bar{q}_{32}\frac{N\zeta_{21}\zeta_{32}V_s\lambda_{21}^{(2)}|A_{31}|}{\hbar}\sin(\varphi_{32} - \varphi_{31}), \tag{45.17b}$$

$$\bar{q}_{32}\left[(\omega_{c32}^2 - \omega_{32}^2)\cos\varphi_{32} + \frac{\omega_{32}^2}{Q_{\Sigma 32}}\sin\varphi_{32}\right] = \bar{q}_{21}\frac{N\zeta_{21}\zeta_{32}V_s\lambda_{32}^{(2)}|A_{31}|}{\hbar}\cos(\varphi_{21} - \varphi_{31}) + \left(\frac{8P_1^{(32)}\omega_{32}}{Q_{e1}^{(32)}}\right)^{1/2}, \tag{45.17c}$$

$$\bar{q}_{32}\left[(\omega_{c32}^2 - \omega_{32}^2)\sin\varphi_{32} - \frac{\omega_{32}^2}{Q_{\Sigma 32}}\cos\varphi_{32}\right] = -\bar{q}_{21}\frac{N\zeta_{21}\zeta_{32}V_s\lambda_{32}^{(2)}|A_{31}|}{\hbar}\sin(\varphi_{21} - \varphi_{31}). \tag{45.17d}$$

The concepts of magnetic and total Q-factors have been introduced in § 44. The form of the magnetic Q-factor of a three-level maser can be found in the process of changing from (45.11) to (45.15):

$$Q_{Mmn} = \frac{\hbar\omega_{mn}^2}{\zeta_{mn}^2 NV_s \operatorname{Im}\lambda_{mn}^{(1)}}. \tag{45.18}$$

The magnetic Q-factor Q_{Mmn} has the same sign as Im $\lambda_{mn}^{(1)}$. Remembering that $\chi''_{mn} = (|\mu_{mn}|^2/\hbar)$ Im $\lambda_{mn}^{(1)} N$ and $aV_s = 4\pi\xi$ [see (44.10)] we can rewrite (45.18) in the well-known form: $Q_{Mmn} = (4\pi\xi\chi''_{mn})^{-1}$.

If the pump power satisfies the condition $|A_{13}|^2\tau^2 \ll 1$, then χ''_{32} describes a line shape very close to Lorentzian (§ 24, Vol. 1).

A two-frequency circuit resonating at the signal and pump frequencies is quite sufficient for the normal functioning of a maser amplifier. In order to change to this case we must make one of the fields equal to zero in equations (45.17), e.g. $P_1^{(21)} = 0$ and $\bar{q}_{21} = 0$. Squaring each of equations (45.17c) and (45.17d) and then adding them we obtain

$$\bar{q}_{32}^2 \left[(\omega_{c32}^2 - \omega_{32}^2)^2 + \frac{\omega_{32}^4}{Q_{\Sigma 32}^2} \right] = \frac{8P_1^{(32)}\omega_{32}}{Q_{e1}^{(32)}}. \tag{45.19}$$

Equation (45.19) does not differ in form from (44.12) and it leads to the same expressions for the amplification factor and the bandwidth as were obtained in § 44.

The value of the magnetic Q-factor Q_{M32} depends on the pump power. It has been shown in § 24, Vol. 1, that the best pumping satisfies the conditions

$$(T\tau)^{-1} \ll |A_{13}|^2 \ll \tau^{-2}, \tag{45.20}$$

if we are dealing with a solid paramagnetic ($T \gg \tau$).

It is in this range that the maximum of |Im $\lambda_{mn}^{(1)}$| is achieved; moreover Q_{M32}, and thus the amplification also, does not alter when the pump power is altered. When the intensity of the pumping passes in either direction outside the limits of (45.20), $|Q_{M32}|$ rises.

Let us examine briefly two special cases:

(a) $T_{12} = T_{23} = T_{13} = T$; for the range (45.20) Table V.2 leads to

$$Q_{M32}^{-1} = NV_s a\, |\mu_{32}|^2\, \tau[2(\sigma_{22}^0 - \sigma_{33}^0) - (\sigma_{11}^0 - \sigma_{33}^0)]/2\hbar. \tag{45.21}$$

The magnetic Q-factor can be negative only if

$$\frac{\sigma_{22}^0 - \sigma_{33}^0}{\sigma_{11}^0 - \sigma_{33}^0} = \frac{\omega_{32}}{\omega_{31}} < \frac{1}{2}. \tag{45.22}$$

This is the case of a large frequency ratio. One of the disadvantages of the working substances under discussion can be seen in the fact that the pump frequency should be greater than double the signal frequency. In the centimetric waveband the problem of providing a pump source of the necessary power does not, in practice, arise, but as we go further into the millimetric band it becomes more and more acute. Let us reduce (45.21) to a form that is more convenient for calculations:

$$Q_{M32}^{-1} = \frac{2\pi\xi\,|\mu_{32}|^2\,\tau N(2\omega_{32} - \omega_{31})}{3kT}. \tag{45.23}$$

Here ξ is the filling factor; N is the total number of paramagnetic particles in the specimen; T is the temperature.

(b) $T_{12} \ll T_{23}, T_{13}$; then from Table V.1 we have

$$Q_{M32}^{-1} = -NV_s a\,|\mu_{32}|^2\,\tau(\sigma_{11}^0 - \sigma_{22}^0)/\hbar. \tag{45.24}$$

In principle the magnetic Q-factor at the signal frequency is negative for any relation between the frequencies of the signal and idler transitions. Quantitatively, however, $|Q_{M32}^{-1}| \sim \sigma_{11}^0 - \sigma_{22}^0 \sim \omega_{21}$, so the frequency of the idler transition should not be too low. When the conditions mentioned above are satisfied (45.24) becomes

$$Q_{M32}^{-1} = -\frac{4\pi\xi\,|\mu_{32}|^2\,\tau\omega_{21}N}{3kT}. \tag{45.25}$$

To make a numerical estimate let us substitute some typical values in (45.25): $|\mu_{32}|^2 = 10^{-40}$, $\tau = 5\times10^{-9}$ sec, $\omega_{21} = 6\times10^{10}$ sec^{-1}, $N = 10^{19}$ cm^{-3}, $T = 4$°K and $\xi = 10^{-1}$. With these figures $|Q_{M32}| \simeq 5\times10^3$. With a higher value for the filling factor the modulus of the magnetic Q-factor may be slightly lower.

We shall not deal in detail with gaseous working substances because no practical use has been made of them in radiofrequency three-level masers. The basic reason for this is that the susceptibility is too low. One interesting feature of gases should be pointed out, however. Since the dominating relaxation mechanism in a gas is connected with collisions between the molecules, all the longitudinal and transverse relaxation times can be considered to be equal: $T = \tau$. In this kind of system a change in the sign of the imaginary part of the susceptibility need not be accompanied by inversion of the populations of the corresponding pair of levels. Moreover, it can be seen from Table V.2 that provided

$$\frac{1}{4} < \frac{\omega_{021}}{\omega_{031}} < \frac{3}{4} \tag{45.26}$$

saturation of the secondary transition ensures negative susceptibility at the frequencies ω_{021} and ω_{032} simultaneously. Amplification at either of these frequencies is therefore possible in principle.

45.3. The Three-frequency Resonator. Frequency Conversion

Let us examine a three-level quantum system placed in a resonator whose eigenfrequencies are close to the frequencies of all three transitions. Let the pumping intensity, as before, satisfy the inequalities (45.20). Let us make the power of one of the external signals equal to zero, let us say $P_1^{(21)} = 0$. Under these conditions power at a frequency ω_{21} appears as a result of the combined action of the pump and the signal $P_1^{(32)}$ through the non-linear properties of the quantum system. Therefore the device may be frequency converter as well as an amplifier. We give the name of conversion factor to the quantity

$$G_{32/21} = P_2^{(21)}/P_1^{(32)}, \tag{45.27}$$

which is introduced in the same sort of way as the transmission coefficient $G_{32/32} = P_2^{(32)}/P_1^{(32)}$.

In order to find the amplification and conversion factors and bandwidths we use equations (45.17), introducing for the sake of brevity the notation

$$s = NV_s\zeta_{21}\zeta_{32}\lambda^{(2)}\,|A_{31}|\,\hbar^{-1}. \tag{45.28}$$

In future we shall omit the suffixes of the resonance components of the susceptibility and the cross-susceptibility, since for a solid under saturation conditions $\lambda_{32}^{(1)} = -\lambda_{21}^{(1)}$ and $\lambda_{32}^{(2)} = \lambda_{21}^{(2)}$ (see Tables V.1 and V.2). The permissible detunings $\Delta = \omega_{32} - \omega_{c32} = \omega_{c21} - \omega_{21}$ are naturally limited by the resonator passband, i.e. we can consider that $|\Delta| \ll \omega_{32}$ and ω_{21}. Let us transform equations (45.17), remembering to eliminate the phase terms. To do this we solve (45.17a) and (45.17b) for $\cos\varphi_{21}$ and $\sin\varphi_{21}$; then we substitute the result in the remaining pair of equations and obtain

$$\bar{q}_{32}\left[2\omega_{32}\Delta\left(1+\frac{s^2}{\omega_{21}\omega_{32}(4\Delta^2+\omega_{21}^2/Q_{\Sigma 21}^2)}\right)\cos\varphi_{32}+\frac{\omega_{32}^2}{Q_{S32}}\sin\varphi_{32}\right]$$
$$=\left(\frac{8P_1^{(32)}\omega_{32}}{Q_{e1}^{(32)}}\right)^{1/2},$$

$$\bar{q}_{32}\left[2\omega_{32}\Delta\left(1+\frac{s^2}{\omega_{21}\omega_{32}(4\Delta^2+\omega_{21}^2/Q_{\Sigma 21}^2)}\right)\sin\varphi_{32}-\frac{\omega_{32}^2}{Q_{S32}}\cos\varphi_{32}\right]=0.$$

The change to the equation in which the phases are absent

$$\bar{q}_{32}^2\left[4\omega_{32}^2\Delta^2\left(1+\frac{s^2}{\omega_{21}\omega_{32}(4\Delta^2+\omega_{21}^2/Q_{\Sigma 21}^2)}\right)^2+\frac{\omega_{32}^4}{Q_{S32}^2}\right]$$
$$=\frac{8P_1^{(32)}\omega_{32}}{Q_{e1}^{(32)}}, \tag{45.29}$$

is quite trivial.

In another variant when $P_1^{(32)} = 0$, $P_1^{(21)} \neq 0$, we obtain by proceeding in exactly the same way:

$$\bar{q}_{32}^2 \left[4\omega_{32}^2 \Delta^2 \left(1 + \frac{s^2}{\omega_{21}\omega_{32}(4\Delta^2 + \omega_{21}^2/Q_{\Sigma 21}^2)} \right)^2 + \frac{\omega_{32}^4}{Q_{S32}^2} \right]$$

$$= \frac{8P_1^{(21)}\omega_{21}}{Q_{e1}^{(21)}} \frac{s^2}{\omega_{21}^2(4\Delta^2 + \omega_{21}^2/Q_{\Sigma 21}^2)}. \tag{45.30}$$

Equations (45.17) are completely symmetrical in the frequencies ω_{32} and ω_{21}. Therefore to find the amplitudes of the oscillations $\bar{q}_{21}$ it is sufficient to make the suffixes 32 and 21 change places in (45.29) and (45.30):

$$\bar{q}_{21}^2 \left[4\omega_{21}^2 \Delta^2 \left(1 + \frac{s^2}{\omega_{21}\omega_{32}(4\Delta^2 + \omega_{32}^2/Q_{\Sigma 32}^2)} \right)^2 + \frac{\omega_{21}^4}{Q_{S21}^2} \right]$$

$$= \frac{8P_1^{(32)}\omega_{32}}{Q_{e1}^{(32)}} \frac{s^2}{\omega_{32}^2(4\Delta^2 + \omega_{32}^2/Q_{\Sigma 32}^2)} \tag{45.31}$$

when $P_1^{(21)} = 0$ and

$$\bar{q}_{21}^2 \left[4\omega_{21}^2 \Delta^2 \left(1 + \frac{s^2}{\omega_{21}\omega_{32}(4\Delta^2 + \omega_{32}^2/Q_{\Sigma 32}^2)} \right)^2 + \frac{\omega_{21}^4}{Q_{S21}^2} \right]$$

$$= \frac{8P_1^{(21)}\omega_{21}}{Q_{e1}^{(21)}} \tag{45.32}$$

when $P_1^{(32)} = 0$.

Let us compare (45.29) with the analogous expression (45.19) which is valid for a two-frequency resonator. Firstly, the place of the total Q-factors $Q_{\Sigma mn}$ is now occupied by the other quantities Q_{Smn}, which are defined by

$$Q_{S21}^{-1} = Q_{\Sigma 21}^{-1} - s^2 \frac{Q_{\Sigma 32}}{\omega_{21}^2\omega_{32}^2} \frac{\omega_{32}^2/Q_{\Sigma 32}^2}{4\Delta^2 + \omega_{32}^2/Q_{\Sigma 32}^2},$$

$$Q_{S32}^{-1} = Q_{\Sigma 32}^{-1} - s^2 \frac{Q_{\Sigma 21}}{\omega_{21}^2\omega_{32}^2} \frac{\omega_{21}^2/Q_{\Sigma 21}^2}{4\Delta^2 + \omega_{21}^2/Q_{\Sigma 21}^2}. \tag{45.33}$$

It is interesting to note that the additional term (we shall denote it by $-Q_{\pi mn}^{-1}$) is essentially negative. The exceptional case of $Q_{\Sigma 21} < 0$ (or $Q_{\Sigma 32} < 0$) relates to the oscillatory mode. Therefore the additional (compared with an ordinary maser) tuning of the resonator to the combination frequency leads to a decrease in the losses at the working frequency. We note that this new regenerative effect is not coupled with a population inversion. The quantity

$Q_{\pi mn}^{-1}$ reaches its maximum with zero detuning. Let us see what the expressions (45.33) look like in this case:

$$Q_{S21}^{-1} = Q_{\Sigma 21}^{-1} - s^2 \frac{Q_{\Sigma 32}}{\omega_{21}^2 \omega_{32}^2};$$
$$Q_{S32}^{-1} = Q_{\Sigma 32}^{-1} - s^2 \frac{Q_{\Sigma 21}}{\omega_{21}^2 \omega_{32}^2}. \tag{45.34}$$

The question arises: how great is the part played by the additional effect and under what conditions is it essential? This can be answered by comparing $Q_{\Sigma mn}^{-1}$ and $Q_{\pi mn}^{-1}$; from (45.34)

$$\frac{Q_{\pi mn}^{-1}}{Q_{\Sigma mn}^{-1}} = s^2 \frac{Q_{\Sigma 21} Q_{\Sigma 32}}{\omega_{21}^2 \omega_{32}^2} = |A_{13}|^2 \tau^2 \frac{Q_{\Sigma 21} Q_{\Sigma 32}}{|Q_{M21} Q_{M32}|}. \tag{45.35}$$

The Q-factor ratio differs little from unity even when there is strong regeneration. For example, for $G_0 = 50$ calculation gives $|Q_{\Sigma 32}/Q_{M32}| = 4$, whilst $Q_{\Sigma 21}/Q_{M21} \leqq 1$. From this we can conclude that with normal amplification factors the presence of a resonator at the difference frequency is significant only when $|A_{13}|^2 \tau^2 \gtrsim 1$. Generally, the pumping power of a paramagnetic maser satisfies the condition $|A_{13}|^2 T\tau \sim 1$, i.e. is T/τ times less. For ruby at the temperature of liquid helium $T/\tau \sim 10^6$. This estimate shows that the phenomenon of cross-susceptibility has no effect at all on the properties of a paramagnetic amplifier.

The situation is essentially different near the self-excitation threshold and in the oscillatory mode. As $|Q_M|$ approaches Q_L, and therefore $Q_\Sigma \to \infty$, the relative contribution of Q_π to the total Q-factor rises. At the oscillation threshold Q_{S32} and Q_{S21} simultaneously become zero since in accordance with (45.34) the corresponding Q-factors are connected by the equation

$$Q_{S32}^{-1}/Q_{\Sigma 32}^{-1} = Q_{S21}^{-1}/Q_{\Sigma 32}^{-1}. \tag{45.36}$$

This means that the oscillation process starts at both frequencies simultaneously (ω_{21} and ω_{32}). This kind of simultaneous appearance of instability in two coupled circuits is characteristic of parametric systems. In addition there are other facts which point to the parametric nature of the effects connected with the cross-susceptibility of a substance. There is a similarity in the structures of the basic equations we have used here and those used to describe parametric systems (Bloom and Chang, 1957). This does not mean that a maser and a parametric amplifier or oscillator are equivalent. The action of a maser is based on the negative susceptibility achieved in one of the transitions. The secondary parametric effect appears only under specific conditions of a three-frequency resonator because of the presence of cross-susceptibility, and does not depend on the sign of the population differences.

Let us return to the questions of conversion and amplification, noting that, in the light of what has been said above, the conversion is parametric. The resonance conversion factor $G_{21/32}$ can be found from (45.30) for $\Delta = 0$ by substituting $\bar{q}_{32}$ in the well-known expression $P_2^{(32)} = \omega_{32}^3 \bar{q}_{32}^2 / 2Q_{e2}^{(32)}$:

$$G_{21/32} = \frac{4/Q_{e1}^{(21)} Q_{e2}^{(32)}}{Q_{S32}^{-2}} \left(s^2 \frac{Q_{\Sigma 21}^2}{\omega_{21}^3 \omega_{32}} \right). \tag{45.37}$$

The conversion band coincides with the amplification band $\Delta\omega = \omega_{32}/Q_{S32} \simeq \omega_{32}/Q_{\Sigma 32}$. In just the same way, using (45.31), we find that

$$G_{32/21} = \frac{4/Q_{e1}^{(32)} Q_{e2}^{(21)}}{Q_{S32}^{-2}} \left(s^2 \frac{Q_{\Sigma 21}^2}{\omega_{21} \omega_{32}^3} \right) \tag{45.38}$$

for the band $\Delta\omega \simeq \omega_{32}/Q_{\Sigma 32}$. The amplification factor is

$$G_{32/32} = \frac{4/Q_{e1}^{(32)} Q_{e2}^{(32)}}{Q_{S32}^{-2}}. \tag{45.39}$$

Therefore in the system under discussion two processes take place simultaneously: maser amplification, and parametric conversion of the signal frequency. The relative efficiency of these two processes can be determined by comparing the coefficients characterizing them:

$$G_{21/32}/G_{32/32} = \frac{Q_{e1}^{(32)}}{Q_{e1}^{(21)}} s^2 \frac{Q_{\Sigma 21}^2}{\omega_{21}^3 \omega_{32}}, \tag{45.40}$$

$$G_{32/21}/G_{32/32} = \frac{Q_{e2}^{(32)}}{Q_{e2}^{(21)}} s^2 \frac{Q_{\Sigma 21}^2}{\omega_{21} \omega_{32}^3}; \tag{45.41}$$

in both cases the conversion efficiency is proportional to $Q_{\Sigma 21}^2$. By improving the quality of the resonator we can ensure that magnetic losses will be predominant,† i.e. $Q_{\Sigma 21} \approx Q_{M21}$. Substituting in (45.40) and (45.41) for Q_{M21} from (45.18) for s^2 from (45.28) and putting $Q_{e1}^{(21)} = Q_{e1}^{(32)}$ and $Q_{e2}^{(21)} = Q_{e2}^{(32)}$ we obtain

$$G_{21/32}/G_{32/32} = \frac{\omega_{32}}{\omega_{21}} |A_{13}|^2 \tau^2, \tag{45.42}$$

$$G_{32/21}/G_{32/32} = \frac{\omega_{21}}{\omega_{32}} |A_{13}|^2 \tau^2. \tag{45.43}$$

The conversion factor, as was to be expected, is proportional to the pumping power. The example discussed above confirms that, under typical conditions for a maser, the conversion effect is practically absent since

† We neglect the dielectric losses in the working substance.

$|A_{13}|^2\tau^2 \sim 10^{-6}$. Without making the amplification factor worse the pumping power can be raised until $|A_{13}|^2\tau^2 \sim 1$. Now the conversion factor is comparable with or even larger than the amplification factor. But from the practical point of view this mode of operation is not very convenient since it requires exceptionally large pumping powers. For example, a ruby maser amplifier needs a pumping power of only a few milliwatts but a converter needs kilowatts.

45.4. The Range of Applicability of the Balance Equations

When constructing the theory of a three-level amplifier we started with the general system of equations for the density matrix (24.5). This was done to enable us to estimate the possible part played by parametric effects. Generally, however, the theory of a paramagnetic maser is built up on the basis of the simpler balance equations (45.1) to which (24.5) reduces with quite definite assumptions. The latter should always be kept in mind when solving actual problems. We can now move on to finding the limits of the applicability of the balance equations.

In § 24, Vol. 1, we found the solutions of (24.5) in the stationary state. Unless we assume a stationary state the amplitude of the off-diagonal elements of the density matrix should obey the following equations:

$$\dot{\lambda}_{31} + \frac{1}{\tau_{31}}(1 - i\Delta_{31})\,\lambda_{31} = iA_{31}(\sigma_{11} - \sigma_{33}) + i(A_{32}\lambda_{21} - A_{21}\lambda_{32}),$$

$$\dot{\lambda}_{32} + \frac{1}{\tau_{32}}(1 - i\Delta_{32})\,\lambda_{32} = iA_{32}(\sigma_{22} - \sigma_{33}) + i(A_{31}\lambda_{12} - A_{12}\lambda_{31}),$$

$$\dot{\lambda}_{21} + \frac{1}{\tau_{21}}(1 - i\Delta_{21})\,\lambda_{21} = iA_{21}(\sigma_{11} - \sigma_{22}) + i(A_{23}\lambda_{31} - A_{31}\lambda_{23}), \quad (45.44)$$

where $\Delta_{mn} = (\omega_{mn} - \omega_{0\,mn})\,\tau_{mn}$.

If the high-frequency fields are limited by the condition $|A_{mn}|^2\tau_{mn}^2 \ll 1$ above, then the cross terms of the type of $A_{23}\lambda_{21}$, $A_{21}\lambda_{32}$, $A_{31}\lambda_{12}$, etc., can be omitted because of their small size. We shall assume further that τ_{mn} is much less than other times characteristic of a change in the state of the system, such as T or the decay time of the resonator $T_c = Q_L/\omega$. Then $|\dot{\lambda}_{mn}/\lambda_{mn}| \ll \tau_{mn}^{-1}$ and the derivatives of λ_{mn} in (45.44) can be neglected. As a result equations (45.44) can be considerably simplified:

$$\lambda_{31} = i\tau_{31}A_{31}(\sigma_{11} - \sigma_{33})/(1 - i\Delta_{31}),$$

$$\lambda_{32} = i\tau_{32}A_{32}(\sigma_{22} - \sigma_{33})/(1 - i\Delta_{32}), \quad (45.45)$$

$$\lambda_{21} = i\tau_{21}A_{21}(\sigma_{11} - \sigma_{22})/(1 - i\Delta_{21}).$$

Substituting (45.45) in the equations defining the change in the diagonal components (24.5) we finally obtain equations that are the same as (45.1):

$$\dot{\sigma}_{nn} = \sum_k (w_{kn}\sigma_{kk} - w_{nk}\sigma_{nn}) + \sum_k W_{kn}(\sigma_{kk} - \sigma_{nn}). \tag{45.46}$$

Here W_{kn} denotes the expressions

$$\frac{|\mu_{kn}|^2 H_{kn}^2 \tau_{kn}}{2\hbar^2(1 + \Delta_{kn}^2)} = \frac{2\,|A_{kn}|^2 \tau_{kn}}{1 + \Delta_{kn}^2}, \tag{45.47}$$

which are the same as the probabilities of stimulated transitions between the k and n levels (see, for example, Bloembergen, 1956). Recapitulating the assumptions that have been made in the process of changing to equations (45.46) we conclude that the latter are valid if

$$\tau \ll T, \quad Q_L/\omega, \quad W^{-1}. \tag{45.48}$$

We notice that the first and second inequalities from (45.48) are used when analysing non-stationary processes. We can firmly state that the balance equations are quite suitable for calculating the stationary state of a solid-state maser amplifier. The inequality $\tau \ll T$ is always satisfied in paramagnetic crystals at low temperatures. Because of this the inequality $\tau \ll W_{31}^{-1}$ is compatible with the saturation condition and in practice is always satisfied. But the balance equations are not suitable for calculations involving three-level amplifiers if the working substance is a gas. For gases $T = \tau$ and with the secondary transition saturation condition the inequality $\tau \ll W_{31}^{-1}$ is not satisfied.

When (45.48) is satisfied the expression (45.18) found above for the magnetic Q-factor can have the population difference explicitly introduced into it

$$Q_{M\,mn} = 1/4\pi\xi\chi''_{mn} = -\frac{\hbar}{4\pi\xi\tau\,|\mu_{mn}|^2(\sigma_{mm} - \sigma_{nn})\,N}. \tag{45.49}$$

For calculating the magnetic Q-factor by using (45.49) it is sufficient to know the populations of the working levels, which can easily be found by solving the balance equations (45.46). For the characteristic curve of a system we often make use not of the population difference but of the inversion factor

$$R_{mn} = -\frac{\sigma_{mm} - \sigma_{nn}}{\sigma_{mm}^0 - \sigma_{nn}^0}. \tag{45.50}$$

Let us calculate the inversion factor R_{32}, assuming that a pumping field ω_{31} and a signal ω_{32} are acting on the paramagnetic. The balance equations allowing for the fact that $w_{mn} = w_{nm}\,e^{\hbar\omega_{mn}/kT} = w_{nm}(1 + \hbar\omega_{mn}/kT)$ are of the form:

$$\dot{\sigma}_{33} = w_{13}\left(\sigma_{11} - \sigma_{33} - \frac{1}{3}\frac{\hbar\omega_{31}}{kT}\right) + w_{23}\left(\sigma_{22} - \sigma_{33} - \frac{1}{3}\frac{\hbar\omega_{32}}{kT}\right)$$

$$+ W_{31}(\sigma_{11} - \sigma_{33}) + W_{32}(\sigma_{22} - \sigma_{33}),$$

$$\dot{\sigma}_{22} = w_{23}\left(\sigma_{33} - \sigma_{22} + \frac{1}{3}\frac{\hbar\omega_{32}}{kT}\right) + w_{12}\left(\sigma_{11} - \sigma_{22} - \frac{1}{3}\frac{\hbar\omega_{21}}{kT}\right)$$

$$+ W_{32}(\sigma_{33} - \sigma_{22}), \tag{45.51}$$

$$\dot{\sigma}_{11} = w_{13}\left(\sigma_{33} - \sigma_{11} + \frac{1}{3}\frac{\hbar\omega_{31}}{kT}\right) + w_{12}\left(\sigma_{22} - \sigma_{11} + \frac{1}{3}\frac{\hbar\omega_{21}}{kT}\right)$$

$$+ W_{31}(\sigma_{33} - \sigma_{11}).$$

In the stationary state with saturation pumping

$$R_{32} = \frac{w_{12}\dfrac{\omega_{31}}{\omega_{32}} - w_2}{w_2 + W_{32}}. \tag{45.52}$$

We use w_2 to denote the probability of a non-radiative transition from level 2 to all the other levels of the system, i.e. in the present case $w_{12} + w_{23}$. If the signal is weak and we can consider that $W_{32} \ll w_2$, the expression can be simplified:

$$R_{32} = \frac{w_{12}}{w_2}\frac{\omega_{31}}{\omega_{32}} - 1. \tag{45.53}$$

Lastly, when all the relaxation transition probabilities are equal the inversion factor

$$R_{32} = \frac{1}{2}\frac{\omega_{21}}{\omega_{31}} - 1 \tag{45.54}$$

is positive only provided that $\omega_{31} > 2\omega_{32}$. We often meet this condition.

46. Four-level Masers

46.1. Some Methods of increasing the Inversion in Four-level Paramagnetics

The minimum number of levels required for the continuous operation of a paramagnetic maser is three. However, the spectra of the working substances used in practice is not as simple as this. For example, Cr^{3+} ions are characterized by fourfold, and Fe^{3+} by sixfold spin degeneracy of the ground state. The presence of spin levels which do not take part in maser operation reduces the efficiency since these levels are populated. But this disadvantage can be turned to good account if one merely slightly complicates the pumping circuit. This idea has been embodied in four-level masers.

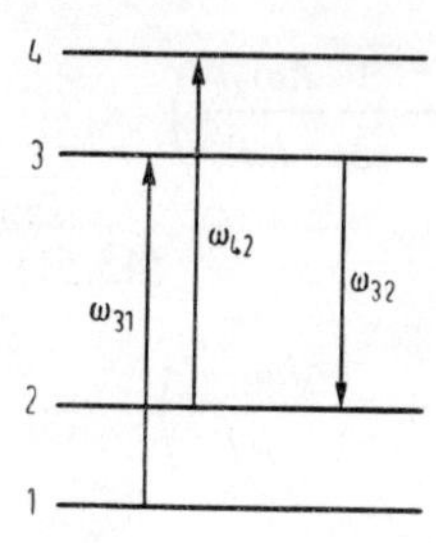

FIG. X.9a. Push–pull pumping of a four-level substance.

The most effective method of pumping in a system with four levels is shown in Fig. X.9a. Amplification takes place at the frequency ω_{32}, whilst pumping is carried out simultaneously at two frequencies: ω_{31} and ω_{42}. With this kind of two-frequency pumping the inversion factor is larger than can be achieved by the simplest three-level circuit. The principle is broadly that saturation of the 3–1 transition increases σ_{33}, whilst saturation of the 4–2 transition reduces σ_{22}. Therefore this circuit has been called a push–pull circuit. In order to find the inversion coefficient we must use equations (45.1), remembering that the suffixes have a series of four values. Of the four equations of the set (45.1) we need only two:

$$\begin{aligned}\dot{\sigma}_{22} &= w_{12}\sigma_{11} + w_{32}\sigma_{33} + w_{42}\sigma_{44} \\ &\quad - (w_{21} + w_{23} + w_{24})\,\sigma_{22} + W_{42}(\sigma_{44} - \sigma_{22}), \\ \dot{\sigma}_{44} &= w_{14}\sigma_{11} + w_{24}\sigma_{22} + w_{34}\sigma_{33} \\ &\quad - (w_{41} + w_{42} + w_{43})\,\sigma_{44} - W_{42}(\sigma_{44} - \sigma_{22}).\end{aligned} \tag{46.1}$$

We consider that both pumping fields produce saturation. This means that we can put $\sigma_{22} = \sigma_{44}$ and $\sigma_{33} = \sigma_{11}$ in all the terms of the equations except $W_{42}(\sigma_{44} - \sigma_{22})$. In the stationary state the inversion coefficient is

$$R_{32} = \frac{\omega_{31}(w_{12} + w_{14}) + \omega_{42}(w_{14} + w_{34})}{\omega_{32}(w_{12} + w_{23} + w_{34} + w_{14})} - 1. \tag{46.2}$$

This expression can be considerably simplified when all the relaxation probabilities are equal

$$R_{32} = \frac{\omega_{31} + \omega_{42}}{2\omega_{32}} - 1. \tag{46.3}$$

The sum of the pumping frequencies must be more than twice the signal frequency to achieve inversion.

The presence of two different pumping frequencies considerably complicates the design of the microwave system of the maser. In practice use is always made of a special case of the scheme shown in Fig. X. 9a; this is the case of symmetrical arrangement of the four levels when $\omega_{31} = \omega_{42} = \omega_p$. The symmetrical arrangement of the levels necessary for operation of a push–pull circuit in this way is achieved by choosing a suitable orientation of the crystal in the magnetizing field. For example, the symmetrical spectrum in a ruby occurs for $\theta = 54{\cdot}74°$ (Fig. A.III.9). The inversion coefficient when the pumping frequencies are the same is

$$R_{32} = \frac{\omega_p}{\omega_s} - 1. \tag{46.4}$$

Inversion occurs provided we can satisfy the single condition

$$\omega_p > \omega_s.$$

Stepwise pumping can be used in a four-level paramagnetic, as an alternative to push–pull pumping. In this case the levels must be arranged as

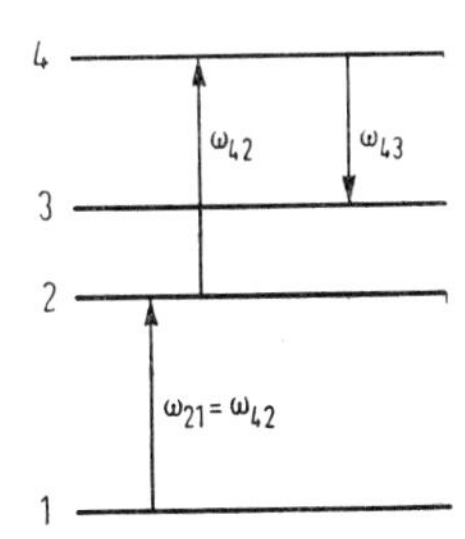

FIG. X.9b. Stepwise pumping of a four-level substance.

shown in Fig. X.9b. Frequencies ω_{21} and ω_{42} are equal, so the pumping simultaneously saturates both the secondary transitions and we can consider that $\sigma_{11} = \sigma_{22} = \sigma_{44}$. In this case the inversion coefficient is

$$R_{43} = \frac{\omega_{42}}{\omega_{43}} \frac{2w_{13} + w_{23}}{w_{13} + w_{23} + w_{34}} - 1 \tag{46.5}$$

or when all the relaxation probabilities are equal

$$R_{43} = \frac{\omega_p}{\omega_s} - 1. \tag{46.6}$$

Stepwise excitation can, of course, be carried out when the pumping frequencies are not the same but there are the same design difficulties as with push–pull pumping. The same kind of result as with stepwise pumping can be achieved by directly saturating the 4–1 transition. The advantage of the first method is that the pumping frequency is half as great. We shall not deal with other possible ways of pumping in four-level paramagnetics (Shteinshleiger *et al.*, 1962) since the calculation in this case does not differ from the above. The advantages of four-level pumping schemes over three-level ones are obvious: a given inversion coefficient can be achieved with a lower ratio ω_p/ω_s; when the pumping and signal frequencies are equal in four-level systems a high inversion coefficient, i.e. a high amplifier efficiency, is achieved.

46.2. The Part played by Cross-relaxation Processes in Masers

Push–pull pumping is one of the ways of increasing the population difference of the working levels of the substance, i.e. increasing the amplification factor and bandwidth of the maser. This justifies its popularity although from the point of view of the simplicity of tuning it is not as good as three-level systems. Another possible way of achieving the same end is increasing the concentration of the paramagnetic ions in the crystal. The second method has certain difficulties, however. First, increasing the concentration leads to a stronger spin–spin interaction which broadens the lines. As a result the saturation threshold is raised and higher pumping powers are required to reach it. But at the same time an even greater difficulty appears: as the concentration is increased the probability of cross-relaxation processes rises and hinders the creation of a negative working level population difference. Maiman (1960a) has shown experimentally the important part played by this effect in ruby crystals. Maiman investigated the dependence of the amplifier parameters on the concentration and temperature using the combined pumping method with $f_p = 23{\cdot}5$ GHz, $f_s = 9{\cdot}3$ GHz. Comparison of the experimental and theoretical data showed that the cross-relaxation processes can be ignored when the chromium ion concentration is not greater than 0·1%. But by 0·6% their effect rises so sharply that it is generally not possible to produce a negative temperature in a specimen cooled with liquid helium.

The efficiency of the cross-relaxation processes between levels 3 and 2 depends on the degree of overlap of the resonance curves corresponding to the pumping and signal transitions. The balance equations, supplemented by the cross-relaxation terms, can be written in the form (Bloembergen *et al.*, 1959)

$$\dot{\sigma}_{22} = w_{12}\left(\sigma_{11} - \sigma_{22} - \frac{1}{4}\frac{\hbar\omega_{21}}{kT}\right) + w_{23}\left(\sigma_{33} - \sigma_{22} + \frac{1}{4}\frac{\hbar\omega_{32}}{kT}\right)$$
$$+ w_{24}\left(\sigma_{44} - \sigma_{22} + \frac{1}{4}\frac{\hbar\omega_{42}}{kT}\right) + \frac{1}{4\tilde{T}_{21}}\left(\sigma_{33} - \sigma_{22} + \frac{1}{4}\frac{\hbar\omega_{32}}{kT}\right)$$
$$+ W_{24}(\sigma_{44} - \sigma_{22}), \tag{46.7}$$

$$\dot{\sigma}_{44} = w_{14}\left(\sigma_{11} - \sigma_{44} - \frac{1}{4}\frac{\hbar\omega_{41}}{kT}\right) + w_{24}\left(\sigma_{22} - \sigma_{44} - \frac{1}{4}\frac{\hbar\omega_{42}}{kT}\right)$$

$$+ w_{34}\left(\sigma_{33} - \sigma_{44} - \frac{1}{4}\frac{\hbar\omega_{43}}{kT}\right) + W_{24}(\sigma_{22} - \sigma_{44}).$$

Solving (46.7) for the case of a stationary state with saturation pumping ($\sigma_{11} = \sigma_{33}$ and $\sigma_{22} = \sigma_{44}$), assuming that all $w_{ik} = 1/(4T_1)$, we obtain

$$R_{32} = \frac{\omega_{31}/\omega_{32} - (1 + T_1/\tilde{T}_{21})}{1 + T_1/\tilde{T}_{21}}. \tag{46.8}$$

Here $\tilde{T}_{21}$ is the cross-relaxation time. In a very dilute crystal the $T_1 \ll \tilde{T}_{21}$ spin–lattice relaxation process predominates and (46.8) becomes (46.4). But if the paramagnetic ion concentration is high and the opposite situation ($T_1 \gg \tilde{T}_{21}$) obtains, then $R_{32} < 0$ whatever the pumping power. This result is not unexpected. In § 24 we have already met the case of $T_{21} \ll T_{32}, T_{31}$ and found that population inversion cannot be achieved in the transition with the lowest relaxation time.

Unlike spin–lattice relaxation, cross-relaxation is not temperature-dependent. At the temperature rises the ratio $T_1/\tilde{T}_{21}$ decreases and one expects that at a certain temperature the concentrated crystal would start to show amplifying properties. The above-mentioned crystal with 0·6 % of chromium started to amplify at 77 °K, although it absorbed at 4·2 °K. From what has been said it is clear that masers can operate at temperatures above that of liquid helium. The loss in bandwidth and amplification caused by the higher temperature is compensated for to a certain extent by the high density of the paramagnetic ions. But it must not be forgotten that as the temperature and concentration rise the relaxation time decreases, and for saturation of the secondary transition more power is required than in ordinary masers with liquid-helium cooling.

We have thus seen that the cross-relaxation processes can introduce large correction factors into the population distribution function of a paramagnetic in a strong pumping field. Above we have discussed only cross-relaxation between the levels of the working transition which is connected with the Kronig–Bouwkamp mechanism (§ 21, Vol. 1) and, as well being concentration-dependent, is strongly dependent on the degree of overlap of the resonance curves of the 2–1 and 3–1 transitions. The overlap increases as the 2 and 3 levels approach each other; thanks to this the cross-relaxation process establishes the low-frequency limit for a maser with a given working substance. When there is a symmetrical arrangement of the levels, as in the combined pumping method, there is yet another cross-relaxation mechanism operative; it consists of a transition from level 2 to level 1 with the simultaneous transition of the neighbouring spin from 3 to 4. This process, however, has no effect on the population difference $\sigma_{33} - \sigma_{22}$ of interest to us, so it was not taken into consideration when deriving (46.8).

In the examples given cross-relaxation either prevented us from obtaining a negative spin temperature at the selected transition or had no effect on it. At the same time a number of cases can be cited when inversion is increased by cross-relaxation. In particular amplification becomes possible at a frequency greater than the pumping frequency (Arams, 1960). The levels of the

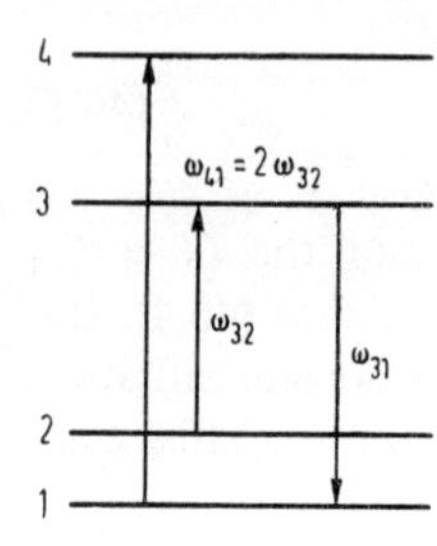

FIG. X. 9c. Arrangements of levels in ruby with $\theta = 90°$, $H_0 =$ 1675 Oe. The pumping frequency $\omega_{32} = \omega_{41}/2$ is less than the working frequency ω_{31}.

chromium ions in ruby with $H = 1675$ 0e, $\theta = 90°$ are arranged (Fig. X.9c) so that $\omega_{32} = \omega_{41}/2$. Assuming that the $3 \to 2$ transition is saturated we can find the population difference $\sigma_{33} - \sigma_{11}$.

With the arrangement shown in Fig. X.9c a cross-relaxation event consists of the transition of one ion from level 1 to level 4 and the simultaneous transition of two neighbouring ions from 3 to 2. The efficiency of this kind of process (Bloembergen *et al.*, 1959; Bogle, 1961) depends on the presence of two paramagnetic ions in the neighbourhood of the one under discussion, i.e. is proportional to the square of the concentration N. In addition these ions must be at level 3, and the probability of this is σ_{33}^2. If a suitable situation is created the transition takes place with a probability w_c and as a result the number of transitions in unit time is†

$$\sigma_{11} w_c N^2 \sigma_{33}^2.$$

Likewise for the number of reverse transitions from 4 to 1 we have

$$\sigma_{44} w_c N^2 \sigma_{22}^2.$$

Allowing for cross-relaxation the balance equations can be generalized as follows:

$$\dot{\sigma}_{11} = \sum_{k \neq 1} \left(\sigma_{kk} - \sigma_{11} + \frac{\hbar \omega_{k1}}{4kT} \right) w_{1k} + w_c N^2 (\sigma_{44}\sigma_{22}^2 - \sigma_{11}\sigma_{33}^2),$$
$$\dot{\sigma}_{44} = -\sum_{k \neq 4} \left(\sigma_{44} - \sigma_{kk} - \frac{\hbar \omega_{4k}}{4kT} \right) w_{k4} - w_c N^2 (\sigma_{44}\sigma_{22}^2 - \sigma_{11}\sigma_{33}^2). \quad (46.9)$$

† A more rigorous approach to cross-relaxation processes is developed in § 21, Vol. 1.

We shall not write out the rest of the equations of the system since their form is obvious and they are not needed for our calculations. In order of magnitude the spin–lattice terms of the right-hand side are not greater than w_{mn}. In conditions where cross-relaxation is dominant ($w_{mn} \ll N^2 w_c$) an estimate in the case of a stationary state shows that

$$|\sigma_{44}\sigma_{22}^2 - \sigma_{11}\sigma_{33}^2| \sim \frac{w_{mn}}{w_c N^2} \ll 1, \tag{46.10}$$

i.e. $\sigma_{44}\sigma_{22}^2 \approx \sigma_{11}\sigma_{33}^2$. Remembering that, due to pumping, $\sigma_{22} = \sigma_{33}$ we come to the conclusion that in addition $\sigma_{11} = \sigma_{44}$. Therefore saturation of the 2–3 transition in the presence of strong cross-relaxation leads to saturation of the $1 \to 4$ transition. It is sometimes said that these two transitions are harmonically coupled.

We can now estimate the conversion coefficient using equations (46.9) and the relations obtained between the populations:

$$R_{31} = \frac{\omega_{32}}{\omega_{31}} \frac{3w_{24} + 2w_{34} + w_{12}}{w_{12} + w_{13} + w_{24} + w_{34}} - 1. \tag{46.11}$$

The sign of the difference depends on the actual values of w_{mn}. When they are all equal

$$R_{31} = \frac{3}{2}\frac{\omega_p}{\omega_s} - 1 \tag{46.12}$$

and the inversion condition is reduced to

$$\frac{\omega_p}{\omega_s} > \frac{2}{3}.$$

We could continue to discuss examples of different kinds of cross-relaxation. In particular, processes accompanied by the re-orientation of four or even five spins may play a certain part (Roberts *et al.*, 1961). It must be pointed out, however, that the higher the order of the process the greater must be the concentration of the spins for which the cross-relaxation process is effective. Calculation of the inversion coefficients is in all cases basically the same as that described.

47. Practical Information on Resonator-type Paramagnetic Amplifiers

The advantage of masers over amplifiers of other types is their extremely low noise level. Masers are therefore used at the input of receivers when the signal being received is so weak that it cannot be distinguished in any other way from the background noise of the receiving equipment. Of course, all the inherent capabilities of a paramagnetic amplifier can be exploited only if its stability is high. The required stability cannot be provided by two-level devices, which in addition are technically more complex than multi-level ones. They have therefore not been put to practical use even in pulse receivers. Because of the practical value of continuously operating masers we shall pay most attention to them when discussing individual components and assemblies. The parameters of some amplifiers are given in Table X.1.

47.1. Paramagnetic Crystals

The paramagnetic crystal is the most important component of a maser. It determines the most important characteristics of the amplifier, such as the working and secondary frequencies and the efficiency. The design of the device, of course, is by no means unimportant, but it must not be forgotten that the best design can do no more than make the maximum use of the possibilities inherent in the working substance. Not every paramagnetic crystal will satisfy the numerous and often contradictory requirements. Let us list the basic ones with multi-level amplifiers in mind:

(a) The spin splitting of the ground orbital level of the paramagnetic ion should be greater than threefold. Therefore only ions with a spin of $S \geqslant 1$ are of use. We should point out that operation at higher orbital levels is impossible since they are not in practice populated.

(b) Transitions corresponding to a chosen signal frequency and a convenient pumping frequency should be possible between the levels. To achieve this the degeneracy of the paramagnetic ion ground state should be partly removed by the crystal field, and the initial splitting should have a value in the required range. The necessary paramagnetic spectrum structure can then be produced by selecting the magnitude and orientation of the external magnetic field. On the other hand, in this kind of situation there is mixing of the different spin states and transitions are possible between each pair of levels. For these reasons we have to exclude from consideration all crystals with cubic symmetry, since they do not have an initial splitting of the ground state. We shall discuss the spectra of paramagnetic crystals in greater detail in Appendix III.

(c) It must be easy to saturate the transitions to be used. Because of this we try to have long relaxation times T and τ. As a rule ions with an isotropic g-factor close to 2·0 have a long spin–lattice relaxation time. Deviations of the g-factor from this value indicate considerable spin–orbit interaction which, in its turn, is incompatible with a long relaxation time. It is for this reason that no one has yet been successful in using Ni^{2+} ions. The best characteristics are shown by Cr^{3+}, Fe^{3+} and Gd^{3+} which have long enough longitudinal relaxation times at liquid-helium temperatures. The transverse

TABLE X.1. PARAMETERS OF SINGLE-RESONATOR PARAMAGNETIC AMPLIFIERS

f_s (MHz)	f_p (MHz)	Working substance	Working temperature (°K)	$G^{1/2}\,\Delta f$ (MHz)	H (Oe)	Crystal orientation	Remarks	References
9,060	17,520	Lanthanum ethylsulphate	4·2		2,850		La (0·2%) is added to reduce relaxation time in idler transition	Scovil *et al.*, 1958; Singer, 1959; Troup, 1959
2,800	9,400	Potassium cobalticyanide (with Cr^{3+})	1·4	1·8	2,200	$\theta = 0°$		McWhorter and Meyer, 1958
1,373	8,000	Potassium cobalticyanide (with Cr^{3+})	< 2		1,000			Artman *et al.*, 1958
1,382	9,070	Potassium cobalticyanide (with Cr^{3+})	1·25	1·85	1,200			Autler and McAvoy, 1958
300	5,300	Potassium cobalticyanide (with Cr^{3+})	1·6	1	60		Signal circuit is superconducting winding round crystal	Kingston, 1958
400	11,800	Ruby	1·7	0·4	70	$\theta = 90°$		Wessel, 1959
1,420	11,270	Ruby	4·2	20	2,000	$\theta = 90°$		Jelley and Cooper, 1961; Jelley, 1962
9,000	24,200	Ruby	4·2	40	4,200	$\theta = 54°$		Kikuchi *et al.*, 1959
9,500	23,000	Ruby	1·4	230	4,200	$\theta = 54°$	Resonator with frequency and coupling tuning	Gianino and Dominick, 1960
9,500	23,500	Ruby	60	3·8	4,200	$\theta = 54°$		Ditchfield and Forrester, 1958; Maiman, 1960a
9,300	23,500	Ruby	78	14	4,200	$\theta = 54°$		Maiman, 1960a
10,590	9,595	Ruby	4·2		1,675	$\theta = 90°$	Cross-relaxation pumping	Arams, 1960
9,000	35,500	Rutile (Cr^{3+})	4·2	24	3,200			Gerritsen and Lewis, 1960
37,000	70,000	Rutile (Fe^{3+})	4·2	10–50	3,900			Foner *et al.*, 1961
36,000	95,000	Rutile (Fe^{3+})	4·2	10–50	13,000			
76,000	135,000	Rutile (Fe^{3+})	4·2	10–50	19,000			
53,000	78,200	Rutile (Fe^{3+})	4·2	40	5,500	$\varphi = 0$; $\theta = 73°$		Carter, 1961
70,000	120,000	Rutile (Fe^{3+})	4·2	100	4,600	$\varphi = 0$; $\theta = 45°$		Hughes and Kremenek, 1963

time τ is not temperature-dependent and it can be increased in magnetically dilute crystals. When the concentration of paramagnetic ions is low ($<1\%$) the ions are separated from each other by the diamagnetic components of the host lattice, thus reducing the spin–spin interaction between them.

(d) The density of the paramagnetic ions should be as high as possible to ensure efficient amplification. This requirement is incompatible with the preceding one and a compromise solution has to be found in each actual case. From the point of view of increasing the number of working spins it is natural to try and have paramagnetic ions with not too great a spin, since the spins at the secondary levels take no part in the operation of the maser. The Gd^{3+} ion, which has a spin of $S = \frac{7}{2}$, has eightfold splitting of the ground level and is not as good as such ions as $Cr^{3+}(S = \frac{3}{2})$ and Fe^{3+} $(S = \frac{5}{2})$.

(e) Each cause of inhomogeneous broadening of the paramagnetic resonance line makes the crystal worse. The fraction of the usefully used spins is equal to the ratio of the homogeneous line broadening $\Delta\omega_0$ to the inhomogeneous broadening $-(\Delta\omega_0)^*$. One of the reasons for inhomogeneous broadening is the hyperfine interaction between the electron and nuclear spins. The best results are therefore obtained when the nuclear spins of the paramagnetic ions themselves and of the different nuclei surrounding them are as small as possible. Inhomogeneous broadening is also caused by various kinds of imperfection in the crystal structure which bring about local variations in the magnitude and direction of the crystal field. This imposes very stringent requirements when growing the crystals, and in the subsequent processing.

(f) The crystal should be mechanically strong and chemically stable to give the maser a long life and to keep its parameters constant. This requirement is not met by crystals grown from aqueous solutions, for example gadolinium ethyl-sulphate.

(g) The dielectric losses in the crystals should be minimal. Otherwise it is not possible to use specimens of large dimensions. There are no limitations on the dielectric constant of the crystals. In a number of cases it is useful to have a large ε since it leads to the possibility of making dielectric resonators from the paramagnetic crystals themselves.

The parameters of some paramagnetic crystals used for amplification are listed in Table X.2. Ruby and rutile are the best.

An additional paramagnetic impurity can be introduced into the crystal as well as the basic one. The additional impurity is selected so that one of its resonance frequencies is close to the frequency of the idler transition of the basic paramagnetic. The cross-relaxation between the two kinds of ion leads to an effective reduction in the longitudinal relaxation time of the idler transition. As a result there is an increase in the inversion coefficient of the populations of the working pair of levels, the dynamic range of the amplifier

TABLE X.2. DATA ON PARAMAGNETIC CRYSTALS FOR MASERS

Crystal	Ion	Initial splitting (MHz)	Optimum concentration	T_1 (sec) at 4·2°K	Line width Δf (MHz)	ε	References
La ethyl-sulphate $La(C_2H_5SO_4)_3 \cdot 9H_2O$	Gd^{3+}		~0·5%	10^{-4}			Scovil *et al.*, 1958
Potassium cobalticyanide $K_3Co(CN)_6$	Cr^{3+}		~0·5%	0·2 (for 1·4°K)	~50		McWhorter and Meyer, 1958
Synthetic ruby Al_2O_3	Cr^{3+}	11·4	0·05–0·1%	2×10^{-3}	~60	10	Kikuchi *et al.*, 1959
Synthetic rutile TiO_2	Cr^{3+}	43·3	~0·1%	$2{\cdot}5 \times 10^{-3}$		130–260	Karlov and Manenkov, 1964
Synthetic rutile TiO_2	Fe^{3+}	43·3 and 81·3	~0·1%	10^{-3}	~200	130–260	Karlov and Manenkov, 1964

is extended, and there is a reduction in the time taken to restore steady amplification in a maser subjected to the action of a saturating pulsed signal. Cerium ions added to the $La(C_2H_5SO_4)_3 \cdot 9H_2O:Gd^{3+}$ crystal are an example of such an additional impurity. The anomalously short relaxation time at the $(-\frac{1}{2}) \rightarrow (-\frac{3}{2})$ transition of the gadolinium ions is caused by the presence of cross-relaxation.

47.2. Resonators

A very important component of a three-level maser is the resonator. It should be chosen so that the value of the parameter $G^{1/2} \Delta f$ is as high as possible with the lowest possible pump power. This imposes definite requirements on the properties of the resonator at the amplification and excitation frequencies. These frequencies are determined by the paramagnetic spectrum of the crystal. The difference between the two is so great that one mode of the resonator cannot be used to fulfil both functions. Pumping and amplification are achieved either with two different modes of a cavity resonator, or by the use of hybrid types of resonator. The optimum polarization of the high-frequency magnetic field to which the maximum probability of a quantum transition corresponds depends on the strength and orientation of the magnetizing field and on the levels between which the transition is occurring. For example, the transition between adjacent levels ($\Delta m = \pm 1$) of the Gd^{3+} ion in $La(C_2H_5SO_4)_3 \cdot 9H_2O:Gd^{3+}$ (Scovil *et al.*, 1958) is stimulated by an alternating field at right angles to the constant one. The transition with $\Delta m = \pm 2$ used as the secondary one is excited by a parallel field. Because of the large initial splitting of the levels in ruby the relationship of the polarization of the signal and pumping fields depends essentially on H_0 and θ. It is, of course, difficult to select a resonator with modes that are optimal for both polarizations—generally it is quite impossible. We therefore look for compromise solutions based on selecting the angle between the constant field and the resonator field. In this case the interaction of the field with the working substance is not the largest possible but the reduction can be made acceptably small. The question of the position of the crystal in the resonator can be solved simply: maximum use should be made of the regions where the antinodes of the magnetic fields of the signal and the pump intersect.

Fairly rigid requirements are imposed on the resonator Q-factor. At the pump frequency the Q-factor is not limited by any critical conditions, but the higher it is the lower the pump power required. The loaded Q-factor of the resonator at the amplification frequency is controlled by the magnetic Q-factor of the working substance. We must have $Q_L \approx -Q_M$. Generally, the magnetic Q-factor reaches 10^2–10^3 and the same value is chosen for the resonator Q-factor. It is best to have as high as possible a natural Q-factor

Q_a, adjusting the value of the loaded Q-factor by changing the coupling to the waveguide.

It is possible to use a simple cavity resonator in a maser if the pump and signal frequencies lie in the same region of the radio spectrum. The maser designed by Morris *et al.* (1959) is an example of this type, amplifying a 3 cm wave with one-centimetre pumping. Here a rectangular resonator tuned by a

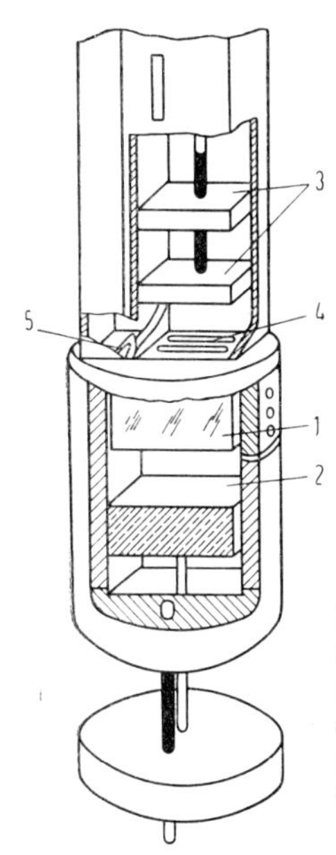

FIG. X.10. Design of paramagnetic amplifier with resonator tuning and adjustable coupling between resonator and waveguide in signal section (Gianino and Dominick, 1960): 1—paramagnetic crystal; 2—plunger for controlling volume of resonator; 3—dielectric plunger for adjusting coupling; 4—holes for coupling to signal waveguide; 5—hole for coupling to pumping waveguide.

plunger is used. The basic drawback of the design is the impossibility of independent tuning for each of the resonator frequencies used. Figure X.10 shows a design with a rectangular resonator in which, as well as resonator tuning, it is possible to adjust the coupling to the waveguide (Gianino and Dominick, 1960). By selecting the optimum position for the dielectric plunger $G^{1/2} \Delta f$ can be raised to 230 MHz with single-frequency push–pull pumping of the ruby crystal.

Cavity resonators of a simple type do not permit frequency control of each of the working modes over a wide enough range. Designs of this kind are therefore complicated to tune. Hybrid resonators have considerable advantages from this point of view. The design of Artman *et al.* (1958), which is shown in Fig. X.11, is a combination of a rectangular cavity resonator and a strip resonator. The first is tuned to the pump frequency, which is in the 3 cm band. The second is the amplifier circuit and resonates at a frequency of 1380 Mc/s. The loop is sited in the resonator so that it does not introduce any distortion of the pump field. Therefore the cavity resonance frequency does not depend upon its length. On the other hand, the movement of the plunger has no effect on the resonance conditions in the strip line formed by the loop and the wall of the resonator.

When the working frequency of the maser is very low the design principle of the amplifying circuit must be altered. For example, in u.h.f. masers

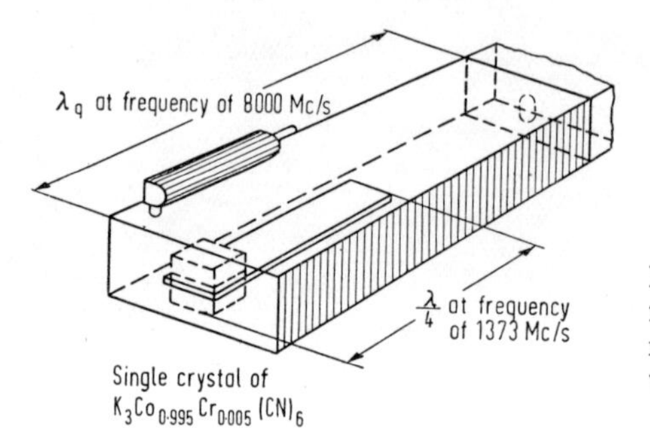

FIG. X.11. Hybrid two-frequency resonator (Artman *et al.*, 1958). The cavity resonates at the pump frequency and the wire strip at the signal frequency.

(Kingston, 1958; Wessel, 1959) the amplifying circuit is formed by the inductance of a winding round the paramagnetic specimen and a certain additional capacitance. The circuit for pumping is once again a cavity resonator. The winding may be covered with a superconducting material in order to raise the Q-factor.

If the dielectric losses in the paramagnetic specimen are small, the resonator filling factor may be as much as 1. In this case the surfaces of the crystal itself are the walls of the resonator. In practice, of course, the resonator cannot be tuned, but there is often no need for this: when operating in high-order modes the eigenfrequency spectrum is fairly dense, and modes can always be found which are close in frequency to the paramagnetic transitions. Using higher-order modes of oscillation does not lead to an excessive increase in the resonator dimensions provided that the crystal dielectric constant is large. Ruby has $\varepsilon \sim 10$ and the use of the crystal as a resonator is possible in centimetric band masers. When we move into the millimetric band it becomes natural to use the higher-order modes. Here designs with crystal resonators are better than all others.

There is no sense in separately discussing resonators for two-level masers since they are ordinary single-mode designs. Crystal resonators are also used.

47.3. Cryostats and Magnets

Liquid helium—and therefore special cryostats—are required to cool amplifying paramagnetic crystals. The laboratory glass cryostat is a double Dewar vessel. The outer vessel is filled with liquid nitrogen to reduce the

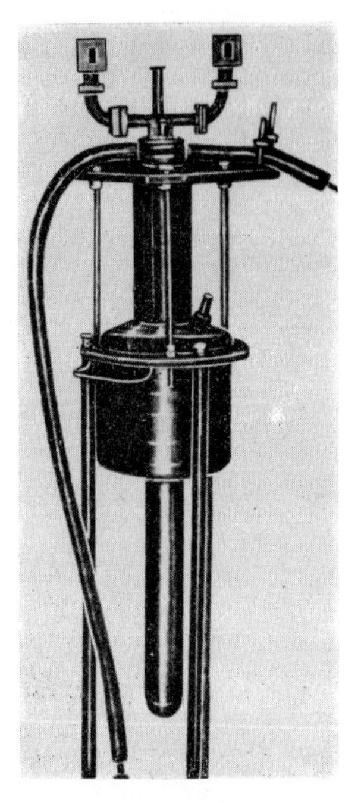

FIG. X.12. Metal cryostat containing the resonator of a paramagnetic maser amplifier (Karlova *et al.*, 1963).

FIG. X.13. Harvard 21 cm radio telescope (Bloembergen, 1961). The paramagnetic maser amplifie is sited near the reflector.

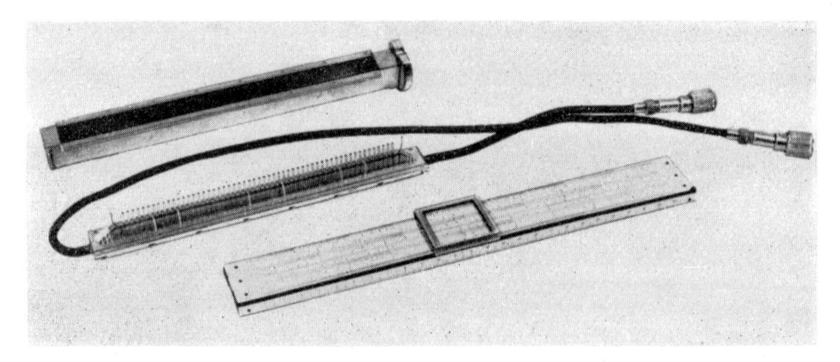

FIG. X.16. Three-centimetre travelling-wave paramagnetic amplifier (Shteinshleiger *et al.*, 1964).

flow of heat from the surroundings to the helium. Likewise the waveguides connecting the cooled components to the rest of the apparatus should be made of materials with low thermal conductivity such as German silver, stainless steel, etc. In a working maser one litre of liquid helium is lost by evaporation in a few hours.

The resonator containing the paramagnetic crystal is placed in the narrower tail section of the cryostat (Fig. X.12). The external diameter of this section determines the size of the gap between the poles of the magnet. The necessary uniformity of the field in the gap can only be provided with a magnet having a large pole diameter (up to 200 mm). This results in a large and heavy installation weighing hundreds of kilogrammes. We should add that special stabilizers are necessary to stabilize the current supplied to the electromagnet windings with an accuracy up to 10^{-5}. Such clumsy and complex devices are clearly unsuitable for devices designed to be used under conditions in the field. For example, it is best to site the maser used as the pre-amplifier of a radio telescope as close as possible to the reflector, i.e. near its focus (Fig. X.13). In this case permanent magnets with small tuning coils and shimming rings are preferable. They weigh considerably less than electromagnets.

The most radical and promising solution of the problem of the magnetic field was the use of superconducting magnets. A new era began in paramagnetic amplifier technique with the appearance of this kind of magnet in 1961 (Kunzler, Buehler, Hsu and Wernick, 1961; Kunzler, Buehler, Hsu, Matthias and Wahl, 1961). Nb_3Sn and NbZr are considered the most promising superconducting materials at present. The first of them remains superconducting to a higher temperature than any other known material (18·05°K) and is suitable for producing fields stronger than 100 kOe. If very strong fields have to be produced, the magnet is made in the form of a solenoid. For moderate fields magnets with iron cores and superconducting windings are used.

The superconducting magnet is placed with the other components of the maser that require cooling in a cryostat having a volume of several litres. The whole system is very compact and in addition requires no constant supply for the magnet windings, thus eliminating the need for field stabilization. Very strong fields are essential to the development of masers for millimetric and shorter waves. To readers who are interested in the questions of superconducting magnets we can recommend the more detailed surveys by Karasik (1962) and Boom and Livingston (1962).

Paramagnetic maser amplifiers have not yet been brought into extensive use, chiefly because of the requirement for liquid helium. Many causes dictate the necessity for cooling the working substance. The population difference is increased with a reduction in temperature, i.e. a high inversion coefficient is obtained. Furthermore, paramagnetic crystals have long spin–

lattice relaxation times at low temperatures and the secondary transition can be saturated with low pumping powers. Lastly, the lower the temperature the lower the level of the natural noise of the system. At present cryogenics is still badly adapted for operation in combination with radio equipment under field conditions, although the position is improving [closed cycle liquifiers (Higa and Wiebe, 1963)]. An investigation of the possibilities of working with more accessible coolants than liquid helium has consequently always been considered to be expedient. The inevitable losses in amplification and bandwidth can be justified by easier equipment operating conditions. These losses can be neutralized to a certain extent by changing to more concentrated paramagnetic crystals. The idea of a maser operating at a high temperature has been investigated experimentally (Maiman, 1960a; Ditchfield and Forrester, 1958). With liquid-nitrogen cooling (78°K) in a ruby with a 0·2% chromium content $G^{1/2}\,\Delta f = 14$ MHz was obtained. But it is more notable that amplification was observed even with $T = 195$°K, when solid carbon dioxide was the coolant.

47.4. Ways of increasing the Efficiency of Maser Amplifiers

The basic users of maser amplifiers in radio astronomy and radar have to deal with broad spectral bands of signals. Therefore a maser must definitely be a broad-band device. Whatever the minimum permissible band Δf_{min}, the use of amplifiers is justified only in the case $G_0^{1/2}\Delta f \gg \Delta f_{min}$. Ways of increasing the efficiency of single-resonator amplifiers can easily be listed. First, there is a limited increase in the resonator filling factor. Much depends on successful selection of the working levels and the method of pumping. The general law is that the inversion coefficient rises as the frequency ω_p/ω_s and relaxation time T_s/T_i ratios increase. The latter can be artificially increased by choosing conditions in which the part played by cross-relaxation processes is important. The push–pull method of pumping is the most efficient of those known. Lastly, much also depends on a suitable resonator design which permits not only the eigenfrequencies but also the coupling with the feeder to be adjusted within the necessary limits. The figure of merit of $G_0^{1/2}\,\Delta f = 230$ MHz (Gianino and Dominick, 1960) is clearly close to the maximum for single-resonator circuits. A further advance in the direction of broadening the bandwidth of maser amplifiers can be made with two-circuit schemes and travelling-wave systems.

48. Multi-resonator Amplifiers and Travelling-wave Amplifiers

48.1. Systems with Coupled Circuits

Because of regeneration the product of the amplification factor and the bandwidth in a single-resonator amplifier is a constant. If $\Delta\omega_0 \gg \Delta\omega_c$ the bandwidth of the amplifier is determined when $G_0 \gg 1$ by the regenerated bandwidth of the resonator, and for a reflex scheme is (see § 44)

$$\Delta\omega = 2\Delta\omega_c/G_0^{1/2}. \tag{48.1}$$

In order to increase the bandwidth whilst preserving the amplification we must reduce the loaded Q-factor of the resonator $Q_L = Q_e$ and at the same time, in the same respect, $|Q_M|$. The maximum value of the bandwidth is therefore determined by the maximum achievable inversion coefficient. If there is any necessity for a further broadening of the bandwidth of the amplifier we must change to a system of coupled resonators (Shteinshleiger, 1959 and 1962; Karlov and Prokhorov, 1963; Aseev, 1955; Cook *et al.*, 1961; Nagy and Friedman, 1963; Higa and Clauss, 1963).

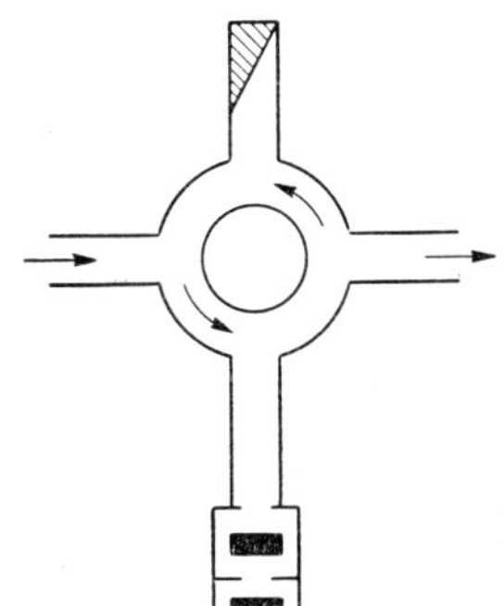

FIG. X.14a. Block diagram of a two-circuit amplifier.

The block diagram of a reflex amplifier with two coupled resonators is shown in Fig. X.14a, and its equivalent circuit in Fig. X.14b. In order to find the parameters of the amplifier it is not necessary to solve the complete set of equations as was done before. We can use the well-known results of circuit theory for this purpose. The reflection coefficient of the resonator is given by (Slater, 1950)

$$G = \left|\frac{1-\beta}{1+\beta}\right|^2, \tag{48.2}$$

in which the quantity $\beta = Z/Z_0$ is the ratio of the input impedance of the resonator to the characteristic impedance of the supply feeder. For a passive resonator

$$\beta = \frac{Q_e^{-1}}{Q_a^{-1} + iy}, \tag{48.3}$$

where y is the relative detuning.

Expression (48.3), when generalized in accordance with the rule (44.32)

$$Q_a^{-1} \to Q_a^{-1} + Q_M^{-1},$$

$$y \to y\left(1 - \frac{\omega T_2}{2Q_M}\right),$$

describes an active resonator. By proceeding as indicated we can obtain the expression for the amplification factor (44.34).

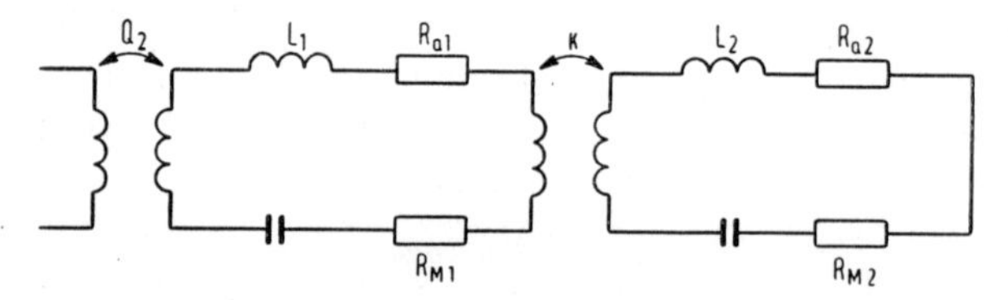

FIG. X.14b. Equivalent diagram for the two-circuit amplifier.

The technique for calculating the amplification factor of a reflex circuit with coupled resonators is similar to that described. In this case Z must be taken to be the input impedance of the succession of coupled circuits. The ratio $Z/Z_0 = \beta$ for any number of stages is given by the continued fraction (Karlov and Prokhorov, 1963)

$$\beta = Q_e^{-1}\left\{Q_{a1}^{-1} + Q_{M1}^{-1} + i\left(y_1 - \frac{Q_0}{Q_{M1}}y_0\right)\right.$$
$$\left. + \frac{k_1^2}{Q_{a2}^{-1} + Q_{M2}^{-1} + i\left(y_2 - \frac{Q_0}{Q_{M2}}y_0\right) + \dfrac{k_2^2}{Q_{a3}^{-1} + Q_{M3}^{-1} + i\left(y_3 - \dfrac{Q_0}{Q_{M3}}y_0\right) + \cdots}}\right\}^{-1} \tag{48.4}$$

Here k_l is the coupling coefficient of the lth resonator with the $(l+1)$th; Q_e is the Q-factor of the coupling of the first resonator with the supply feeder, Q_0 is the Q-factor of the resonance line of the working substance.

In particular when $k_1 = 0$ expression (48.4) gives the input impedance of a single circuit. Let us now move on to analyse a two-circuit scheme (Fig. X.14b). We shall immediately take the losses inside each of the resonators to be negligibly small and put $Q_{a1} = Q_{a2} = \infty$, which is in good agreement with actual conditions in a maser. Assuming all the circuits to be tuned to the same frequency ($y_1 = y_2 = y_0 = y$), let us examine the three possible cases that differ in the distribution of the working substance between the resonators.

(a) The input circuit contains no working substance ($Q_{M1} = \infty$). In this case the amplification factor is given by an expression that follows from (48.2) and (48.4):

$$G = \{[k^2 + Q_e^{-1}\,|Q_M|^{-1} - y^2(1+Q_0/|Q_M|)]^2 + y^2[Q_e^{-1}(1 + Q_0/|Q_M|) + |Q_M|^{-1}]^2\} \times \{[-k^2 + Q_e^{-1}|Q_M|^{-1} + y^2(1 + Q_0/|Q_M|)]^2 + y^2[Q_e^{-1}(1 + Q_0/|Q_M|) - |Q_M|^{-1}]^2\}^{-1}. \qquad (48.5)$$

In particular with zero detuning

$$G_0 = \left(\frac{k^2 + Q_e^{-1}\,|Q_M|^{-1}}{-k^2 + Q_e^{-1}\,|Q_M|^{-1}}\right)^2. \qquad (48.6)$$

It is well known that the form of the frequency response depends on the value of the coupling between the circuits (Aseev, 1955). It follows from the equation $dG/dy = 0$ that the extremes of the response curve are located at the points

$$y = 0$$

and

$$y = \pm\left[\frac{2k^2(1 + Q_0/|Q_M|) - Q_e^{-2}(1 + Q_0/|Q_M|)^2 - Q_M^{-2}}{2(1 + Q_0/|Q_M|)}\right]^{1/2}. \qquad (48.7)$$

There are three extrema, i.e. the frequency curve has two humps, if

$$k^2 \geqq k_c^2 = \frac{Q_e^{-2}(1 + Q_0/|Q_M|)^2 + Q_M^{-2}}{2(1 + Q_0/|Q_M|)}. \qquad (48.8)$$

The value $k = k_c$ is called critical or optimum coupling. With coupling equal to or less than critical the frequency curve has a maximum when $y = 0$ whose value is given by (48.6).

The amplifier is stable if

$$k^2 < Q_e^{-1}\,|Q_M|^{-1}. \tag{48.9}$$

For example, if $k^2 = Q_M^{-2}$ the stability condition is of the form $|Q_M| > Q_e$ The case of $k = k_c$ corresponds to a flat-topped resonance curve. Let us find the bandwidth of an amplifier in a case which is both simple to calculate and close to reality:

$$Q_e^{-2}(1 + Q_0/|Q_M|)^2 \simeq Q_M^{-2}.$$

The value of the critical coupling in this case is

$$k_c^2 = 1/Q_M^2(1 + Q_0/|Q_M|) \tag{48.10}$$

and with high amplification the bandwidth is

$$\Delta\omega = \omega_c\left[\left(-1+\sqrt{2}\right)\frac{k^2 - Q_e^{-1}\,|Q_M|^{-1}}{1 + Q_0/|Q_M|}\right]^{1/2}. \tag{48.11}$$

The constant quantity now, in the case of coupled circuits, is not $G_0^{1/2}\Delta\omega$, as for a single-circuit amplifier, but $G_0^{1/4}\Delta\omega$. From (48.6), (48.9) and (48.10) we obtain

$$G_0^{1/4}\Delta\omega = \frac{\omega_c}{Q_0 + |Q_M|}\left(-2 + 2\sqrt{2}\right)^{1/2} \simeq \frac{\omega_c}{Q_0 + |Q_M|}. \tag{48.12}$$

Let us now compare single-circuit and two-circuit schemes, considering their magnetic Q-factors to be equal and $|Q_M| \gg Q_0$. Then it follows from (44.36) and (48.12) that $(G_0^{1/4}\Delta\omega)_{\text{two-circ}} = \frac{1}{2}(G_0^{1/2}\Delta\omega)_{\text{single circ}}$, i.e. when the amplification factors are equal the bandwidth of a two-circuit amplifier is $0{\cdot}5\,G_0^{1/4}$ times greater.

(b) Both resonators contain a working substance, so $Q_{M1} = Q_{M2} = Q_M$. The amplification factor of this kind of device is

$$\begin{aligned} G = {} & \{[k^2 + |Q_M|^{-1}Q_e^{-1} + Q_M^{-2} - y^2(1 + Q_0/|Q_M|)^2]^2 \\ & + y^2(1 + Q_0/|Q_M|)^2\,(Q_e^{-1} + 2|Q_M|^{-1})^2\} \\ & \times \{[k^2 - |Q_M|^{-1}Q_e^{-1} + Q_M^{-2} - y^2(1 + Q_0/|Q_M|)^2]^2 \\ & + y^2(1 + Q_0/|Q_M|)^2\,(Q_e^{-1} - 2|Q_M|^{-1})^2\}^{-1}. \end{aligned} \tag{48.13}$$

The amplification at resonance is

$$G_0 = \left(\frac{k^2 + |Q_M|^{-1}\,Q_e^{-1} + Q_M^{-2}}{k^2 - |Q_M|^{-1}\,Q_e^{-1} + Q_M^{-2}}\right)^2, \tag{48.14}$$

and the stability condition is

$$|Q_M| > Q_e(1 + k^2 Q_M^2). \tag{48.15}$$

As will be shown later (§ 49), a coupling $k = |Q_M|^{-1}$ is best from the point of view of stability. With this coupling and when $G_0 \gg 1$ the parameter $G_0^{1/4} \Delta\omega$ is

$$G_0^{1/4} \Delta\omega = \frac{2\omega}{|Q_M| + Q_0} \left(\sqrt{2} - 1\right)^{1/2}. \tag{48.16}$$

This is $\sqrt{2}$ times greater than in an amplifier with a passive input resonator.

(c) The working substance is situated only in the input resonator. The amplifier has a two-maximum frequency curve, and each of the maxima has a bandwidth

$$\Delta\omega = \frac{\omega}{Q_0 + |Q_M|} G_0^{-1/2}, \tag{48.17}$$

if the coupling coefficient is $k^2 = 1/Q_M^2(1 + Q_0/|Q_M|)$. An arrangement with an active input resonator is the least efficient of all those discussed.

We have investigated arrangements of two identically tuned circuits in series. Amplifiers with untuned resonators can be similarly investigated. Data relating to some of them are given in Table X.3. (Cook *et al.*, 1961; Nagy and Friedman, 1963.) The amplifier whose design is shown in Fig. X.15

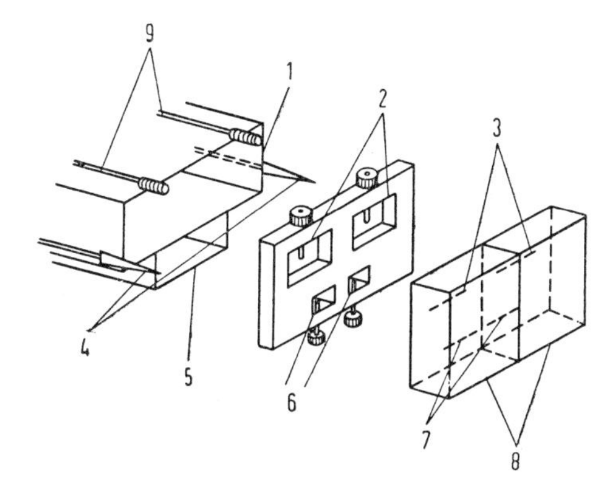

FIG. X.15. Parallel connection of two-circuit amplifiers with passive input circuits. Signal section: 1—waveguide; 2—adjustable diaphragms; 3—coupling slots. Pumping section: 5—waveguide; 6—adjustable diaphragms; 7—coupling slots. Other components: 4—iron cones for altering magnetic field; 8—ruby crystals; 9—adjustment drives.

TABLE X.3. DATA OF SOME MULTI-RESONATOR AMPLIFIERS AND TRAVELLING WAVE AMPLIFIERS

Working frequency (Mc/s)	Pump frequency (Mc/s)	Working substance	T (°K)	G (db)	Δf (Mc/s)	H (Oe)	Orientation	Remarks	References
Multi-resonator amplifiers									
8,720	22,300	Ruby	4·2	20	20	3860	$\theta = 54°$	Passive input resonator	Cook, Cross *et al.*, 1961
								Two amplifiers in parallel with passive input	Nagy and Friedman, 1963
2,388		Ruby	4·2	34	2·5			Two amplifiers in series	Higa and Clauss, 1963
Travelling-wave amplifiers									
9,650	24,000	Ruby	4·2	30	35	4100	$\theta = 55°$		Haddad and Rowe, 1962
2,390	13,000	Ruby	1·8	35	13	2530	$\theta = 90°$		DeGrasse *et al.*, 1959
10,000	23,000	Ruby	1·8	20	20	4300	$\theta = 54°$		Karlova *et al.*, 1963
6,000	23,000	Ruby	4·2	20	20	4000	$\theta = 90°$		Shteinshleiger *et al.*, 1962
2,190	12,600	Ruby	1·8	30	20	2450	$\theta = 90°$		Okwit *et al.*, 1961

has a very high efficiency (Nagy and Friedman, 1963). It has a pair of two-circuit units with a passive input resonator. These amplifying units are connected in parallel to the signal and pumping transmission lines. The difference between them is 20–30 Mc/s. In this kind of amplifier $G_0^{1/2}\Delta f \sim$ 1700 MHz has been obtained. Finally, the efficiency of an amplifier can be increased by connecting up single-resonator masers in series (Higa and Clauss, 1963). There is no difference between the calculations for this kind of arrangement and those for multi-cascade single-circuit valve amplifiers. No practical attempts have been made to construct a multi-cascade maser. This can be explained by its complexity of design and the considerable inconvenience in use.

48.2. Travelling-wave Amplifiers

The possibilities of broadening the band of a maser amplifier are not limited merely to multi-resonator arrangements. Even better results can be achieved in travelling-wave amplifiers (DeGrasse *et al.*, 1959; Butcher, 1958; DeGrasse *et al.*, 1961; Misezhnikov and Shteinshleiger, 1961; Kontorovich, 1960; Karlova *et al.*, 1963; Haddad and Rowe, 1962; Okwit and Smith, 1961; Okwit *et al.*, 1961; Tabor, 1963). The reason for the narrowing of the bandwidth of a resonator amplifier is its regenerative nature. High amplification can be achieved in travelling-wave systems thanks to the large effective interaction length; positive feedback is a harmful effect leading to self-excitation and special measures are taken to eliminate it. The bandwidth of a travelling-wave amplifier is therefore chiefly determined by the pass band of the delay system $\Delta\omega_{ds}$ and the width of the absorption line of the working substance. Generally, $\Delta\omega_0 < \Delta\omega_{ds}$, which makes it possible to tune the maser comparatively simply within the limits of $\Delta\omega_{ds}$.

A general view of a delay system for the three-centimetre band is shown in Fig. X.16.† Delay systems of the comb type are used most often since they have suitable polarization of the magnetic field and sufficient delay. The magnetic field of a wave in this kind of system is elliptically polarized in a plane at right angles to the teeth of the comb and is concentrated near its base. The amplifying crystal may therefore have a small diameter. The plane of the comb breaks up the straight wave into regions with circular polarization of a different sign. This can easily be confirmed by successively following the rotation of the vector of the magnetic field at symmetrically fixed points. For a reverse wave these regions change places. Therefore by placing the amplifying crystal on one side of the comb and an absorbing one (a non-inverted paramagnetic or ferrite) on the other amplification in only one direction can be achieved and self-excitation of the amplifier thus prevented.

† The authors are grateful for the kindness of V. B. Shteinshleiger in making Figs. X.16 and X.17a available to them.

It is quite simple to obtain expressions for the amplification factor and the bandwidth of a travelling-wave amplifier in general form. The intensity of a wave travelling along the delay system varies exponentially. Therefore the power P_0 at the section $z = z_0$ and its change in a distance D are connected by the relation

$$P_D = P_0(1 - e^{-2\alpha D}). \tag{48.18}$$

If the distance D is small, then

$$P_D \simeq 2\alpha D P_0, \tag{48.19}$$

where P_D is the power absorbed by an element of the system of length D (we shall consider that D is the period of the delay structure).

The power P_0 is the energy per unit length multiplied by the group velocity v_g. If we introduce $W_D = W_1.D$ (the energy per period of the structure), then $P_0 = (W_D/D)\, v_g$ and from (48.19) we can write

$$\alpha = P_D/2W_D v_g = \omega/2Q_\Sigma v_g. \tag{48.20}$$

The quantity Q_Σ that appears here is in effect the Q-factor for the period of the delay structure. Just as in a resonator, the Q-factor is made up of two parts: Q_a and Q_M. The first is connected with the ohmic losses and the second allows for the absorption (or amplification) caused by the working substance.

If the length of the system is l, the power amplification factor is

$$G = e^{-2\alpha l} = e^{\omega l/|Q_M| v_g} e^{-\omega l/Q_a v_g}. \tag{48.21}$$

Expressing this formula in terms of the parameters $m = c/v_g$ (the group velocity index), $n = l/\lambda$ (the length of the system expressed in wavelengths in free space), and L (the total attenuation due to ohmic losses) and changing to a logarithmic scale we obtain

$$G_{(\mathrm{db})} = 27{\cdot}3\, \frac{mn}{|Q_M|} - L_{(\mathrm{db})}. \tag{48.22}$$

The magnetic Q-factor of a cell in the delay system can be expressed in terms of the susceptibility of the working substance and the filling factor in the form (45.49)

$$Q_M = (4\pi\chi''\xi)^{-1}. \tag{48.23}$$

When the quantities χ'' and ξ are calculated we must take into consideration all the distinctive features of the system such as the actual polarization of the wave field and the vector of the matrix element of the dipole moment (see, for example, Misezhnikov and Shteinshleiger, 1961).

When the shape of the line is Lorentzian the magnetic Q-factor depends on the frequency difference as follows:

$$Q_M^{-1} = Q_{M0}^{-1} \frac{1}{1 + y^2 Q_0^2}. \tag{48.24}$$

Substituting (48.24) in (48.22) we obtain for the bandwidth at the $G_0/2$ (i.e. 3 db) level

$$\Delta\omega = \Delta\omega_0 [3/(G_{0(\mathrm{db})} + L_{(\mathrm{db})} - 3)]^{1/2}. \tag{48.25}$$

It should be noted that the use of travelling-wave amplifiers is most reasonable in the decimetric waveband where we always have $\Delta\omega_c \ll \Delta\omega_0$. There is hardly any alteration in the Q-factors of the resonators and the widths of the resonance lines as the frequency rises. This leads to the bandwidths of the resonator and the substance becoming comparable in the millimetric band. Travelling-wave amplifiers have no special advantages here.

Of the many possible types of delay systems periodic structures are the ones used in masers. They include disconnected pin combs (Figs. X.16 and X.17) and Karp systems which take the form of a series of pins connected at both ends. A certain amount of additional delay can be introduced by the

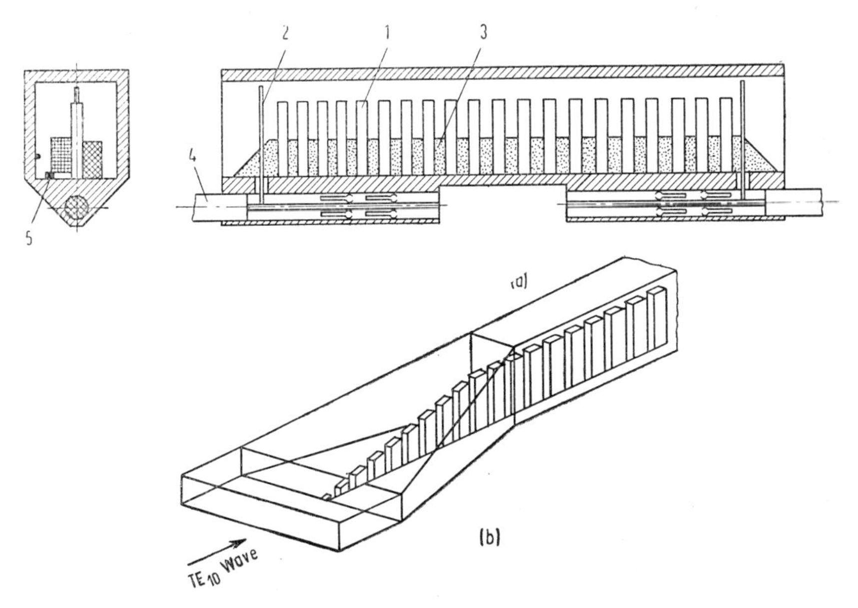

FIG. X.17. Methods of exciting a maser delay system. (a) Excitation by a probe from a coaxial line (Shteinshleiger *et al.*, 1964); 1—comb; 2—coupling probes; 3—ruby; 4—coaxial lines; 5—ferrite. (b) Waveguide excitation (Karlova *et al.*, 1963).

active material itself if it has a large ε. Spiral systems are inconvenient because the waves propagated in them do not have constant polarization.

The delay system together with the working substance and the material that acts as an isolator is put inside a section of waveguide, which is resonant (although its Q-factor is low) for the pump field. Excitation of the delay system can be achieved by a probe which is a continuation of the coaxial line, or by the method shown in Fig. X.17b. The reader will find further detailed information on the design of travelling-wave masers, and on the calculation of the delay structures, in the references quoted.

49. Non-linear and Non-stationary Phenomena in Amplifiers

49.1. *Saturation of Amplifiers*

An important practical feature of a maser amplifier is its dynamic range. Although this kind of amplifier is best used only for the reception of weak signals, it is not free from being saturated by a random strong pulse. The question arises: at what input signal power does saturation of the amplifier appear and how quickly is the normal state restored after the action of the saturating signal is over?

Saturation of the working substance by the signal being amplified occurs, as can be seen from (45.52), if

$$W_{32} \gtrsim w_2 . \tag{49.1}$$

This criterion permits us to make a direct estimate only of the strength of the saturating field in the resonator. We, however, are interested in the corresponding signal power at the input to the amplifier, which depends on the amplification. Let us investigate the effect of saturating a reflex amplifier with a high amplification $\sqrt{G_0} \gg 1$. In order to distinguish the quantities relating to the saturated amplifier we shall mark them with primes.

In the case of $|Q_M| \simeq Q_e \ll Q_a$ (44.35) gives

$$\sqrt{G_0'} = \frac{2Q_e^{-1}}{Q_e^{-1} - |Q_M'|^{-1}} . \tag{49.2}$$

We now turn to the relations (45.49) and (45.52) from which we conclude that

$$Q_M' = Q_M \frac{w_2 + W_{32}}{w_2} . \tag{49.3}$$

Substituting (49.3) in (49.2) and expressing Q_M in terms of G_0 we arrive at the equation

$$\frac{W_{32}}{w_2} + 1 = \frac{2/\sqrt{G_0} - 1}{2/\sqrt{G'_0} - 1},$$

from which it follows that

$$W_{32} = 2w_2\left(\frac{1}{\sqrt{G'_0}} - \frac{1}{\sqrt{G_0}}\right). \tag{49.4}$$

We shall express the probability of a stimulated transition W_{32} by P_1.

For $G \gg 1$ equation (44.30) becomes

$$G = -P_a/P_1.$$

If we allow for the fact that the power P_a depends on the energy of the field in the resonator W_c by the relation

$$P_a = \frac{\omega_{32}W_c}{Q_M} = -\frac{\omega_{32}W_c}{Q_e},$$

then the energy can be expressed in terms of P_1 as

$$W_c = \frac{Q_e G}{\omega_{32}} P_1. \tag{49.5}$$

On the other hand, the energy W_c depends or the amplitude of the field within the paramagnetic crystal:

$$W_c = \frac{1}{8\pi\xi}\int_{Vs} H_{32}^2\, dV \simeq \frac{H_{32}^2 V_s}{8\pi\xi}. \tag{49.6}$$

As a result (49.5) and (49.6) give the expression for H_{32}^2:

$$H_{32}^2 = \frac{8\pi Q_e}{\omega_{32}}\frac{\xi}{V_s} GP_1, \tag{49.7}$$

which must then be substituted in (45.47)

$$W_{32} = \frac{4\pi\xi Q_e |\mu_{32}|^2 \tau G'}{\hbar^2 \omega_{32} V_s} P_1. \tag{49.8}$$

Now from (49.4) and (49.8) we can write the final result

$$\frac{P_1}{b} = \left(\frac{G_0}{G'_0}\right)^{3/2} - \frac{G_0}{G'_0}. \tag{49.9}$$

If we decide upon the permissible drop in the amplification factor, we can use this expression to find the corresponding power of the saturating input signal.† A graph of G_0'/G_0 as a function of P_1/b is shown in Fig. X.18. The numerical value of the coefficient b, which it is natural to call the saturation parameter, is determined by the parameters of the amplifier under investigation in accordance with

$$b = \frac{\hbar^2 \omega_{32} V_s w_2}{2\pi\xi \, |\mu_{32}|^2 \, \tau Q_e G_0^{3/2}}. \tag{49.10}$$

For example, with the values $f_{32} = 2{\cdot}8$ GHz, $Q_e = 10^3$, $|\mu_{32}| = 10^{-20}$, $\tau = 5 \times 10^{-9}$ sec, $w_2 = 10^2$ sec^{-1}, $V_s = 1$ cm^3, $\xi = 0{\cdot}1$ and $G_0 = 100$ we have $b \simeq 5 \times 10^{-7}$ W. The power amplification factor is half its normal value when there is a power at the input of $P_1 = 4 \times 10^{-7}$ W.

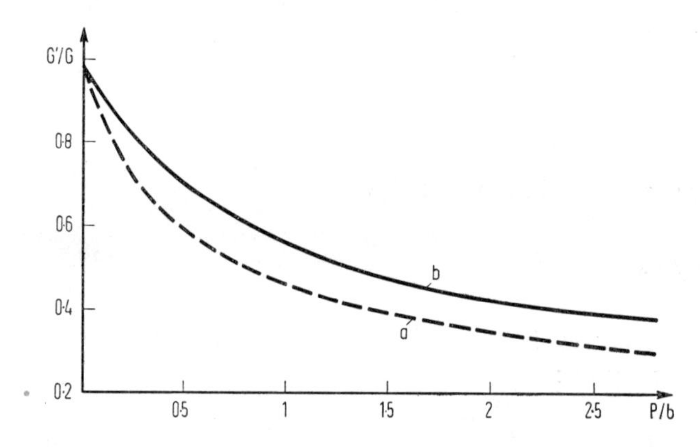

FIG. X.18. Saturation curves of maser amplifiers: *a*—resonator amplifier; *b*—travelling-wave amplifier.

The value of the saturating power, as we have seen, is proportional to $w_2 = w_{12} + w_{23}$. From the point of view of extending the dynamic range of the amplifier it is desirable to have a substance which has a high w_2. But this should not be achieved at the expense of large values of the relaxation probabilities between the working levels since in this case the inversion coefficient is low [see (45.52)]. Working substances which are best from the point of view of saturability of the amplifier are therefore those that have a short idler transition relaxation time: $T_i \ll T_s$; T_p.

† An expression similar to (49.9) is derived in Butcher's paper (1958). Its derivation, however, like the general style of his paper, is unjustifiably complicated. Karlov and Manenkov (1964) also discuss the question of the saturation of maser amplifiers.

In a travelling-wave amplifier the change in the energy flux is given by

$$\frac{dP_0}{dz} = 2\alpha' P_0 = \frac{\omega}{|Q_\Sigma|\, v_g} P_0. \tag{49.11}$$

Neglecting ohmic losses we shall put $Q = Q_M$. When there is saturation Q_M is given here by (49.3) and the problem is reduced to finding W_{32}. To do this we make use, on one hand, of (49.6) and on the other hand, of the relation $W_D = P_0 D/v_g$ [see (48.2)]. As a result, instead of (49.11), we obtain

$$\frac{dP_0}{dz} = \frac{2\alpha P_0}{1 + P_0 S}. \tag{49.12}$$

The solution of this equation is obviously

$$\ln \frac{P_0(z)}{P_1} + S[P(z) - P_1] = 2\alpha z. \tag{49.13}$$

Putting $z = l$ and noting that $P_0(l)/P_1 = G_0'$ and $2\alpha l = \ln G_0$ we obtain for $G' \gg 1$

$$\frac{P_1}{b_t} = \frac{G_0}{G_0'} \ln \frac{G_0}{G_0'}. \tag{49.14}$$

The saturation parameter b_t is given by

$$b_t = \frac{1}{SG_0} = \frac{\hbar^2 v_g A w_2}{4\pi\xi\, |\mu_{32}|^2\, \tau G_0}, \tag{49.15}$$

where $A = V_s/D$ is the cross-sectional area of the working substance.

The corresponding curve is shown in Fig. X.18. It is still difficult to compare a resonator amplifier and a travelling wave amplifier by considering (49.9) and (49.14) or the graphs corresponding to them (Fig. X.18); to do this we must know the value of the saturation parameter b which is determined by quite definite data. In general all that can be said is that when the saturation parameters b and b_t are equal a travelling-wave amplifier is saturated with stronger signals than a resonator amplifier.

By proceeding in similar manner we can find the dynamic range of an amplifier with finite ohmic losses. The difference lies in the form of the saturation parameter:

$$b_t = \frac{|Q_M|}{|Q_M| - Q_a} \frac{\hbar^2 v_g A w_2}{4\pi\xi\, |\mu_{32}|^2\, \tau G_0}. \tag{49.16}$$

49.2. The Amplification Recovery Time in Three-level Masers

Normal amplification is not restored in a maser immediately after the action of a saturation signal has ceased. A certain time is required for the relaxation processes to return the system to the stationary state. The population difference of the working levels of an out-of-balance system returns to equilibrium as

$$D = D_0 + (D_s - D_0)\,e^{-t/T_1}. \tag{49.17}$$

Here $D = (\sigma_{33} - \sigma_{22})\,N$; D_0 is the steady-state population difference; D_s is the population difference produced by the saturating signal.

The magnetic Q-factor, as is shown by (45.49), is inversely proportional to D. Substituting (45.49) in (44.35) we obtain

$$G_t^{1/2} = \frac{1 + KD}{1 - KD}, \tag{49.18}$$

in which $K = 4\pi\xi\tau\,|\mu_{32}|^2/\hbar Q_e$. Then replacing D in (49.18) by its value according to (49.17) and assuming only that the normal amplification is $G_0 \gg 1$ we obtain at a time t after the end of the saturating signal

$$G_t^{-1/2} = \frac{2G_0^{-1/2} - [2G_0^{-1/2} - (1 - KD_s)]\,e^{-t/T_1}}{2 + [G_0^{-1/2} - (1 - KD_s)]\,e^{-t/T_1}}. \tag{49.19}$$

From this the time taken to restore the amplification to the value G_t is

$$t = T_1 \ln \frac{[1 + G_t^{-1/2}]\,(1 - KD_s - 2G_0^{-1/2})}{2(G_t^{-1/2} - G_0^{-1/2})}. \tag{49.20}$$

If the saturation is complete, so that it can be taken that $D_s = 0$, then the time for restoration to a value $G_{t_\alpha}^{1/2} = \alpha G_0^{1/2}$ such that $G_{t_\alpha}^{1/2} \gg 1$ is

$$t_\alpha = T_1 \ln \left(\frac{1}{2}\,\frac{\alpha}{1 - \alpha}\,G_0^{1/2} \right). \tag{49.21}$$

Taking, for example, $\alpha = 0{\cdot}9$ and $G_0 = 100$ we obtain $t_\alpha = 3{\cdot}8T_1$.

In the case when the amplifier has been subjected to a comparatively small amount of saturation, i.e. $KD_s \approx 1$ and $1 - KD_s = 2G_s^{-1/2} \ll 1$, the time taken for restoration from the value $G_s^{1/2} = \beta G_0^{1/2}$ to $G_{t_\alpha}^{1/2} = \alpha G_0^{1/2}$ is

$$t_\alpha = T_1 \ln \frac{\alpha(1 - \beta)}{\beta(1 - \alpha)}. \tag{49.22}$$

For example, with $\alpha = 0{\cdot}9$ and $\beta = 0{\cdot}3$ (49.22) gives $t_\alpha = 3T_1$.

When calculating the restoration time of a travelling-wave amplifier we shall proceed from (48.21), in which we substitute $|Q_M|^{-1} = KD$ and allow for (49.17):

$$G_t = e^{\frac{\omega l}{v_g} K[D_0 + (D_s - D_0)\exp(-t/T_1)]}. \tag{49.23}$$

It is not difficult to change from (49.23) to the expression for the restoration time

$$t = T_1 \ln \frac{\ln G_0/G_s}{\ln G_0/G_t} \tag{49.24}$$

or, considering that $G_t = \alpha^2 G_0$ and $G_s = \beta^2 G_0$, we can write

$$t_\alpha = T_1 \ln \frac{\ln 1/\beta^2}{\ln 1/\alpha^2}. \tag{49.25}$$

To compare this with a resonator amplifier let us look at the same example: $\beta = 1/G_0^{1/2} = 0{\cdot}1$ and $\alpha = 0{\cdot}9$ leads to $t_\alpha = 3{\cdot}1 T_1$: $\beta = 0{\cdot}3$ and $\alpha = 0{\cdot}9$ corresponds to $t_\alpha = 2{\cdot}5 T_1$. From the example discussed it can be seen that a travelling-wave amplifier restores its properties slightly more quickly than a resonator amplifier with the same T_1 and G_0.

The relaxation time of the population difference of the working pair of levels (T_1) is determined by all the probabilities w_{mn} and by the pumping intensity. This question will be discussed in detail in § 51. With saturation pumping $T_1 = 2/(3w_2)$.

Therefore a reduction in the restoration time of a maser amplifier can be achieved in the same way as an increase in its dynamic range: the use of working substances with $T_i \ll T_s$ and T_p. This kind of situation can be created artificially by introducing a special impurity into the crystal to ensure intense cross-relaxation between the necessary levels.

49.3. The Stability of the Amplification Factor

We became acquainted with an obvious cause of instability of amplification in the section devoted to two-level masers. The nature of the instability in this case is trivial: the inverted system relaxes to an equilibrium state. Let us find out how the amplification factor changes in this case. In § 49.2 we discussed a similar problem in connection with a system with different equilibrium properties; in that case it had a high amplification. We now deal with a paramagnetic in a state of thermodynamic equilibrium.

If we calculate the change in the amplification factor without going beyond the bounds of the condition $G^{1/2} \gg 1$ we can put $KD \simeq 1$ in (49.18). Within these limits the quantity $G^{-1/2}$ relaxes in accordance with the same law as D. The time t_β in which the amplification factor decreases from $G_0^{1/2}$ to $\beta G_0^{1/2}$ is

$$t_\beta = T_1 \ln \frac{(1 - KD_0) - 2G_0^{-1/2}}{(1 - KD_0) - 2G_0^{-1/2}/\beta}. \tag{49.26}$$

Putting $KD_0 = -1$ ($D_0 < 0$ by definition) and $G_0 = 100$ we find that the amplification factor G_0 drops by a factor of four ($\beta = 0{\cdot}5$) in a time $t_\beta = 0{\cdot}1T$. We can also calculate the time in which the amplification drops to unity. In this case we proceed from (49.18) and (49.17), putting $G_t = 1$, which gives

$$t'_\beta = T_1 \ln \left| \frac{1 - KD_0 - 2G_0^{-1/2}}{KD_0} \right|. \tag{49.27}$$

With $G_0 \gg 1$ the time t'_β hardly depends at all on G_0; putting $KD_0 = -1$ we obtain $t'_\beta = T_1 \ln 2 = 0{\cdot}7T_1$.

In amplifiers that operate continuously the amplification is inherently constant with time. However, fluctuations of various parameters such as the pumping power and frequency, the strength of the magnetic field and the temperature cause variations in the value of the magnetic Q-factor and thus in the amplification factor. The stability of an amplifier with respect to variations of Q_M depends above all on its physical form. For example the relative instability of a single resonator amplifier calculated from (49.2) is

$$\frac{\delta G_0}{G_0} = G_0^{1/2} \left| \frac{\delta Q_M}{Q_M} \right|. \tag{49.28}$$

A two-circuit arrangement with a passive input resonator has the same instability. The instability of an arrangement with two coupled active resonators is expressed by the formula

$$\frac{\delta G_0}{G_0} = G_0^{1/2} \left| \frac{1 - k^2 Q_M^2}{1 + k^2 Q_M^2} \frac{\delta Q_M}{Q_M} \right| \tag{49.29}$$

obtained from (48.14). With $k^2 = Q_M^{-2}$ the amplifier is absolutely stable. A travelling-wave amplifier has good stability because of its low regeneration. Its relative instability is

$$\frac{\delta G_0}{G_0} = \ln G_0 \left| \frac{\delta Q_M}{Q_M} \right| \tag{49.30}$$

which is less than with a resonator amplifier.

A comparison of different amplifier arrangements according to their saturability, amplification recovery times and stability definitely indicates the advantages of travelling-wave amplifiers over resonator amplifiers. Preference should be given to working substances for which the condition $T_s/T_i \gg 1$ can be satisfied. This can be achieved, for example, by introducing an additional paramagnetic doping which interacts by cross-relaxation with the basic doping.

50. Noise in Maser Amplifiers

The continued interest displayed in maser amplifiers is chiefly due to their low noise level. We shall see a little later that the noise in maser amplifiers, at least formally, can be reduced to thermal noise; therefore by reducing the temperature at which masers operate the noise level can be made quite low. The serious disadvantage of maser amplifiers, the necessity of low temperatures for the operation, has thus been turned into an important advantage over types of amplifier. The point is that a thermionic vacuum tube (or transistor) cannot operate at a very low temperature† since the temperature determines their conductivity. Furthermore, the shot noise and the partition noise inherent in vacuum tube and transistor amplifiers are completely absent in solid-state paramagnetic maser amplifiers.‡ It is true that in the latter the pump signal is a possible source of noise. But, as Weber (1957) has shown, the pump signal that saturates one of the transitions in a three-level system has hardly any effect on the amplifier noise.

In this section we shall use the results of Chapters II and VIII of Vol 1 to determine the spectral and integral noise characteristics of maser amplifiers (the noise energy, the noise factor, the effective noise temperature). Here we shall use the so-called two-level idealization of a maser amplifier. In the case of a three-level paramagnetic amplifier we shall consider that the pump signal acting between one of the working levels and the third subsidiary one is absent, and the system of two-level molecules (spins) is in therm alequilibrium with the corresponding dissipative system (for example, the ion lattice), but at a negative temperature. This temperature T_M is called the molecular (spin) temperature. Just as in the case of a two-level molecular beam oscillator we shall not worry about the mechanism for selecting the molecules at the higher level and shall replace the beam of excited molecules passing through the resonator by a system of molecules that are in thermal equilibrium with a certain dissipative system, also at a negative temperature.

† Parametric amplifiers (e.g. with crystal diodes), which have a fairly low noise level, are now being widely used.

‡ In a molecular (e.g. ammonia molecular beam) amplifier shot noise is possible in principle. It can, however, be neglected (Gordon *et al.*, 1955; Shimoda *et al.*, 1956; Troitskii, 1958a and 1959).

50.1. The Mean Value of the Noise Energy in a Resonator

Let us calculate the energy of the electromagnetic field in the resonator of a maser amplifier in which there are excited molecules. For the sake of simplicity we shall carry out the whole analysis here, and in future, on the assumption that there is no signal to be amplified. This should not introduce any significant error since we are interested in the noise properties of amplifiers for *weak* signals.

In the calculation of the energy of the electromagnetic field in the resonator of a maser amplifier we shall make use of a set of equations for the squared quantities that is a generalization of equations (35.10)–(35.13). When there is no amplified signal the set of equations (35.5)–(35.6) has zero as its solution. The generalization of the equations mentioned above involves allowing for the dissipation of the electromagnetic field energy in the resonator in the presence of several dissipative subsystems which are in thermal equilibrium—each at its own temperature T_i. The necessity for this procedure is caused by the fact that in a maser amplifier the input and output loads and the resonator are at different temperatures.

We shall first generalize equation (35.3), which describes the time-dependent behaviour of the mean value of a certain quantity taking into account dissipation processes. We shall do this for the case of two dissipative subsystems (i = 1, 2). By analogy with equation (11.4), which defines the coefficient $\Phi_\nu^\pm$, we have

$$\Phi_\nu^\pm = \frac{\pi}{\hbar^2}\sum_{\alpha\alpha'} \delta(\mp\omega_\nu + \omega_{\alpha'\alpha}) \langle\alpha\,|F_\nu^{(1)}|\,\alpha'\rangle \langle\alpha'\,|F_\nu^{(1)}|\,\alpha\rangle\, \overset{(1)}{p}_{\alpha_1^{(1)}}\, \overset{(2)}{p}_{\alpha_1^{(2)}}$$

$$+ \frac{\pi}{\hbar^2}\sum_{\alpha\alpha'} \delta(\mp\omega_\nu + \omega_{\alpha'\alpha}) \langle\alpha\,|F_\nu^{(2)}|\,\alpha'\rangle \langle\alpha'\,|F_\nu^{(2)}|\,\alpha\rangle\, \overset{(1)}{p}_{\alpha_1^{(1)}}\, \overset{(2)}{p}_{\alpha_1^{(2)}}$$

$$+ \frac{2\pi}{\hbar^2}\sum_{\alpha\alpha'} \delta(\mp\omega_\nu + \omega_{\alpha'\alpha}) \langle\alpha\,|F_\nu^{(1)}|\,\alpha'\rangle \langle\alpha'\,|F_\nu^{(2)}|\,\alpha\rangle\, \overset{(1)}{p}_{\alpha_1^{(1)}}\, \overset{(2)}{p}_{\alpha_1^{(2)}}. \tag{50.1}$$

Here $F_\nu^{(1)}$ and $F_\nu^{(2)}$ are the operators of the dissipative subsystems whose interaction with the radiation field of the νth mode is of the form†

$$\hat{V}_\nu = -(\hat{a} + \hat{a}^+)(F_\nu^{(1)} + F_\nu^{(2)}). \tag{50.2}$$

Furthermore, $\overset{(1)}{p}_{\alpha_1^{(1)}} . \overset{(2)}{p}_{\alpha_1^{(2)}}$ is the density matrix of a dissipative system in a state α', its individual parts not interacting with each other. This means that

† We are interested in the attenuation of only the νth mode and are not discussing the case when we have to take the mutual resistance into consideration (Shteinshleiger, 1955). In this case $\Delta(\omega_0 - \omega_{\nu'}) = 1$.

we neglect the region of the dissipative system in which there is a temperature gradient.

It is easy to see that the third term in (50.1) is equal to zero since the matrix element product

$$\langle\alpha\,|F_\nu^{(1)}|\,\alpha'\rangle\,\langle\alpha'\,|F_\nu^{(2)}|\,\alpha\rangle$$

$$= \langle\alpha^{(1)}\,|F_\nu^{(1)}|\,\alpha_1^{(1)}\rangle\,\delta_{\alpha^{(2)}\alpha_1^{(2)}}\cdot\langle\alpha_1^{(2)}\,|F_\nu^{(2)}|\,\alpha^{(2)}\rangle\,\delta_{\alpha_1^{(1)}\alpha^{(1)}} \tag{50.3}$$

is finite only in the case $\alpha^{(1)} = \alpha_1^{(1)}$ and $\alpha^{(2)} = \alpha_1^{(2)}$, so that $\omega_{\alpha'\alpha} = 0$ and $\delta(\pm\omega_\nu + \omega_{\alpha'\alpha})$ is zero everywhere except at the point $\omega_\nu = 0$.

The coefficient $\Phi_\nu^\pm$ therefore consists of the sum of the terms $\Phi_i^\pm$ each of which describes the interaction of the radiation field with only a single dissipative subsystem. It can be shown that this also holds in the case of a large number of dissipative subsystems. Naturally, between the coefficients Φ_i^+ and Φ_i^- in thermal equilibrium there is a relation similar to (11.5):

$$\Phi_i^+ = \Phi_i^-\,e^{-\frac{\hbar\omega}{kT}}. \tag{50.4}$$

Proceeding from the above, the generalization of (35.3) is reduced to increasing the number of terms defining the interaction of the electromagnetic field with the dissipative subsystems. From the equation thus obtained it is easy to write a set of equations for the quadratic quantities. In the case of resonance $\omega_\nu = \omega_0$ (ω_0 is the molecular transition frequency) and, in the absence of a signal to be amplified, this set is of the form

$$\dot{R}_3 = U - T_1^{-1}(R_3 - R_3^0), \tag{50.5}$$

$$\dot{n} = -U - 2\gamma_0(n - n_0) - 2\gamma_1(n - n_1) - 2\gamma_2(n - n_2), \tag{50.6}$$

$$\dot{U} = -\frac{\alpha^2}{2}\,[\overline{R_1^2} + \overline{R_2^2} + R_3 + 2\langle nR_3\rangle] - (\gamma_L + T_2^{-1})\,U, \tag{50.7}$$

$$\frac{d}{dt}\,(\overline{R_1^2} + \overline{R_2^2}) = -\,\langle R_3U + UR_3\rangle\, - 2T_2^{-1}\left[\overline{R_1^2} + \overline{R_2^2} - \frac{N}{2}\right]. \tag{50.8}$$

Here $n = \langle\hat{a}^+\hat{a}\rangle$ is the mean number of photons in the νth mode; n_0, n_1, n_2 are the equilibrium values of n corresponding to the different dissipative subsystems (T_0, T_1, T_2):

$$n_i = \frac{1}{e^{\frac{\hbar\omega_0}{kT_i}} - 1}, \tag{50.9}$$

$\gamma_L = \gamma_0 + \gamma_1 + \gamma_2$ is the total attenuation coefficient of the electromagnetic field,† which is equal to $\gamma_L = \sum_{0,1,2} (\Phi_i^- - \Phi_i^+)$. This quantity can be expressed in terms of the corresponding Q-factors

$$\gamma_L = \frac{\omega_0}{2Q_L} = \frac{\omega_0}{2}(Q_a^{-1} + Q_{e1}^{-1} + Q_{e2}^{-1}).$$

Further R_1, R_2, R_3 are the components of the energy spin of the molecular system; R_3^0 is the equilibrium value of R_3. In the amplification mode this quantity will be positive ($T_M < 0$):

$$R_3^0 = -\frac{N}{2}\frac{e^{\frac{\hbar\omega_0}{kT_M}} - 1}{e^{\frac{\hbar\omega_0}{kT_M}} + 1}, \tag{50.10}$$

T_1, T_2 are the longitudinal and transverse relaxation times and, lastly, U denotes the quantity

$$U = \frac{i\alpha}{2}\langle\overline{R_+ a} - \overline{R_- a^+}\rangle, \tag{50.11}$$

$$R_\pm = R_1 \pm iR_2. \tag{50.12}$$

We recall that the bar on top denotes averaging over the time $2\pi/\omega_0$.

Let us find the stationary solution of the set of equations (50.5)–(50.8). We introduce the notation

$$n_{0\Sigma} = \frac{\gamma_0}{\gamma_L} n_0 + \frac{\gamma_1}{\gamma_L} n_1 + \frac{\gamma_2}{\gamma_L} n_2 \tag{50.13}$$

and note the fact that

$$\eta_0 = \frac{\alpha^2 R_3^0 T_2}{2\gamma_L}. \tag{50.14}$$

We put the derivatives in the set (50.5)–(50.8) equal to zero and neglect correlations of the kind $\langle \Delta n \Delta R_3 \rangle$. Using the same transformations as in § 36.3 we obtain

$$U = -2\frac{\eta_0}{1-\eta_0}\frac{n_{0\Sigma} + \frac{1}{2} + \frac{N}{4R_3^0}}{(\gamma_L^{-1} + T_2)}, \tag{50.15}$$

† The set of equations (35.5)–(35.6) for the linear quantities contains the total attenuation γ_L.

$$n = n_{0\Sigma} + \frac{\eta_0}{1-\eta_0} \frac{n_{0\Sigma} + \dfrac{1}{2} + \dfrac{N}{4R_3^0}}{1+\gamma_L T_2}, \tag{50.16}$$

$$R_3 = R_3^0 - \frac{\eta_0}{1-\eta_0} \frac{2\gamma_L T_1 \left(n_{0\Sigma} + \dfrac{1}{2} + \dfrac{N}{4R_3^0}\right)}{1+\gamma_L T_2}, \tag{50.17}$$

$$\overline{R_1^2} + \overline{R_2^2} = \frac{N}{2} + \frac{\eta_0}{1-\eta_0} \frac{2\gamma_L T_2 R_3^0 \left(n_{0\Sigma} + \dfrac{1}{2} + \dfrac{N}{4R_3^0}\right)}{1+\gamma_L T_2}. \tag{50.18}$$

In the expressions (50.15)–(50.18) the minimum value of $(1 - \eta_0)$ is determined by the condition $T_1 U \sim R_3^0$.

In thermal equilibrium, when the temperatures of the subsystems are equal ($T_0 = T_1 = T_2 = T_M$), it follows from the equations (50.16)–(50.18) that

$$n = n_0, \quad R_3 = R_3^0, \quad \overline{R_1^2} + \overline{R_2^2} = \frac{N}{2}. \tag{50.19}$$

This can easily be confirmed by using the expressions for n_i and R_3^0 (see (50.9) and (50.10)); it turns out that

$$n_{0\Sigma} = n_0 \quad \text{and} \quad n_{0\Sigma} + \frac{1}{2} + \frac{N}{4R_3^0} = 0. \tag{50.20}$$

We recall that the quantities n_0, R_3^0, $\overline{R_1^2} + \overline{R_2^2} = N/2$ relate to the case when there is no interaction between the field and the molecules ($\alpha^2 = 0$). Therefore when there is thermal equilibrium the interaction of the radiation field with the molecules has no effect.

In conclusion let us analyse the expression for the energy of the field in a resonator that follows from (50.16):

$$H_0 = \left(n + \frac{1}{2}\right)\hbar\omega_0 = \left(n_{0\Sigma} + \frac{1}{2}\right)\hbar\omega_0 + \frac{\eta_0}{1-\eta_0} \frac{n_{0\Sigma} + \dfrac{1}{2} + \dfrac{N}{4R_3^0}}{1+\gamma_L T_2}\hbar\omega_0. \tag{50.21}$$

The expression obtained gives the noise energy in the resonator when there is no signal. The first two terms are the energy of the thermal fluctuations of the radiation field and the energy of the zero-point oscillations of the vacuum in the empty resonator. When there is a molecular system in the resonator the thermal and zero-point fluctuations cause stimulated emission of the molecules which together with spontaneous emission makes up the third term of (50.21).

As has already been pointed out, the factor $1/(1 - \eta_0)$ characterizes the amplification of a maser amplifier† (connected, of course, with the stimulated emission). In actual amplifiers this factor reaches quite high values. Therefore in the case of a solid-state paramagnetic amplifier ($\gamma_L T_2 \ll 1$) the amplified noise is much greater than the non-amplified. On the other hand in the case of an ammonia beam molecular amplifier ($\gamma_L T_2 \gg 1$) with the not very large amplification $[1 - \eta_0]\, \gamma_L T_2 > 1$ the amplified noise is less than the non-amplified. This is because the thermal noise is amplified over a very narrow band compared with the resonator bandwidth in which, as we shall see a little later, the unamplified noise in concentrated.‡

Let us see how the change in the molecular temperature T_M affects the noise energy of the field in a resonator. When the whole system is in thermal equilibrium ($T_0 = T_1 = T_2 = T_M$), as has already been pointed out, $n = n_0$ and $H_0 = (n_0 + \frac{1}{2})\, \hbar\omega_0$. In this case the quantity U characterizing the correlation of the field coordinate $q_\nu = \sqrt{\hbar/2\omega_0}(a + a^+)$ and of the transverse component of the energy spin R_2,

$$U = -\alpha \sqrt{\frac{2\omega_0}{\hbar}} \langle R_2 q_\nu \rangle = -\frac{(e_1 \cdot A_1)}{\hbar} \langle R_2 q_\nu \rangle, \tag{50.22}$$

is equal to zero. When there is no equilibrium, e.g. in the simplest case $T_0 = T_1 = T_2 \neq T_M$, the correlation (50.22) of the states of the field and the molecules is no longer equal to zero and the energy of the field in the resonator is not $(n_0 + \frac{1}{2})\, \hbar\omega_0$. Figure X.19 shows the quantity $[\eta_0/(1 - \eta_0)] \times [n_0 + \frac{1}{2} + (N/4R_3^0)]$ defining H_0 and U as a function of the temperature T_M.

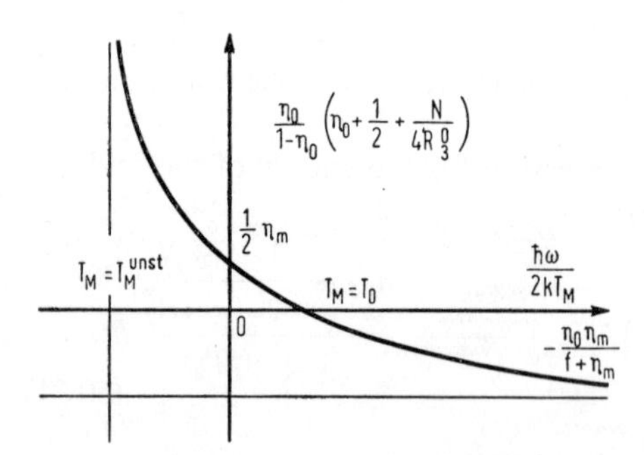

FIG. X.19. Dependence of the quantity $[\eta_0/(1 - \eta_0)]\, [n_0 + \frac{1}{2} + (N/4R_3^0)]$ on the molecular temperature T_M. T_M^{unst} is the value of T_M for which the amplifier becomes unstable, $\eta_m = \alpha^2 T_2 N/4\gamma_L$.

† An expression containing the quantity R_3, and not R_3^0 will, of course, be a more accurate one for the amplification factor of a field in a resonator. We shall, however, use the parameter η_0 defined in terms of R_3^0 since the distinction between R_3 and R_3^0 is very small.

‡ The pass band of a molecular amplifier ($\gamma_L T_2 \gg 1$) is $T_2^{-1}(1 - \eta_0)$ (see, for example, Gordon and White, 1958).

If the molecular temperature T_M is lower than the resonator temperature T_0 the correlation of the field and the molecules, in accordance with (50.22) and (50.15), is negative and $H_0 < (n_0 + \frac{1}{2})\, \hbar\omega_0$. Therefore in this case, as a result of the interaction of the molecular system and the radiation field, the latter is "cooled". In the case of $T_M > T_0$, or for $T_M < 0$, the correlation of the states is positive and the field as a result is "heated": $H_0 > (n_0 + \frac{1}{2})\, \hbar\omega_0$.

50.2. The Spectral Distribution of the Noise Energy in the Resonator of a Maser Amplifier

Let us determine the spectral distribution of the energy found in § 50.1. The mean value of the noise energy does not provide a complete description of the noise properties of maser amplifiers.

To find the noise spectrum we need the corresponding correlation function whose spectrum gives the required spectral energy distribution. The autocorrelation function Ψ_{qq} is such a function since the field energy can be expressed in terms of q_ν^2:

$$H_\omega = \tfrac{1}{2}(\omega^2 + \omega_0^2)\,(q^2)_\omega \approx \omega_0^2(q^2)_\omega, \quad \langle H_\nu \rangle = \omega_\nu^2 \langle q_\nu^2 \rangle. \tag{50.23}$$

We use the results of Chapter II, in which it is stated that the correlation function Ψ_{qq} obeys the same equations as the quantity q itself. The latter, together with the other variables p_ν, R_1, R_2, is defined by equations (35.5) and (35.6) in which the variables a^+, a must be replaced by p_ν and q_ν.†

Therefore using (35.5)–(35.7) we have:

$$\frac{dq}{dt} = -\gamma_L q + p, \tag{50.24}$$

$$\frac{dp}{dt} = -\omega_0^2 q - \gamma_L p + (e_1 \cdot A_0) R_1, \tag{50.25}$$

$$\frac{dR_1}{dt} = -\omega_0 R_2 - T_2^{-1} R_1, \tag{50.26}$$

$$\frac{dR_2}{dt} = \frac{(e_1 \cdot A_0)}{\hbar} R_3 q + \omega_0 R_1 - T_2^{-1} R_2. \tag{50.27}$$

Here, just as before, we are discussing the case of resonance, $\omega_\nu = \omega_0$. The set of equations (50.24)–(50.27) will be soluble if the quantity R_3 in it is

† We have changed from the annihilation and creation operators to the coordinate and momentum operators for the sake of convenience.

considered as known: $R_3 = R_3^0$. The general solution of this set, which is similar to (36.14), for the field coordinate is of the form

$$q = A\cos(\omega_0 t + \psi_1)\,e^{\beta_1 t} + B\cos(\omega_0 t + \psi_2)\,e^{\beta_2 t}, \tag{50.28}$$

where A, B, ψ_1, ψ_2 are constants determined from the initial conditions and β_1 and β_2 are defined by (36.15).

Therefore the autocorrelation function is of the form†

$$\Psi_{qq} = C_1\cos(\omega_0|\tau| + \varphi_1)\,e^{\beta_1|\tau|} + C_2\cos(\omega_0|\tau| + \varphi_2)\,e^{\beta_2|\tau|}. \tag{50.29}$$

The constants C_1, C_2, φ_1, φ_2 are determined by the "initial" values (for $\tau = 0$) of the function Ψ_{qq} and its derivatives $\Psi_{qq}(0)$, $\dot{\Psi}_{qq}(0)$, $\ddot{\Psi}_{qq}(0)$, $\dddot{\Psi}_{qq}(0)$:

$$C_1\cos\varphi_1 = \frac{2\omega_0^2\Psi_{qq}(0)\,\beta_2 + \omega_0^2\dot{\Psi}_{qq}(0) + \dddot{\Psi}_{qq}(0)}{-2\omega_0^2(\beta_1 - \beta_2)}, \tag{50.30}$$

$$C_2\cos\varphi_2 = -\frac{2\omega_0^2\Psi_{qq}(0)\,\beta_1 + \omega_0^2\dot{\Psi}_{qq}(0) + \dddot{\Psi}_{qq}(0)}{-2\omega_0^2(\beta_1 - \beta_2)}, \tag{50.31}$$

$$C_1\sin\varphi_1 = \frac{\omega_0^3\Psi_{qq}(0) - 3\omega_0\beta_2\dot{\Psi}_{qq}(0) + \omega_0\ddot{\Psi}_{qq}(0) - \beta_2\omega_0^{-1}\dddot{\Psi}_{qq}(0)}{-2\omega_0^2(\beta_1 - \beta_2)}, \tag{50.32}$$

$$C_2\sin\varphi_2 = -\frac{\omega_0^3\Psi_{qq}(0) - 3\omega_0\beta_1\dot{\Psi}_{qq}(0) + \omega_0\ddot{\Psi}_{qq}(0) - \beta_1\omega_0^{-1}\dddot{\Psi}_{qq}(0)^{\ddagger}}{-2\omega_0^2(\beta_1 - \beta_2)}. \tag{50.33}$$

The initial values of the derivatives of the correlation function Ψ_{qq} are determined from equations similar to (50.24)–(50.27):

$$\begin{aligned}
\dot{\Psi}_{qq} &= -\gamma_L\Psi_{qq} + \Psi_{pq},\\
\dot{\Psi}_{pq} &= -\omega_0^2\Psi_{qq} - \gamma_L\Psi_{pq} + (e_1\cdot A_1)\Psi_{R_1q},\\
\dot{\Psi}_{R_1q} &= -\omega_0\Psi_{R_2q} - T_2^{-1}\Psi_{R_1q}.
\end{aligned} \tag{50.34}$$

Differentiating the first equation of (50.34) several times and eliminating each time the derivatives of the other correlation functions for $\tau = 0$ we obtain:

$$\begin{aligned}
\Psi_{qq}(0) &= \langle q^2\rangle, \quad \dot{\Psi}_{qq}(0) = -\gamma_L\langle q^2\rangle + \langle pq\rangle,\\
\ddot{\Psi}_{qq}(0) &= -\omega_0^2\langle q^2\rangle + (e_1\cdot A_0)\langle R_1q\rangle - 2\gamma_L\langle pq\rangle,\\
\dddot{\Psi}_{qq}(0) &= 3\omega_0^2\gamma_L\langle q^2\rangle - \omega_0^2\langle pq\rangle - (e_1\cdot A_0)(2\gamma_L + T_2^{-1})\langle R_1q\rangle\\
&\quad - (e_1\cdot A_0)\omega_0\langle R_2q\rangle.
\end{aligned} \tag{50.35}$$

† The modulus of the time τ is taken since the correlation function is even.

‡ In the derivation of these expressions we have omitted terms in $\ddot{\psi}(0)$ and $\dddot{\psi}(0)$ which are ω_0^2/β_i^2 times smaller than those which have been included.

Here we have made use of the definition of the mean value of the product of the random quantities $\Psi_{xy}(0) = \langle xy \rangle$ and omitted a series of terms of the order of

$$\frac{\gamma_L^2}{\omega_0^2} \ll 1.$$

Substituting the expressions obtained for the derivatives of the correlation function (50.35) in (50.30)–(50.33) we obtain the final expression for the constants that determine the correlation function (50.29):

$$C_1 \cos \varphi_1 = \frac{(\gamma_L + \beta_2)\langle q^2 \rangle - (\boldsymbol{e}_1 \cdot \boldsymbol{A}_0)/2\omega_0 \langle R_2 q \rangle}{-(\beta_1 - \beta_2)},$$

$$C_2 \cos \varphi_2 = -\frac{(\gamma_L + \beta_1)\langle q^2 \rangle - (\boldsymbol{e}_1 \cdot \boldsymbol{A}_0)/2\omega_0 \langle R_2 q \rangle}{-(\beta_1 - \beta_2)}, \tag{50.36}$$

$$C_1 \sin \varphi_1 = \frac{-(\gamma_L + \beta_2)\langle pq \rangle + \frac{1}{2}(\boldsymbol{e}_1 \cdot \boldsymbol{A}_0)\langle R_1 q \rangle + \beta_2 (\boldsymbol{e}_1 \cdot \boldsymbol{A}_0)/2\omega_0 \langle R_2 q \rangle}{-\omega_0(\beta_1 - \beta_2)},$$

$$C_2 \sin \varphi_2 = -\frac{-(\gamma_L + \beta_1)\langle pq \rangle + \frac{1}{2}(\boldsymbol{e}_1 \cdot \boldsymbol{A}_0)\langle R_1 q \rangle + \beta_1 (\boldsymbol{e}_1 \cdot \boldsymbol{A}_0)/2\omega_0 \langle R_2 q \rangle}{-\omega_0(\beta_1 - \beta_2)}. \tag{50.37}$$

We notice that in the derivation of (50.36) we have neglected the term containing $(2\gamma_L + T_2^{-1})\langle R_1 q \rangle$ since it is small by comparison with the term containing $\omega_0 \langle R_2 q \rangle$ [see below, (50.39)], and in the derivation of (50.37) we have used the condition $\beta_{1,2}(2\gamma_L + T_2^{-1}) \ll \omega_0^2$.

The expression found for the correlation function (50.29), (50.36) and (50.37) allows us to calculate the spectral distribution of the energy in a resonator (positive frequencies)

$$\begin{aligned} 2(q^2)_\omega &= \frac{1}{\pi} \int_{-\infty}^{\infty} \Psi_{qq}(|\tau|)\, e^{i\omega\tau}\, d\tau = \frac{1}{\pi} \frac{-\beta_1 C_1 \cos \varphi_1}{(\omega_0 - \omega)^2 + \beta_1^2} \\ &+ \frac{1}{\pi} \frac{-C_1 \sin \varphi_1 (\omega_0 - \omega)}{(\omega_0 - \omega)^2 + \beta_1^2} + \frac{1}{\pi} \frac{-\beta_2 C_2 \cos \varphi_2}{(\omega_0 - \omega)^2 + \beta_2^2} \\ &+ \frac{1}{\pi} \frac{-C_2 \sin \varphi_2 (\omega - \omega_0)}{(\omega_0 - \omega)^2 + \beta_2^2}. \end{aligned} \tag{50.38}$$

In the derivation of (50.38) we have used the evenness of the correlation function $\Psi_{qq}(\tau)$ and have neglected the non-resonance terms.

In the spectrum (50.38) the second and fourth terms can be omitted since they are γ_L/ω_0 times less than the rest. This can easily be confirmed. From

(35.3) we can calculate the stationary correlations $\langle pq \rangle$ and $\langle R_1 q \rangle$ (averaged over the period $2\pi/\omega_0$):

$$\langle pq \rangle = \frac{(e_1 \cdot A_0)}{2\omega_0} \langle R_2 q \rangle,$$

$$\langle R_1 q \rangle = \frac{\eta_0 \gamma_L}{\omega_0 (1 + \gamma_L T_2)} \langle R_2 q \rangle, \tag{50.39}$$

and substitute them in (50.37). As a result we obtain

$$C_1 \sin \varphi_1 = -C_2 \sin \varphi_2 = \frac{\gamma_L}{\omega_0} \frac{(1 - \eta_0) + \gamma_L T_2}{(\beta_1 - \beta_2)(1 + \gamma_L T_2)} \frac{(e_1 \cdot A_0)}{2\omega_0} \langle R_2 q \rangle. \tag{50.40}$$

Let us now compare (50.40) and (50.36). It is easy to see that the terms defined by (50.40) in the frequency band $(\omega_0 - \omega) \sim \beta_1, \beta_2$ are really γ_L/ω_0 times smaller than the terms defined by (50.36). We notice that there is no sense in discussing greater detunings than β_1 and β_2 since the whole of our analysis is valid in the range

$$(\omega_0 - \omega) \sim \gamma_L \text{ (or } T_2^{-1}).$$

Before moving on to discuss the noise spectra of maser amplifiers let us examine the case when there is no interaction between the molecules and the radiation field. It follows from (50.38), in which all the terms except for the third equal zero ($\langle R_2 q \rangle = 0$ and $\beta_1 = -T_2^{-1}$, $\beta_2 = -\gamma_L$), that

$$2(q^2)_\omega = \frac{1}{\pi} \frac{\gamma_L \langle q^2 \rangle}{(\omega_0 - \omega)^2 + \gamma_L^2}, \tag{50.41}$$

where $\langle q^2 \rangle$ is defined from (50.21) and (50.23) as

$$\langle q^2 \rangle = \left(n_{0\Sigma} + \frac{1}{2} \right) \frac{\hbar}{\omega_0}.$$

Finally, let us examine the spectral density of the noise energy in the resonator of a maser amplifier. For the sake of simplicity we shall limit ourselves to the two extreme values of the parameter $\gamma_L T_2$: $\gamma_L T_2 \ll 1$ (a solid-state paramagnetic amplifier), and $\gamma_L T_2 \gg 1$ (a molecular beam amplifier).

In these two cases the roots β_1, β_2 of the characteristic equation take the simple form [see (36.14)†]:

$$\begin{aligned} \beta_2 &= -(T_2^{-1} + \gamma_L) + \frac{\gamma_L (1 - \eta_0)}{1 + \gamma_L T_2}, \\ \beta_1 &= -\frac{\gamma_L (1 - \eta_0)}{1 + \gamma_L T_2}. \end{aligned} \tag{50.42}$$

† We recall that for an amplifier $\eta_0 \leq 1$.

Substituting in (50.38) the values calculated from (50.21), (50.23), (50.15), (50.22) and (50.42) of the mean values and coefficients

$$\gamma_L + \beta_1 = \frac{\gamma_L(\eta_0 + \gamma_L T_2)}{1 + \gamma_L T_2},$$
$$\gamma_L + \beta_2 = -T_2^{-1} + \frac{\gamma_L(1 - \eta_0)}{1 + \gamma_L T_2}, \tag{50.43}$$

$$\frac{(e_1 \cdot A_0)}{2\omega_0}\langle R_2 q\rangle = \frac{\hbar}{\omega_0}\frac{\eta_0}{1 - \eta_0}\frac{\gamma_L\left(n_{0\Sigma} + \frac{1}{2} + \frac{N}{4R_3^0}\right)}{1 + \gamma_L T_2}, \tag{50.44}$$

$$\langle q^2\rangle = \frac{\hbar}{\omega_0}\left(n_{0\Sigma} + \frac{1}{2} + \frac{\eta_0}{1 - \eta_0}\frac{n_{0\Sigma} + \frac{1}{2} + \frac{N}{4R_3^0}}{1 + \gamma_L T_2}\right), \tag{50.45}$$

we obtain the spectral distribution of the energy in the case of a paramagnetic amplifier ($\gamma_L T_2 \ll 1$)

$$\frac{H_\omega}{\omega_0^2} = 2(q^2)_\omega = \frac{1}{\pi}\frac{\gamma_L(1 - \eta_0)\langle q^2\rangle}{(\omega_0 - \omega)^2 + \gamma_L^2(1 - \eta_0)^2} \tag{50.46}$$
$$= \frac{1}{\pi}\frac{\gamma_L(1 - \eta_0)\left[n_{0\Sigma} + \frac{1}{2} + \frac{\eta_0}{1 - \eta_0}\left(n_{0\Sigma} + \frac{1}{2} + \frac{N}{4R_3^0}\right)\right]\frac{\hbar}{\omega_0}}{(\omega_0 - \omega)^2 + \gamma_L^2(1 - \eta_0)^2}.$$

In this we have omitted the third term of (50.38) since it is at least $|\gamma_L T_2(1 - \eta_0)|^{-2}$ times less than the first. In the second case, $\gamma_L T_2 \gg 1$, the spectrum will consist of two parts:

$$\frac{H_\omega}{\omega_0^2} = 2(q^2)_\omega = \frac{1}{\pi}\frac{\gamma_L\left(n_{0\Sigma} + \frac{1}{2}\right)\frac{\hbar}{\omega_0}}{(\omega_0 - \omega)^2 + \gamma_L^2}$$
$$+ \frac{1}{\pi}\frac{T_2^{-1}(1 - \eta_0)\left[\left(n_{0\Sigma} + \frac{1}{2}\right)\frac{1 - (1 - \eta_0)^2}{1 - \eta_0} + \frac{\eta_0}{1 - \eta_0}\frac{N}{4R_3^0}\right]\gamma_L^{-1}T_2^{-1}\frac{\hbar}{\omega_0}}{(\omega_0 - \omega)^2 + T_2^{-2}(1 - \eta_0)^2}. \tag{50.47}$$

By direct integration of (50.46) and (50.47) it is easy to check that

$$2\int_0^\infty (q^2)_\omega\, d\omega = \langle q^2\rangle,$$

where for the case of $\gamma_L T_2 \gg 1$, the latter is satisfied with an accuracy up to terms of the order of $\eta_0/\gamma_L T_2 \ll 1$.

We have thus found the spectral distribution of the noise energy in a resonator of a maser amplifier. In the first case when the resonator pass band is narrower than the width of the molecular line

$$\gamma_L \ll T_2^{-1},$$

the regeneration effect holds for the whole of the resonator band so the latter narrows to values of $\gamma_L(1 - \eta_0)$ (see Gordon and White, 1958). It is in this frequency range, of course, that the noise energy is distributed. For thermal equilibrium, when $n_{0\Sigma} + \frac{1}{2} + (N/4R_3^0) = 0$, the molecular system is not excited ($\eta_0 < 0$), and the bandwith of a filled resonator broadens to $\gamma_L(1 + |\eta_0|)$. The energy spectrum will be of the form [(50.46) is valid, as before]

$$2(q^2)_\omega = \frac{1}{\pi} \frac{\gamma_L(1 + |\eta_0|)\left(n_0 + \dfrac{1}{2}\right)}{(\omega_0 - \omega)^2 + \gamma_L^2(1 + |\eta_0|)^2}. \tag{50.48}$$

When there is no interaction (50.48) becomes (50.41). On the other hand, in the case of $\gamma_L T_2 \gg 1$ the width of the molecular line is narrower than the resonator pass band. Regeneration will occur only in the band $T_2^{-1}(1 - \eta_0)$ located about the centre of the resonator band $\omega_\nu = \omega_0$. Therefore [for $(1 - \eta_0)\gamma_L T_2 > 1$] most of the noise energy [see the first two terms in (50.45)] whose spectrum is described by the first term in (50.47) lies outside the bandwith of a molecular amplifier and, of course, does not affect the noise properties of the latter.

50.3. The Noise Factor and Effective Noise Temperature of a Resonator Maser Amplifier

The noise properties of a maser amplifier are generally stated in terms of the effective noise temperature and the noise factor. The noise factor of any amplifier is defined by

$$F = \frac{(P_{\text{sig}}/P_{\text{noise}})_1}{(P_{\text{sig}}/P_{\text{noise}})_2}. \tag{50.49}$$

Here $(P_{\text{sig}}/P_{\text{noise}})_{1,2}$ is the ratio of the powers of the signal and the noise at the input and output of the system. It is clear from the definition that if the amplifier does not introduce any additional noise of its own the signal–noise ratios at the input and output are equal and the noise factor of such an ideal amplifier is unity. A little later we shall see that a maser amplifier in this sense is almost an ideal device.

Definition (50.49) can be written slightly differently if we introduce the power amplification factor G of the amplifier:

$$F = \frac{1}{G}\frac{(P_{\text{noise}})_2}{(P_{\text{noise}})_1}. \tag{50.50}$$

Then the noise power at the output of the amplifier can be represented as the sum of the amplified input noise and the amplifier's own noise

$$(P_{\text{noise}})_2 = G(P_{\text{noise}})_1 + (P_{\text{noise}})_0. \tag{50.51}$$

The noise factor of the amplifier can then be written as follows:

$$F = 1 + \frac{(P_{\text{noise}})_0}{G(P_{\text{noise}})_1}. \tag{50.52}$$

If at the input to the amplifier the load has a temperature T_1, then, as we shall show a little later [see formula (50.57)], the noise power at the input in the band df, for $kT \gg \hbar\omega$, will be†

$$(P_{\text{noise}})_1 = kT_1\, df. \tag{50.53}$$

Therefore the noise factor depends not only on the noise properties of the amplifier itself but also on the temperature of the input device. Sometimes this description of the noise is inconvenient (particularly when the temperature of the input device is quite high). We therefore introduce yet another

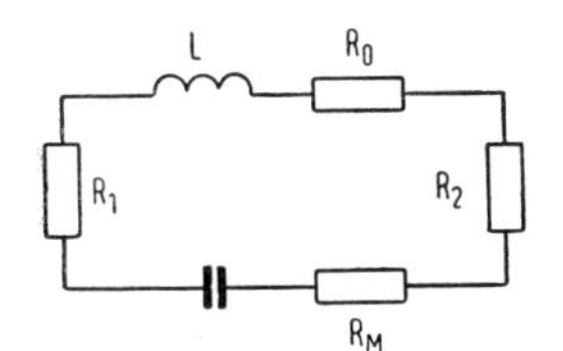

FIG. X.20. Equivalent (series) circuit of a maser amplifier.

measure of the amplifier noise: the effective noise temperature

$$(P_{\text{noise}})_0 = GkT_{\text{ef}}\, df. \tag{50.54}$$

From (50.52), (50.53) and (50.54) we have the relation between the noise factor and the effective noise temperature

$$T_{\text{ef}} = (F-1)\,T_1. \tag{50.55}$$

† If we represent a maser amplifier by an equivalent circuit (see Fig. X.20) expression (50.53) can be written at once according to Nyquist's theorem, as the nominal noise power into a matched load.

Let us move on to an actual calculation of the noise factor and the effective noise temperature of a straight-through resonator maser amplifier (see § 44), a block diagram of which is given in Fig. X.7a. On the input and output sides the resonator is generally protected by unidirectional lines or isolators (non-reciprocal elements based, for example, on the rotation of the plane of polarization of a wave propagated in a waveguide filled with ferrite). These isolators serve to protect the maser from the influence of reflected signals and, in addition, the output isolator protects the amplifier from the noise of the following components, e.g. a crystal mixer whose effective noise temperature may be very high. Let us look in greater detail at the role of an isolator.

Let a noise power $P_{\text{noise}}(T_{\text{ef}\,M})$ arrive at the isolator from the maser amplifier. Then at the output of the isolator there will be the part of $P_{\text{noise}}(T_{\text{ef}\,M})$ that has passed through it, and the power radiated by the isolator itself (which has losses):

$$P_{\text{noise}}(T_{\text{ef}\,M}) \cdot (1 - \alpha_{\rightarrow}) + P_{\text{noise}}(T_2)\,\alpha_{\rightarrow} \simeq P_{\text{noise}}(T_{\text{ef}\,M}).$$

Here $T_{\text{ef}\,M}$ is the effective noise temperature of the maser; T_2 is the isolator temperature; $\alpha_{\rightarrow}$ is the attenuation of the isolator for a signal coming from the maser ($\alpha_{\rightarrow} \to 0$).

At the output of the isolator there will be $P_{\text{noise}}(T_{\text{ef}\,M})$, because of the smallness of $\alpha_{\rightarrow}$. The situation is different for a noise signal coming to the maser from the crystal mixer, whose effective noise temperature can be very high ($T_{\text{ef}\,d} \gg T_2$):

$$P_{\text{noise}}(T_{\text{ef}\,d}) \cdot (1 - \alpha_{\leftarrow}) + \alpha_{\leftarrow} P_{\text{noise}}(T_2) \simeq P_{\text{noise}}(T_2) \ll P_{\text{noise}}(T_{\text{ef}\,d}).$$

Here $\alpha_{\leftarrow}$ is the attenuation coefficient of the isolator for a wave coming towards the maser ($\alpha_{\leftarrow} \leqslant 1$).

Therefore the isolator noise reaches the maser amplifier, instead of the crystal mixer noise. The isolator behaves as a black body (a matched source) with a temperature

$$T_2 \ll T_{\text{ef}\,d}.$$

To calculate the noise characteristics of a maser amplifier we need an expression for the noise power arriving at the input of the maser from the antenna (or signal generator) which is the input load of the maser amplifier (on the maser side the input isolator has a very small attenuation $\alpha_{\rightarrow} \to 0$ and it need not be taken into consideration). We shall assume that the antenna (or signal generator) is matched to the input waveguide, i.e. is an absolutely black body over the appropriate range of frequencies.

The total power incident on the maser amplifier through the cross-section of the waveguide can be written as (Landau and Lifshitz, 1957)

$$(P_{\text{noise}})_1 = H \cdot u_{\text{gr}}. \tag{50.56}$$

Here H is the density of the noise energy in the waveguide multiplied by the cross-sectional area of the waveguide and u_{gr} is the group velocity in the waveguide.

Using the expression for the average number of photons in one mode of the field and the expression for the state (mode) density in a waveguide that follows from (3.58) we obtain†

$$(P_{\text{noise}})_1 = \frac{\hbar\omega}{e^{\hbar\omega/kT_1} - 1}\, \varrho(f)\, df\, u_{\text{gr}} = \hbar\omega n_1\, df. \tag{50.57}$$

Here $\varrho(f)\, df = df/u_{\text{gr}}$ is the number of states per unit length of the waveguide in the spectral range df; T_1 is the temperature of the input load and $f = \omega/2\pi$.

The total noise power at the output of a quantum amplifier is also easy to calculate. First we write in a convenient form the spectral energy density of the noise in the resonator of a paramagnetic amplifier ($\gamma_L T_2 \ll 1$):

$$\begin{aligned} H_f = {} & \left\{ \frac{4Q_\Sigma^2}{2\pi f_0 Q_{e1}} \frac{\hbar\omega_0}{2} \coth \frac{\hbar\omega_0}{2kT_1} + \frac{4Q_\Sigma^2}{2\pi f_0 Q_a} \frac{\hbar\omega_0}{2} \coth \frac{\hbar\omega_0}{2kT_0} \right. \\ & \left. + \frac{4Q_\Sigma^2}{2\pi f_0 Q_{e2}} \frac{\hbar\omega_0}{2} \coth \frac{\hbar\omega_0}{2kT_2} + \frac{4Q_\Sigma^2}{2\pi f_0 Q_M} \frac{\hbar\omega_0}{2} \coth \frac{\hbar\omega_0}{2kT_M} \right\} \\ & \times \frac{1}{1 + \dfrac{4(\omega_0 - \omega)^2}{\omega_0^2 Q_\Sigma^{-2}}}. \end{aligned} \tag{50.58}$$

When writing (50.58) we used the relations (50.9), (50.10), (50.13), (50.46), the definition of the molecular Q-factor Q_M and the total Q-factor Q_Σ:

$$-Q_M^{-1} = Q_L^{-1}\eta_0 = \frac{\alpha^2 R_3^0 T_2}{2\gamma_L Q_L}$$

[see the footnote after equation (44.19)],

$$Q_\Sigma^{-1} = Q_L^{-1} + Q_M^{-1} = \frac{2\gamma_L}{\omega_0}(1 - \eta_0). \tag{50.59}$$

Therefore the noise energy in a resonator consists of the following: the thermal noise due to the input and output loads and due to the resonator's

† The energy $\hbar\omega/2$ of the vacuum oscillation make no contribution to the power flux (Fain, 1966).

own losses and, lastly, the noise of the spontaneous emission from the molecular system, the last being the same as the thermal noise in form. Expression (50.58) is essentially basis for the equivalent circuit shown in Fig. X.20 for the noise of a maser amplifier. It follows from this expression that the noise due to spontaneous emission from the molecules is characterized by the noise resistance R_M which is determined by the molecular Q-factor and has the temperature T_M. Let us derive (50.58) using Nyqu ist'stheorem (Landau and Lifshitz, 1957) for each resistance separately (including $R_M = \omega_0 L/Q_M$†). The noise generated by the resistance R_i creates the stored energy in the circuit (at resonance):

$$L(I^2)_f = L\cdot\frac{4R_i\,\dfrac{\hbar\omega}{2}\coth\dfrac{\hbar\omega}{2kT_i}}{(R_1+R_2+R_0-R_M)^2} = \frac{4Q_\Sigma^2}{2\pi f_0 Q_i}\,\frac{\hbar\omega}{2}\coth\frac{\hbar\omega}{2kT_i} \tag{50.60}$$

($Q_i = \omega_0 L/R_i$, where L is the equivalent circuit inductance).

Since the noise generated in the different resistances is not coherent (50.58) is obtained by adding expressions such as (50.60).

Let us return to the calculation of the noise factor. It is easy to write the expression for the noise power at the output of a paramagnetic maser amplifier. We shall do this, first having eliminated the noise from the output load

$$(P_{\text{noise}})_2' = H_f\,df\,\frac{2\pi f_0}{Q_{e2}} = \frac{4Q_\Sigma^2}{Q_{e2}}\{Q_{e1}^{-1}\hbar\omega_0 n_1 + Q_a^{-1}\hbar\omega_0 n_0 + Q_M^{-1}\hbar\omega_0 n_M\}$$

$$\times\frac{df}{1+\dfrac{4(\omega_0-\omega)^2}{\omega_0^2 Q_\Sigma^{-2}}}, \tag{50.61}$$

where $n_M = 1/(e^{\hbar\omega_0/kT_M}-1)$.

The total power at the output will also include the noise generated by the output load (isolator) reflected from the output of the maser amplifier:

$$(P_{\text{noise}})_2 = (P_{\text{noise}})_2' + \Gamma_2\hbar\omega_2 \tag{50.62}$$

where Γ_2 is the power reflection coefficient at the output of the maser amplifier.

Since the output isolator is matched to the output waveguide we can use a relation similar to (50.57) when writing (50.62).

† We apply Nyquist's theorem to an excited system ($T_M < 0$). This is perfectly correct since at the very beginning of the section we assumed that the molecular system is in equilibrium with the corresponding system, although it is at negative temperature.

We substitute the value of the reflection coefficient expressed in terms of the Q-factor (Slater, 1950)

$$\Gamma_2 = 1 + \frac{4Q_\Sigma^2}{Q_{e2}} \frac{(Q_0^{-1} + Q_M^{-1} + Q_{e1}^{-1})}{\left(1 + \dfrac{4(\omega_0 - \omega)^2}{\omega_0^2 Q_\Sigma^{-2}}\right)} \tag{50.63}$$

in the expression for the total output noise (50.62). In addition we take into consideration (50.61) and (50.62), and also (50.57) for the input noise, in the definition of the noise factor (50.50) and, lastly, we use the expression for the amplification factor of a straight-through maser amplifier (see § 44)

$$G(\omega) = \frac{4Q_\Sigma^2}{Q_{e1}Q_{e2}} \frac{1}{1 + \dfrac{4(\omega_0 - \omega)^2}{\omega_0^2 Q_\Sigma^{-2}}}. \tag{50.64}$$

As a result we have

$$F(\omega) = 1 + \frac{Q_{e1}}{Q_a}\frac{\omega_0 n_0}{\omega n_1} + \frac{Q_{e1}}{Q_M}\frac{\omega_0 n_M}{\omega n_1} + \left[\frac{1}{G(\omega_0)} + \frac{Q_{e1}}{Q_{e2}} - \frac{2}{\sqrt{G(\omega_0)}}\sqrt{\frac{Q_{e1}}{Q_{e2}}\frac{n_2}{n_1}}\right]. \tag{50.65}$$

If the temperature of the input load is high enough ($\hbar\omega \ll kT_1$) (50.65) can be slightly simplified and for the effective noise temperature of a maser amplifier in accordance with the definition (50.55) we have the following expression:

$$kT_{ef}(\omega) = \frac{Q_{e1}}{Q_a}\hbar\omega_0 n_0 + \frac{Q_{e1}}{Q_M}\hbar\omega_0 n_M + \hbar\omega n_2\left[\frac{1}{G(\omega_0)} + \frac{Q_{e1}}{Q_{e2}} - \frac{2}{\sqrt{G(\omega_0)}}\sqrt{\frac{Q_{e1}}{Q_{e2}}}\right]. \tag{50.66}$$

We have thus found the expressions for the noise factor and the effective noise temperature of a straight-through maser amplifier. It can be seen from the relations obtained that F and T_{ef} are very weakly frequency-dependent. The amplification of a maser is often quite large enough for the condition $G(\omega_0)^{1/2} \gg (Q_{e1}/Q_{e2})^{1/2}$ to be satisfied. When the temperatures are equal ($T_0 = T_1 = T_2 \gg \hbar\omega/k$) instead of (50.65) we have

$$F(\omega_0) = \frac{Q_a^{-1} + Q_{e1}^{-1} + Q_{e2}^{-1}}{Q_{e1}^{-1}} + \frac{Q_M^{-1}}{Q_{e1}^{-1}}\frac{\hbar\omega n_{0M}}{kT_1}. \tag{50.67}$$

Then, once again taking into consideration the fact that the amplification is sufficiently great $(-Q_M^{-1} \approx Q_a^{-1} + Q_{e1}^{-1} + Q_{e2}^{-1})$, we finally have

$$F = [1 + Q_{e1}(Q_a^{-1} + Q_{e2}^{-1})]\left(1 - \frac{\hbar\omega_0 n_M}{kT_1}\right). \tag{50.68}$$

In real amplifiers the unloaded Q-factor is generally very high ($Q_a \gg Q_{e1}$).† If, however, a very large output Q-factor $Q_{e2} \gg Q_{e1}$ is used the minimum values of the noise factor and the effective noise temperature will be

$$F = 1 - \frac{\hbar\omega_0 n_M}{kT_1}, \tag{50.69}$$

$$T_{\text{ef}} = -\frac{\hbar\omega_0 n_M}{k}. \tag{50.70}$$

With a high enough molecular temperature ($k|T_M| \gg \hbar\omega$) the maser's effective noise temperature T_{ef} is the same as $|T_M|$. In the opposite case $k|T_M| \ll \hbar\omega$, $T_{\text{ef}} = \hbar\omega_0/k$.

The effective noise temperature of actual paramagnetic amplifiers is of the order of a few degrees Kelvin [2° in the amplifier of McWhorter and Meyer (1958)]. Let us estimate T_{ef} in the case of $k|T_M| \gg \hbar\omega$. For a three-level paramagnetic amplifier in which the pumping is at a frequency ν_{31} and the amplification at a frequency $\nu_{32} < \nu_{21}$ the molecular temperature is (when the relaxation probabilities are equal)

$$|T_M| \approx \frac{2T_0\nu_{32}}{\nu_{21} - \nu_{32}} \approx 2T_0, \quad \text{if} \quad \nu_{32} \leqq \frac{\nu_{21}}{2}.$$

Therefore $T_{\text{ef}} \sim 2T_0$. Since a paramagnetic amplifier generally operates at liquid helium temperature the effective noise temperature really is a few degrees. The procedure for determining the effective noise temperature of a maser experimentally is described in Singer's book (1959).

To conclude let us briefly examine the noise characteristics of a molecular beam oscillator ($\gamma_L T_2 \gg 1$). Within the limits of the band $T_2^{-1}(1 - \eta_0) \ll \gamma_L$ we have from (50.47) for the spectral energy density

$$\frac{H\omega}{\omega_0^2} = \frac{1}{\pi} \frac{\gamma_L\left(n_{0\Sigma} + \frac{1}{2} + \eta_0 \frac{N}{4R_3^0}\right)}{(1 - \eta_0)^2 \gamma_L^2} \frac{\hbar}{\omega} \frac{1}{1 + \frac{(\omega_0 - \omega)^2}{T_2^{-2}(1 - \eta_0)^2}}. \tag{50.71}$$

† This condition eliminates the inaccuracy which we introduced by taking the resonator temperature to be high by assuming that $T_0 = T_2 = T_1$; T_0 is generally a few degrees Kelvin.

Comparing the spectrum (50.71) with the spectrum (50.46) of the noise energy in the resonator of a paramagnetic amplifier ($\gamma_L T_2 \ll 1$) we find that they have the same form. The only difference lies in the frequency bands $\Delta\omega_{1/2}$ in which the noise is contained:

$$\gamma_L T_2 \ll 1, \quad \tfrac{1}{2}\Delta\omega_{1/2} = \gamma_L(1 - \eta_0), \tag{50.72}$$

$$\gamma_L T_2 \gg 1, \quad \tfrac{1}{2}\Delta\omega_{1/2} = T_2^{-1}(1 - \eta_0). \tag{50.73}$$

We notice that (50.72) and (50.73) are respectively the pass bands of a solid-state paramagnetic amplifier and an ammonia molecular beam oscillator (Gordon and White, 1958).

Therefore all the relations derived for the noise factor and the effective noise temperature of a solid-state paramagnetic amplifier hold in the pass band of a molecular amplifier.

CHAPTER XI

Maser Oscillators for the Microwave Range

IN THE preceding chapter we showed that a resonator containing an active medium is an amplifier provided that the radiation from the working substance is sufficient to compensate for the ohmic losses. If we increase the number of active molecules until the power emitted by them compensates for all the sources of losses, including the coupling of the resonator with an external load, the device starts to oscillate independently of the presence of an input signal. The self-excitation condition which was formulated was of the form

$$Q_L^{-1} \leqq -Q_M^{-1}.$$

Several different forms of microwave maser oscillators are at present in operation. Of these, however, only molecular oscillators† have been put to practical use. Because of their high stability they are used as precise frequency standards. Since paramagnetic oscillators are not especially stable they have no advantage in this respect over electronic devices, but from the point of view of physics a paramagnetic oscillator is interesting because of some features of the emission process. One of these is the phenomenon of the self-modulation of the oscillations; this has been observed, for example, at several pumping modes in a ruby maser. Outwardly this effect is very like the spiking of the output of lasers. It should be pointed out in this connexion that microwave oscillators, because of their single-mode nature, are essentially simpler devices than lasers. Therefore a thorough investigation of paramagnetic microwave oscillators may be of great importance in solving one of the main problems in the field of maser oscillators: the problem of the non-stationary nature of the oscillations.

† The term "molecular oscillator" is sometimes used as a synonym for "maser oscillator". We use it to denote the particular class of oscillating devices operating by means of gaseous molecular beams.

(Andronov *et al.*, 1959) can be applied to it. We represent H_{32} in the form of an oscillation with a slowly changing amplitude and phase

$$H_{32} = h_{32}(t) \cos [\omega t - \varphi(t)]. \tag{51.3}$$

We can then find M_{32} by using (45.10) which, taking (45.45) into consideration, becomes

$$M_{32} = \frac{D_{32}\tau \, |\mu_{32}|^2 \, h_{32}}{\hbar(1 + \Delta_{32}^2) \, V_s} [\Delta_{32} \cos (\omega t - \varphi) - \sin (\omega t - \varphi)]. \tag{51.4}$$

Here $D_{32} = N(\sigma_{33} - \sigma_{22})V_s$ is the total number of active molecules in the resonator. We now substitute (51.3) and (51.4) in (51.2) and average it for the time that appears explicitly. This leads to

$$2\dot{h}_{32} + \frac{\omega_c}{Q_L} h_{32} = \frac{4\pi\xi\omega\tau \, |\mu_{32}|^2}{\hbar(1 + \Delta_{32}^2) \, V_s} D_{32}h_{32}, \tag{51.5}$$

which can easily be brought into the form of (51.1b). To do this it is sufficient to multiply both sides of (51.5) by h_{32} and use the fact that the energy of the field in a resonator is $W_c = h_{32}^2 V_s/8\pi\xi$; therefore $m_{32} = h_{32}^2 V_s/8\pi\xi\hbar\omega$ and

$$\dot{m}_{32} = \frac{4\pi\xi\omega\tau \, |\mu_{32}|^2}{\hbar(1 + \Delta_{32}^2) \, V_s} D_{32}m_{32} - \frac{1}{T_c} m_{32}. \tag{51.6}$$

The coefficient B, mentioned above, is

$$B = \frac{4\pi\xi\omega\tau \, |\mu_{32}|^2}{(1 + \Delta_{32}^2) \, \hbar V_s}. \tag{51.7}$$

From the above and using expression (45.47) for the probability of a stimulated transition, it can be shown that $Bm_{32} = W_{32}$. It should be pointed out that until the frequency ω of the oscillations is found (in the form (51.3), giving the solution ω = const.) the quantity Δ_{32}, and therefore B as well, remains unknown. The only exception is the resonance case $\omega_c = \omega_0$, when $\Delta_{32} = 0$.

It now remains to find the additional assumptions necessary for the set of balance equations (45.1) to reduce to (51.1a). Changing from the variables σ_{nn} to the population differences we obtain

$$D_{32} = NV_s(\sigma_{33} - \sigma_{22}),$$

$$D_{31} = NV_s(\sigma_{33} - \sigma_{11}).$$

We can write the initial equations in the form

$$\dot{D}_{32} = -D_{32}(w_{12} + 2w_{23} + 2W_{32}) - D_{31}(w_{13} - w_{12} + W_{31})$$

$$- \frac{\hbar N V_s}{3kT}(\omega_{31}w_{13} - \omega_{21}w_{12} + 2\omega_{32}w_{23}), \tag{51.8a}$$

$$\dot{D}_{31} = -D_{31}(w_{12} + 2w_{13} + 2W_{31}) - D_{32}(w_{23} - w_{12} + W_{32})$$

$$- \frac{\hbar N V_s}{3kT}(\omega_{32}w_{23} + \omega_{21}w_{12} + 2\omega_{31}w_{13}). \tag{51.8b}$$

These equations can be written more concisely as

$$\dot{D}_{32} = -2BD_{32}m_{32} - (D_{32} - D'_{32})(w_{12} + 2w_{23})$$

$$-(D_{31} - D'_{31})(w_{13} - w_{12} + W_{31}), \tag{51.9a}$$

$$\dot{D}_{31} = -(D_{31} - D'_{31})(w_{12} + 2w_{13} + 2W_{31})$$

$$-(D_{32} - D'_{32})(w_{23} - w_{12}) - BD_{32}m_{32}, \tag{51.9b}$$

using primes to denote the values of quantities when there is no oscillation. In the common case when all the relaxation probabilities are equal to each other the equations (51.9) become:

$$\dot{D}_{32} = -2BD_{32}m_{32} - 3w(D_{32} - D'_{32}) - W_{31}(D_{31} - D'_{31}); \tag{51.10a}$$

$$\dot{D}_{31} = -(3w + 2W_{31})(D_{31} - D'_{31}) - BD_{32}m_{32}. \tag{51.10b}$$

The qualitative difference between the sets of equations (51.7), (51.9) and (51.1) consists of the difference in their orders. To change to (51.1) we must postulate that D_{31} = const. and then $D_{31} - D'_{31}$ can be eliminated from the equations. In particular, when all the w_{mn} are equal this leads to the equation

$$\dot{D}_{32} = -3\frac{2w + W_{31}}{3w + 2W_{31}} BD_{32}m_{32} - 3w(D_{32} - D'_{32}), \tag{51.11}$$

which agrees with (51.1a) if we consider that

$$\alpha = 3\frac{2w + W_{31}}{3w + 2W_{31}}, \tag{51.12}$$

$$T_1 = \frac{1}{3w}. \tag{51.13}$$

An analysis of the motions described by both the equations (51.1) and (51.6), (51.10) is carried out in the next section.

51.3. Qualitative Analysis of the Balance Equations

The equations (51.1) are non-linear and their rigorous analytical solutions cannot be found. Nevertheless the basic features of the behaviour of the physical model they describe can be revealed by the methods of qualitative analysis developed in the book by Andronov *et al.* (1959) and applied to the present system by Bespalov and Gaponov (1965).

We are primarily interested in the equilibrium states of the system and their stability. In other words, we must find the singularities in the phase plane of the differential equations (51.1) and analyse the phase trajectories in their vicinity. In order to simplify the form of notation of the equations we shall change, as is usual in the theory of oscillations, to some new dimensionless quantities:

$$t_{\text{new}} = t_{\text{old}}/T_1, \quad n = D_{32}/\overline{D}_{32}, \quad m = m_{32}/\bar{m},$$
$$n' = D'_{32}/\overline{D}_{32}, \quad K_1 = T_1/T_c. \tag{51.14}$$

Taking

$$\overline{D}_{32} = 1/T_cB, \quad \bar{m} = 1/\alpha BT_1 \tag{51.15}$$

as the transformation coefficients, the simplest form of the equations is

$$\dot{m} = K_1(n-1)\,m, \quad \dot{n} = n' - (m+1)\,n. \tag{51.16}$$

The equations (51.16) define parametrically (with time as the parameter) the integral curves on the phase plane. It suffices to divide the first of equations (51.16) by the second to obtain the explicit form of the phase trajectory equations; this gives

$$\frac{dm}{dn} = K_1\,\frac{(n-1)\,m}{n' - (m+1)\,n}. \tag{51.17}$$

The singularities in the phase plane found from the condition that the numerator and denominator of the right-hand side of (51.17) are zero, are at the same time the equilibrium positions of the system, since here $\dot{m} = 0$ and $\dot{n} = 0$. The system has two equilibrium positions:

$$n_1 = n', \quad m_1 = 0;$$
$$n_2 = 1, \quad m_2 = n' - 1. \tag{51.18}$$

In order to determine the type of singularity we must investigate the structure of the phase plane in its immediate neighbourhood; for this purpose we linearize equations (51.16), changing the variables to

$$n = n_{1,2} + \xi,$$

$$m = m_{1,2} + \eta, \tag{51.19}$$

and neglecting higher-order terms that are non-linear in ξ and η. As a result there are two systems of linear differential equations each of which is valid near the selected centre of expansion.

(a) The motion of the mapping point near the equilibrium position denoted by the suffix 1 is determined by the equations

$$\dot{\eta} = K_1(n' - 1)\,\eta, \quad \dot{\xi} = -\xi - n'\eta. \tag{51.20}$$

The solution of these is of the form

$$C_1\, e^{p_1 t} + C_2\, e^{p_2 t},$$

where p_1 and p_2 are the roots of the characteristic equation

$$(p + 1)\,[p - K_1(n' - 1)] = 0.$$

Knowing the roots we can easily find the solutions

$$\eta = \eta_0\, e^{K_1(n'-1)t},$$

$$\xi = \xi_0\, e^{-t} - \frac{n'}{K_1(n' - 1) + 1}\,\eta_0\, e^{K_1(n'-1)t}. \tag{51.21}$$

The topology of the section of the phase plane being investigated depends on the magnitude of the pumping parameter n'. If $n' < 1$, then the singularity is a stable node. When $n' > 1$ the singularity becomes a saddle point and the equilibrium position corresponding to it is unstable.

(b) Near the second equilibrium position the motion of the mapping point obeys the equations

$$\dot{\eta} = K_1(n' - 1)\,\xi, \quad \dot{\xi} = -n'\xi - \eta. \tag{51.22}$$

The characteristic equation

$$p^2 + n'p + K_1(n' - 1) = 0$$

has the roots

$$p_{1,2} = -\tfrac{1}{2}n' \pm \sqrt{\tfrac{1}{4}n'^2 - K_1(n' - 1)}. \tag{51.23}$$

The nature of the singularity at the point in question once again depends on the quantity n'. As long as $n' < 1$ the singularity is of the saddle point type. But this singularity has no physical meaning since it is located in the negative energy half-plane ($m < 0$). Increasing the pumping to values where $n' > 1$ moves the singularity into the upper half-plane. Its type depends on the form of the roots (51.23). Two possibilities arise:

(1) when the condition

$$\tfrac{1}{4}n'^2 > K_1(n' - 1)$$

is satisfied both roots are real and negative. This condition corresponds to a singularity of the stable node type.

(2) In the opposite case the roots are complex. The singularity is now a focus. Its stability is determined by the negative nature of the real part of the roots. For $n' - 1 \sim 1$ and $K_1 \gg 1$ the frequency Ω of the oscillations in the linear approximation is $\sqrt{K_1(n' - 1)}/T_1$. To make an estimate of this we put $T_1 = 10^{-2}$ sec, $T_c = 10^{-7}$ sec, i.e. $K_1 = 10^5$ and $n' = 2$, we obtain $\Omega \simeq 3 \times 10^4\ \text{sec}^{-1}$, which is in poor agreement with the data given by Kikuchi *et al.* (1959) but is much closer to the results of Manenkov *et al.* (1964).

Having elucidated the nature of the singularities for different values of the coefficients of the system we can obtain a qualitative idea of the form of the whole of the phase plane of the system. Figure XI.3a shows a division by trajectories of the phase plane of a system which has not reached the self-excitation threshold ($n' < 1$); Fig. XI.3b relates to a system excited above the threshold ($n' > 1$).

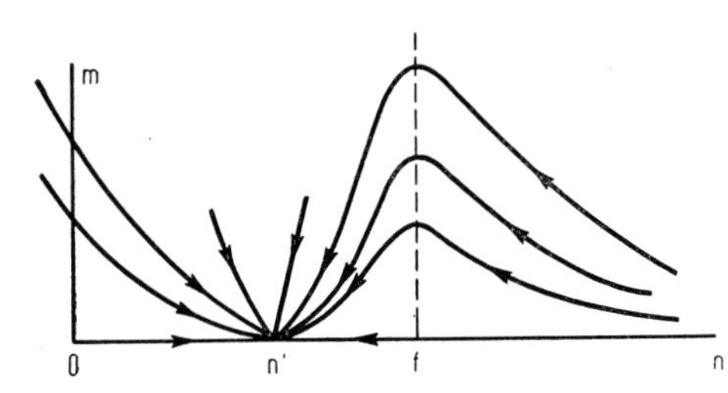

FIG. XI.3a. Phase plane of equations (51.16) for $n' < 1$.

A characteristic feature of the phase plane of the equations (51.16) is the absence of closed trajectories (extreme cycles). From this fact we should conclude that the simple two-level model is unable to explain the undamped peak mode observed in experiment (Kikuchi *et al.*, 1959).

Let us now proceed directly to an analysis of the equations of a three-level system (51.6) and (51.9).

In the dimensionless form defined by equations (51.14), to which we must add

$$r = (D_{31} - D'_{31})/\overline{D}_{32}; \quad \beta = T_1 W_{31}, \tag{51.24}$$

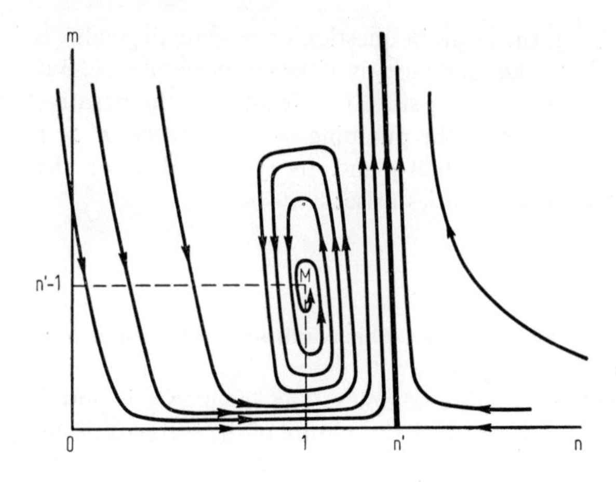

FIG. XI.3b. Phase plane of equations (51.16) for $n' > 1$.

the equations (51.6), (51.10) become

$$\dot{n} = -\frac{2}{\alpha} mn - n + n' - \beta \tag{51.25a}$$

$$\dot{r} = -(1 + 2\beta)\, r - \frac{1}{\alpha} mn \tag{51.25b}$$

$$\dot{m} = K_1(n - 1)\, m. \tag{51.25c}$$

The set (51.25) has two equilibrium positions:

$$n_1 = n'; \quad r_1 = 0; \quad m_1 = 0$$

$$n_2 = 1; \quad r_2 = -(n' - 1)/(2 + 3\beta); \quad m_2 = n' - 1. \tag{51.26}$$

Just as in the two-level model K_1 is a large parameter and when $n \neq 1$, as can be seen from (51.25c), rapid motions take place in the system. The derivative $\dot{m}$ is small and the motion of the mapping point is at low velocity only near the plane $n = 1$. In the actual $n = 1$ plane there lies the asymptotic straight line of the slow motions which, in accordance with (51.25a), is defined by the equation

$$\frac{2}{\alpha} m + \beta r = n' - 1. \tag{51.27}$$

We introduce a new system of coordinates so that one of its axes (the x axis) remains parallel to the n axis and the other (the y axis) runs along the straight line (51.27), whilst the origin of the coordinates is transferred at the same time to the point n_2, r_2, m_2. The old and new coordinates are connected by the relations

$$n = x + 1$$

$$r = y \cos \theta + z \sin \theta - [(n' - 1)/(2 + 3\beta)]$$

$$m = -y \sin \theta + z \cos \theta + n' - 1 \tag{51.28}$$

$$\cos \theta = \left(1 + \frac{\alpha^2\beta^2}{4}\right)^{-1/2}.$$

Substitution of the variables in accordance with (51.28) transforms (51.25) into

$$\dot{x} = -\frac{2}{\alpha}(x + 1)(-y \sin \theta + z \cos \theta) - \left[\frac{2(n' - 1)}{\alpha} + 1\right] x$$

$$- \beta(y \cos \theta + z \sin \theta); \tag{51.29a}$$

$$\dot{y} \cos \theta + \dot{z} \sin \theta = -(1 + 2\beta)(y \cos \theta + z \sin \theta)$$

$$- \frac{1}{\alpha}(x + 1)(-y \sin \theta + z \cos \theta) - \frac{n' - 1}{\alpha} x; \tag{51.29b}$$

$$-\dot{y} \sin \theta + \dot{z} \cos \theta = K_1 x(-y \sin \theta + z \cos \theta + n' - 1). \tag{51.29c}$$

To find the motion along the asymptotic line it is sufficient to put $x = z = 0$, which gives

$$\dot{y} = -\frac{2 + 3\beta}{2} y = \lambda_1 y. \tag{51.30}$$

The mapping point along the asymptotic line moves exponentially with a period

$$\tau_1 = 2T_1/(2 + 3\beta). \tag{51.31}$$

Moving slowly in the direction of the y axis towards the equilibrium position the mapping point at the same time describes a more rapid motion in the plane at right angles. Finding this motion is made easier by the circumstance that $\dot{y}$ can be eliminated from the set (51.29) and by considering y to be a parameter (51.29) can be reduced to a second-order set. However, the problem is still rather complex and we shall limit ourselves to examining a small region around the equilibrium position, i.e. $y = 0$. The characteristic equation has the roots

$$\lambda_{2,3} \simeq -\left\{\frac{2(n'-1)}{\alpha} + 1 + \frac{\alpha\beta\cos^2\theta}{2}\left[\frac{\alpha\beta(1+2\beta)}{2} + \frac{1}{\alpha}\right]\right\}$$

$$\pm i\sqrt{\frac{2K_1}{\alpha}(n'-1)}. \quad (51.32)$$

On the basis of what has been said we can get an idea of the form of the phase trajectories (Fig. XI.3c). It corresponds to the time curve of the emission shown in Fig. XI.1c, i.e. a mode of damped transient oscillations with a certain drift of the centre of the oscillations. As for the undamped peak mode, the question of its nature remains open. The nature of the roots λ_1, λ_2 and λ_3 leaves no doubt in the stability of the oscillation mode. The thought remains that some assumptions taken as the basis of the model are not always valid. There are foundations for such doubts.

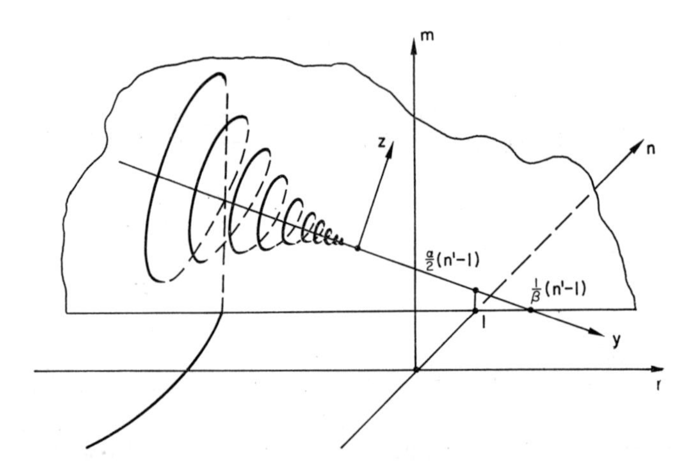

FIG. XI.3c

In fact we have been assuming all the time that we are dealing with homogeneously broadened resonance lines of paramagnetics and have identified $2/\tau$ with the observed line width. However, the assumptions that there is no inhomogeneous broadening can in no way be justified. Moreover, Makhov *et al.* (1960) have sucessfully affected the amplification factor and oscillation mode of a ruby maser by saturating the quadrupole transition in the spectrum of Al^{27} nuclei. This indicates that there is interaction betweem them and the Cr ions. There are therefore sufficient effects to account for the failure of the above theory. To indicate precisely the reason, however, and how we must alter the theory, is not yet possible on the basis of the available experimental material.

52. The Molecular Beam Oscillator

52.1. The State Separation of Gas Molecules

Population inversion in a gaseous medium can be achieved by the same means as in a solid, i.e. by using pump radiation of a sufficiently high power. Whilst in the case of solids this is the only practical possibility, we can use other methods, which are restricted to gases, for obtaining states with negative temperatures. We refer to the separation of the states of molecules in a molecular beam.

In a paper of 1954 Basov and Prokhorov pointed out the particular ways in which molecular beams could be applied to the problems of radiofrequency

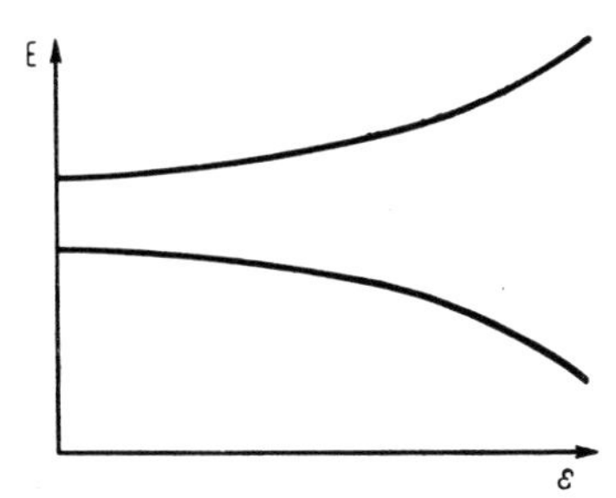

FIG. XI.4. Stark effect in inversion levels of ammonia.

spectroscopy. By using molecular beam techniques two aims are achieved simultaneously: the resolution is increased and the sensitivity of existing methods of experimental radiofrequency spectroscopy is increased. The observed width of the spectral lines in gases is chiefly caused by collisions between the molecules, and by the Doppler effect. The number of collisions can be reduced to a minimum in a molecular beam moving in a sufficiently high vacuum. As a result the observed form of the spectral line turns out to

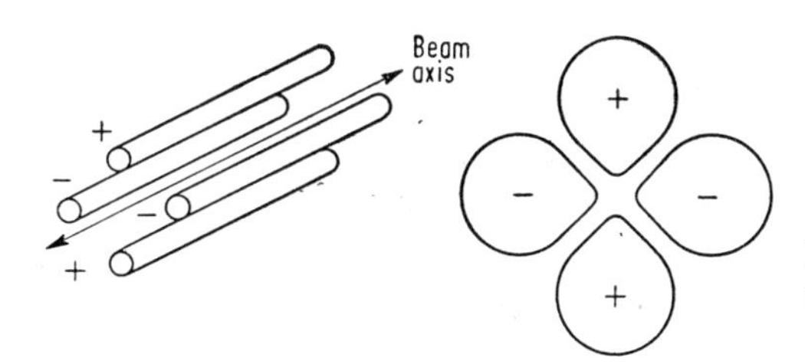

FIG. XI.5. Quadrupole capacitor-type separation system.

be connected only with inhomogeneous broadening mechanisms and the finite time of flight through the region where there is interaction with the high-frequency field (i.e. through the resonator). With a long resonator the time of flight is large, the individual components of the molecular spectrum are narrow, and lines which are very close together can be distinguished. In particular, certain components of the hyperfine structure can be resolved. We can ignore Doppler broadening if we work with a resonator mode which has an infinite wave velocity in the direction of motion of the molecules.

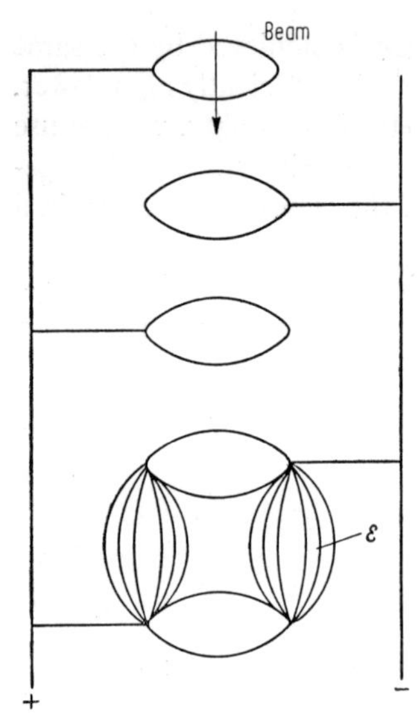

FIG. XI.6. Ring-type sorting system.

The field corresponding to oscillations of this type must not change along the direction of motion for this to be so. In practice, of course, it is impossible to eliminate completely the Doppler frequency shift, because of end effects, but it can be significantly reduced in magnitude by working in the TE_{011} mode of a rectangular resonator, or the TH_{001} mode of a resonator of cylindrical form.

The possibility of physically separating the molecules of a beam in different states proved to be decisive in designing a molecular oscillator. The principle of this kind of separation is based on the use of the Stark or Zeeman effects (depending on the nature of the levels to be divided). As an example we shall take a molecule of the NH_3 type which shows the Stark effect. The energies of different states of the molecule do not depend in the same way on the strength of the applied electric field (Fig. XI.4). The trajectories of the molecules are consequently different in a region where there is a non-uniform electric field. Figure XI.5 shows the arrangement and cross-section of a quadrupole capacitor which is often used as the separation system of a molecular oscillator. Near the axis of the capacitor the strength of the electric field is $|\mathscr{E}_0| \sim r$. It is obvious that as they move in the non-uniform field the molecules are deflected in the direction in which their energy decreases. The force acting on the molecules in it is $-\partial E_i/\partial r$. Because of this the molecules in state 1 will be deflected away from the axis of the system whilst the molecules in state 2 will be focused towards the axis if their energy depends on the field strength in accordance with the law shown qualitatively in Fig. XI.4. As a result the system separates the unexcited and excited molecules spatially, focusing the latter into a narrow axial beam which can be selected by an aperture and allowed to enter the resonator. If the density of the active molecules in the beam is large enough, and the resonator is tuned to the frequency of the molecular transition, we can achieve either amplification or the generation of electromagnetic oscillations of the corresponding frequency.

We can also separate molecules in different states if their energies depend in different ways on the strength of a magnetic field. To do this they should be passed through a non-uniform magnetic field.

52.2. The Theory of a Molecular Oscillator

In the range of variation of the parameters that is accessible to the experimenter a molecular oscillator behaves as a stable self-oscillatory system. Therefore the majority of theoretical investigations do not generally go beyond the stationary mode. Amongst these is the work of Basov and Prokhorov (1955b and 1956), Gordon *et al.* (1954), Shimoda *et al.* (1956). We shall deal with the more general theory that also includes non-stationary processes in a molecular oscillator.

The theoretical investigation of a molecular oscillator can be based on the discussion of its equivalent circuit, which describes the oscillator as two coupled resonant systems: one of them is the resonator and the other is the molecular beam. This kind of system can be described mathematically by the set of equations (19.42)–(19.45).† For a single-mode device, such as the molecular oscillator, the equations can be simplified:

$$\begin{aligned} \dot{s}_1 + \omega_0 s_2 + \frac{1}{\tau} s_1 &= -\frac{1}{\hbar}(\boldsymbol{A}(\boldsymbol{r}) \cdot \boldsymbol{e}_2) s_3 q_s, \\ \dot{s}_2 - \omega_0 s_1 + \frac{1}{\tau} s_2 &= \frac{1}{\hbar}(\boldsymbol{A}(\boldsymbol{r}) \cdot \boldsymbol{e}_1) s_3 q_s, \\ \dot{s}_3 + \frac{1}{\tau}(s_3 - s_3') &= -\frac{1}{\hbar}(\boldsymbol{A}(\boldsymbol{r}) \cdot (\boldsymbol{e}_1 s_2 - \boldsymbol{e}_2 s_1))\, q_s, \\ \ddot{q}_s + \frac{\omega_c}{Q_L}\dot{q}_s + \omega_c^2 q_s &= \int_{V_c} (\boldsymbol{A}(\boldsymbol{r}) \cdot (\boldsymbol{e}_1 s_1 + \boldsymbol{e}_2 s_2))\, dV. \end{aligned} \tag{52.1}$$

† Similar equations were first obtained by Fain (1957) and Oraevskii (1959). They were written in a slightly different form since the variables were the electric field $\mathscr{E}$, the polariza-

Here τ is the mean time of flight of the beam molecules through the resonator, which is the same as both the longitudinal and transverse relaxation times.

First we shall make some simplifying assumptions of a minor nature. The molecular constants $\boldsymbol{e}_1$ and $\boldsymbol{e}_2$ are connected with the off-diagonal elements of the dipole moment matrix by the simple relations [see (19.13)]:

$$\boldsymbol{e}_1 \pm i\boldsymbol{e}_2 = \pm(2i\omega_0/c)\,\boldsymbol{\mu}_{\substack{21\\12}} \tag{52.2}$$

We shall confine ourselves to the case where the matrix element μ_{21} is real, which is equivalent to the constant e_1 being equal to zero. In addition we shall consider that the resonator field in the region occupied by the beam is uniform. Then equations (52.1) become:

$$\ddot{s}_2 + \frac{2}{\tau}\dot{s}_2 + (\omega_0^2 + \tau^{-2})\,s_2 = -\alpha\omega_0 s_3 q_s, \tag{52.3a}$$

$$\dot{s}_3 + \frac{1}{\tau}s_3 = \frac{1}{\tau}s_3' + \frac{\alpha}{\omega_0}\left(\dot{s}_2 + \frac{1}{\tau}s_2\right)q_s, \tag{52.3b}$$

$$\ddot{q}_s + \frac{\omega_c}{Q_L}\dot{q}_s + \omega_c^2 q_s = \hbar\alpha s_2 V_s, \tag{52.3c}$$

where $\alpha = (\boldsymbol{e}\cdot\boldsymbol{A}(\boldsymbol{r}))/\hbar$.

As in the preceding case (see § 51.2) we have recourse to the van der Pol method. We shall try to find the solution of equations (52.1) in the form of quasi-harmonic oscillations with slowly changes amplitudes and phases:

$$\begin{aligned} q_s &= q(t)\sin\Psi_1, \\ s_2 &= s(t)\sin\Psi_2, \end{aligned} \tag{52.4}$$

remembering that

$$\begin{aligned} \Psi_1 &= \omega t + \psi_1(t), \\ \Psi_2 &= \omega t + \psi_2(t), \\ \psi_1(t) - \psi_2(t) &= \Phi(t). \end{aligned} \tag{52.5}$$

---

tion P of the medium and the number of active molecules D. In the case of a uniform field the equations for a molecular oscillator (Oraevskii, 1959) are of the form

$$\ddot{\mathscr{E}} + 4\pi\ddot{P} + \frac{\omega_c}{Q_L}(\dot{\mathscr{E}} + 4\pi\dot{P}) + \omega_0^2\mathscr{E} = 0,$$

$$\ddot{P} + \frac{2}{\tau}\dot{P} + (\omega_0^2 + \tau^{-2})\,P = -2D\mathscr{E}\,|\mu_{12}|^2\,\omega_0/\hbar,$$

$$\dot{D} + \frac{1}{\tau}(D - D') = \frac{2}{\hbar\omega_0}\mathscr{E}\left(\dot{P} + \frac{1}{\tau}P\right).$$

Proceeding by the van der Pol method we obtain the following set of abbreviated equations:

$$\dot{s} + \frac{1}{\tau} s = -\frac{1}{2} \alpha s_3 q \sin \Phi, \tag{52.6a}$$

$$\dot{s}_3 + \frac{1}{\tau} s_3 = \frac{1}{\tau} s_3' + \frac{1}{2} \alpha s q \sin \Phi, \tag{52.6b}$$

$$\dot{q} + \frac{1}{2T_c} q = -\frac{1}{2\omega} \hbar \alpha s V_s \sin \Phi, \tag{52.6c}$$

$$s(\dot{\Psi}_2 - \omega_0) = \tfrac{1}{2} \alpha s_3 q \cos \Phi, \tag{52.7a}$$

$$q(\dot{\Psi}_1 - \omega_c) = -\frac{1}{2\omega_c} \hbar \alpha s V_s \cos \Phi. \tag{52.7b}$$

We can subtract one equation in (52.7) from the other and deduce the following:

$$\dot{\Phi} = -(\omega_0 - \omega_c) - \frac{1}{2}\left(\frac{\hbar \alpha V_s}{\omega_c} \frac{s}{q} + \alpha s_3 \frac{q}{s}\right) \cos \Phi. \tag{52.8}$$

We now change to dimensionless variables and coefficients, introducing them as follows:

$$\begin{gathered} t_{\text{new}} = t_{\text{old}}/\tau, \quad n = s_3/\bar{s}, \quad n' = s_3'/\bar{s}, \quad v = s/\bar{s}, \\ x = \tfrac{1}{2}\alpha \tau q, \quad K_1 = \tau/T_c, \quad \Delta = (\omega_0 - \omega_c)\tau. \end{gathered} \tag{52.9}$$

The transformation coefficient selected is

$$\bar{s} = \frac{2\omega}{\tau T_c \hbar \alpha^2 V_s}. \tag{52.10}$$

In dimensionless form the set of equations (52.6) and (52.8) can be written as

$$\dot{v} + v = -xn \sin \Phi, \tag{52.11a}$$

$$\dot{n} + n = n' + xv \sin \Phi, \tag{52.11b}$$

$$\dot{x} + \tfrac{1}{2}K_1 x = -\tfrac{1}{2}K_1 v \sin \Phi, \tag{52.11c}$$

$$\dot{\Phi} = -\Delta - \left(\frac{1}{2} K_1 \frac{v}{x} + \frac{x}{v} n\right) \cos \Phi. \tag{52.11d}$$

The simplest problem is to find the equilibrium conditions of a system described by equations (52.11). Putting all the derivatives equal to zero we obtain:

$$v = -xn \sin \Phi, \tag{52.12a}$$

$$n = n' + xv \sin \Phi, \tag{52.12b}$$

$$x = -x \sin \Phi, \tag{52.12c}$$

$$\Delta = -\left(\frac{1}{2} K_1 \frac{v}{x} + \frac{x}{v} n\right) \cos \Phi. \tag{52.12d}$$

From equations (52.12) we first of all find the stationary phase

$$\sin^2 \bar{\Phi} = \frac{(K_1 + 2)^2}{4\Delta^2 + (K_1 + 2)^2}. \tag{52.13}$$

Then by eliminating x from (52.12a) and (52.12c) we obtain

$$v(1 - n \sin^2 \bar{\Phi}) = 0, \tag{52.14}$$

from which there follow two possibilities:

$$v_1 = 0 \quad \text{and} \quad n_2 = \frac{1}{\sin^2 \bar{\Phi}}.$$

The set of equations (52.12) defines all the coordinates of the two possible equilibrium positions:

$$n_1 = n', \quad v_1 = 0, \quad x_1 = 0 \tag{52.15}$$

and

$$n_2 = \frac{1}{\sin^2 \bar{\Phi}}, \quad v_2^2 = (n' - n_2) \frac{1}{\sin^2 \bar{\Phi}}, \quad x_2^2 = n' - n_2. \tag{52.16}$$

The equilibrium positions found correspond to two different physical states of the system. An absence of oscillations is characteristic of the first, whilst the second is the self-oscillatory mode of the oscillator. But the second equilibrium position is not always realized, and then only when the self-excitation condition is satisfied:

$$n' \geqq \frac{1}{\sin^2 \bar{\Phi}}. \tag{52.17}$$

This condition means that the beam intensity must give an adequate initial density of active molecules in the resonator since in its expanded form the criterion (52.17) is

$$s_3' \geqq \frac{2\omega}{\tau T_c \hbar \alpha^2 V_s} \frac{4\Delta^2 + (K_1 + 2)^2}{(K_1 + 2)^2}. \tag{52.18}$$

Generally in molecular oscillators $\tau \sim 10^{-4}$ sec, $Q_L \sim 10^4$. For a frequency of $\omega \sim 10^{11}$ sec^{-1} this gives $T_c \sim 10^{-7}$ sec and $K_1 \sim 10^3$. On the other hand the frequency difference is $|\omega_0 - \omega_c| \lessdot \tau^{-1}$, i.e. $\Delta^2 \lessdot 1$, and inequality (52.18) can be simplified by putting it into the frequently used form

$$2s_3' = N' \geqq \frac{\hbar}{4\pi |\mu_{12}|^2 \tau Q_L \xi}. \tag{52.19}$$

We obtain (52.19) by remembering the normalizing condition for the eigenfunctions of the vector-potential (3.61), and the connection between $\boldsymbol{e}_2$ and $\boldsymbol{\mu}_{12}$, (52.2), which gives

$$\alpha^2 = \frac{16\pi\omega_0^2 |\mu_{12}|^2}{\hbar^2 V_c}, \tag{52.20}$$

which must be substituted in (52.18).

The density of the molecules in the resonator (N) is connected with the number of molecules entering the resonator in unit time (N_0) by the relation

$$N = \frac{N_0}{A_s \bar{v}} = \frac{N_0 \tau}{A_s l}, \tag{52.21}$$

in which A_s is the beam cross-section; $\bar{v}$ is the mean velocity of the molecules; l is the length of the resonator; $A_s l = V_s$.

Replacing N' in (52.19) by (52.21) we arrive at another form of the self-excitation condition, which is also often used:

$$N_0' \geqq \frac{\hbar V_s}{4\pi |\mu_{12}|^2 \tau^2 Q_L \xi}. \tag{52.22}$$

With $|\mu_{12}| = 10^{-18}$ CGSe (1 Debye), $\tau = 10^{-4}$ sec, $Q_L = 10^4$, $\xi = 1$, $V_s = 10$ cm^3, we obtain for the starting flux the value $N_0' \simeq 10^{13}$ mol/sec.

Knowing the coordinates of the singular point (52.16) it is not difficult to calculate the power of a molecular oscillator. If the frequency difference is small, and therefore $\sin^2\bar{\Phi} \simeq 1$, then the energy of the field in the resonator is

$$W_c = \frac{q^2\omega_c^2}{2} = \frac{2\omega_c^2(n'-1)}{\alpha^2\tau^2}. \tag{52.23}$$

Hence the power emitted by the molecular beam is

$$P_0 = \frac{\omega_c W_c}{Q_L} = \frac{2\omega_c^3(n'-1)}{\alpha^2\tau^2 Q_L}$$

or in final form

$$P_0 = \frac{2\omega_c^3}{\alpha^2\tau^2 Q_L}\left(\frac{N_0'\tau^2 Q_L \hbar\alpha^2}{4\omega_c^2} - 1\right). \tag{52.24}$$

The power extracted from the resonator is slightly less:

$$P_2 = P_0\frac{Q_L}{Q_e}. \tag{52.25}$$

When the molecular beam is so intense that the oscillation threshold is considerably exceeded and the bracket can be neglected, then

$$(P_0)_{\max} = \tfrac{1}{2}N_0'\hbar\omega. \tag{52.26}$$

The physical meaning of this result is clear. Roughly speaking, when there is a strong field in the resonator, as each molecule passes through the resonator it has time to move from one level to another several times, and on the average half the molecules leave the resonator still in an excited state. To be more precise, the state of the molecular beam leaving the resonator is characterized by the superposition of the wave functions relating to the upper and lower levels. From (52.16) it can be stated that

$$v_2/n_2 = \tan\theta_0 = \sqrt{n'-1}. \tag{52.27}$$

The angle θ_0 gives the position of the energy spin vector in the energy space. When $n' \to \infty$, $\theta_0 \to \pi/2$ simultaneously. The power of an ammonia molecular oscillator estimated from (52.25) and (52.26) for $N_0' = 10^{14}$ mol/sec and $Q_L = Q_e/2$ is 5×10^{-10} W.

Let us find the frequency of the steady-state oscillations of a molecular oscillator. Equation (52.7a) for the stationary condition reduces to

$$\omega - \omega_0 = \frac{1}{2}\alpha\frac{s_3}{s}q\cos\bar{\Phi} = \frac{1}{\tau}\frac{n_2}{v_2}x_2\frac{\cos\bar{\Phi}}{\sin\bar{\Phi}}. \tag{52.28}$$

Generally, $K_1 \gg 1$ and we can take $\sin^2 \bar{\Phi} = 1$, $\cos \Phi = -2\Delta/K_1$, and obtain to a first approximation

$$\omega - \omega_0 = (\omega_c - \omega_0)\frac{\Delta\omega_0}{\Delta\omega_c} = (\omega_c - \omega_0)\frac{2Q_L}{\omega\tau}. \tag{52.29}$$

The greater the time of flight, and the greater the losses in the resonator, the closer is the frequency of the signal generated to the molecular transition frequency.

The analysis of the dynamic model of a molecular oscillator cannot be considered complete until we have investigated the dependence of the stability of its equilibrium positions on the values of the parameters. This problem has been discussed in papers by Khaldre and Khokhlov (1958) and Gurtovnik (1958). The phase space topology of the set (52.11) is essentially dependent on the magnitude of the parameter n'. With $n' < 1$ there is one singularity with coordinates (52.15). With $n' > 1$ there are two singularities, (52.15) now being unstable. The stability of (52.16) depends on the values of the parameters as a whole. Let us linearize equations (52.11) in the vicinity of the singularity (52.16), introducing the variables

$$\begin{aligned} \xi &= n - n_2, \\ \zeta &= x - x_2, \\ \eta &= v - v_2, \\ \varphi &= \Phi - \bar{\Phi}. \end{aligned} \tag{52.30}$$

The linearization leads to the following set of equations:

$$\begin{aligned} \dot{\eta} + \eta + \xi x_2 \sin \bar{\Phi} + \zeta \sin \bar{\Phi} + \varphi x_2 \cos \bar{\Phi} &= 0, \\ \dot{\xi} + \xi - \eta x_2 \sin \bar{\Phi} - \zeta v_2 \sin \bar{\Phi} - \varphi x_2 v_2 \cos \bar{\Phi} &= 0, \\ \dot{\zeta} + \tfrac{1}{2}K_1\zeta + \tfrac{1}{2}\eta K_1 \sin \bar{\Phi} + \tfrac{1}{2}\varphi K_1 v_2 \cos \bar{\Phi} &= 0, \\ x_2\dot{\varphi} + \zeta\Delta + \tfrac{1}{2}\eta K_1 \cos \bar{\Phi} - \tfrac{1}{2}\varphi K_1 v_2 \sin \bar{\Phi} &= 0. \end{aligned} \tag{52.31}$$

In order not to encumber future calculations with superfluous details we shall write the characteristic equation for the case $\Delta = 0$, when $\sin^2 \bar{\Phi} = 1$ and $\cos \bar{\Phi} = 0$:

$$(p + \tfrac{1}{2}K_1)\begin{vmatrix} p+1 & -v_2 & 1 \\ v_2 & p+1 & -v_2 \\ \tfrac{1}{2}K_1 & 0 & p + \tfrac{1}{2}K_1 \end{vmatrix}. \tag{52.32}$$

The problem consists of determining the signs of the real part of each of the roots of the characteristic equation. The presence of even one root with a positive real part indicates that the singularity under investigation is unstable. Since $p_1 = -\frac{1}{2}K_1$ we must examine the roots of the cubic equation

$$p^3 + p^2(2 + \tfrac{1}{2}K_1) + p(n' + \tfrac{1}{2}K_1) + K_1(n' - 1) = 0. \qquad (52.33)$$

Without solving the actual equation we can use the Raus–Hurwitz criterion (Neimark, 1949). The number of roots with a positive real part is the same as the number of changes of sign in the series

$$1, \quad D_1, \quad D_2/D_1, \quad D_3/D_2, \qquad (52.34)$$

in which D_l denotes the lth rank minors of the determinant of the coefficients of the equation (52.33)

$$\begin{vmatrix} 2 + \frac{1}{2}K_1 & K_1(n' - 1) & 0 \\ 1 & n' + \frac{1}{2}K_1 & 0 \\ 0 & 2 + \frac{1}{2}K_1 & K_1(n' - 1) \end{vmatrix}.$$

Writing out the terms of the series (52.34) in explicit form

$$1; 2 + \tfrac{1}{2}K_1; \quad \frac{n'(2 - \frac{1}{2}K_1) + K_1(2 + \frac{1}{4}K_1)}{2 + \frac{1}{2}K_1}; \quad K_1(n' - 1),$$

we notice that three of its four terms are known to be positive. For the oscillator to be stable all we have to do is to satisfy the condition

$$n'(2 - \tfrac{1}{2}K_1) + K_1(2 + \tfrac{1}{4}K_1) > 0. \qquad (52.35)$$

When $K_1 < 4$ inequality (52.35) is satisfied unconditionally. We recall, however, that in reality $K_1 \gg 1$. Because of this, stability is ensured only for values of the molecular concentration limited by the condition

$$n' < \frac{K_1(\frac{1}{4}K_1 + 2)}{\frac{1}{2}K_1 - 2} \simeq \tfrac{1}{2}K_1. \qquad (52.36)$$

Instability appears when the intensity of the flux of the molecules entering the resonator is $\frac{1}{2}K_1$ times greater than the starting value ($n' = 1$). Experimentally, one is limited to far less intense beams, and it is not surprising that the boundary of the stable region has not yet been reached in practice.

The results obtained permit us to make a purely qualitative statement about the nature of the motions in the system when both the singularities (n_1, v_1, x_1) and (n_2, x_2, v_2) are unstable. A stable trajectory should pass in the phase space between them. This means that the oscillations of a molecular oscillator are amplitude-modulated. We have already met the phenomenon of self-modulation when discussing a three-level paramagnetic oscillator (§ 51).

The velocity of the mapping point near the equilibrium position (n_2, v_2, x_2) is determined by the magnitude of the roots of the characteristic equation (52.32). An estimate leads to the values $|p| \sim K_1$. The possible motions of the system can be divided according to their characteristic times into fast $(\Omega \sim T_c^{-1})$ and slow $(\Omega \sim \tau^{-1})$. When a molecular oscillator moves into an unstable region fast motions should appear.

It is interesting to compare molecular and paramagnetic oscillators. The mathematical models of solid-state and molecular beam quantum oscillators that have been described differ in the relation between the longitudinal and transverse relaxation times. In a molecular beam $T_1 = \tau \gg T_c$, in a solid paramagnetic $\tau \ll T_c \ll T_1$. As a consequence a molecular oscillator has, in a certain range of its parameters, instability of the oscillatory equilibrium position, whilst a paramagnetic oscillator (or, to be more precise, its model) is stable for all values of the parameters. This comparison leads us to believe that when there is inhomogeneous broadening of the paramagnetic line $T_1 \geqslant \tau \gg T_c$. In this case a region of instability appears. But we cannot help seeing that this instability does not agree with the experimental data. The theory gives a high threshold for the self-modulation mode. In experiments with ruby this threshold is the same as the self-excitation threshold and, on the contrary, when the pumping increases the oscillation becomes stable.

52.3. The Spectra of Gases used as Working Substances

The creators of the first molecular oscillator used ammonia, $N^{14}H_3$, as the working substance. Preference has been given to this gas in the majority of subsequent designs. The popularity of ammonia is due above all to the intensity of its spectral lines, which lie conveniently in the centimetric waveband. Such qualities as cheapness and low toxicity, chemical stability and a high freezing point are not unimportant in practice. Long before the invention of molecular oscillators ammonia had become a classical subject for radiofrequency spectroscopy and has been investigated far more than other gases.

Certain component of the inversion spectrum of NH_3 are used for obtaining oscillation. The transitions corresponding to them are in the electric dipole category. Without going deeply into the theory of inversion spectra, which the reader can find in Gordy *et al.* (1953), Vuylsteke (1960), and Ginzburg (1947), we shall give merely some information of a general nature relevant to the ensuing discussion.

The NH_3 molecule is a molecule of the symmetrical top type. Three atoms of hydrogen and one atom of nitrogen are located at the apexes of a right pyramid. A number of different oscillatory and rotational motions are possible, in particular oscillations of the nitrogen at right angles to the plane of the hydrogen atoms. The potential energy of the molecule as a function of the distance between the nitrogen and the plane of the hydrogen atoms has the form of a symmetrical curve with two minima. If we neglect the tunnel effect, then each of the lower oscillation levels is doubly degenerate since there are two equivalent possibilities for positioning the nitrogen on one side or the other of the potential barrier. The presence of the tunnel effect eliminates the degeneracy. Since the splitting is connected with the passage of the nitrogen through the potential barrier, its inversion relative to the plane of the hydrogen atoms, it has been given the name of inversion splitting. The analogue of the inversion doublet is the splitting of the frequencies of the oscillations of two identical circuits when there is a weak coupling between them. The magnitude of the inversion splitting rises as we move to higher oscillation states.

As an example we point out the order of magnitude of the splitting in the lowest vibrational energy state characterized by the quantum number $v = 0$, and in two excited ones with $v = 1$ and $v = 2$. This will be 2×10^{10}, 10^{12} and 10^{13} Hz respectively. The inversion states are connected by an electric dipole transition and appear as intense lines in the absorption spectrum of ammonia. Under ordinary conditions the only lines observed are ones relating to the lowest vibrational state since the excited levels are hardly populated at all.

The fine structure of the inversion spectrum is caused by the rotation of the molecule. The nature of the connection between the oscillatory and rotational motions of the molecule is sufficiently obvious. The distance between the atoms, and therefore the form of the potential curve, depends upon the centrifugal forces due to the rotation of the molecule. For example, when the molecule rotates around the axis of symmetry the centrifugal forces try to drag the hydrogen atoms apart, resulting in a decrease in the potential barrier and an increase in the inversion frequency. Each component of the fine structure corresponds to a pair of rotational quantum numbers: J and K. The first determines the total rotational moment of the molecule and the second its projection onto the axis of symmetry. Rotational states with $J = K$ have the greatest statistical weight and therefore the inversion spectrum lines corresponding to it are more intense. Amongst them we can distinguish the line with $v = 0$, $J = 3$, $K = 3$ (abbreviated as the 3,3 line) which has a frequency of 23,870 MHz. This is the one most frequently used in molecular oscillators. The 3,3 line has a complicated hyperfine structure, due to various intermolecular interactions. The electric quadrupole interaction of the N^{14} nucleus with the molecular field splits the levels by several megacycles. The spin–spin interaction of the hydrogen and nitrogen nuclei and the interaction of the spins with the magnetic field of the rotating molecule give splittings up to 100 kHz. For a detailed description of the hyperfine structure of the inversion spectrum we must use the quantum numbers I_H, I_N, J, F, F_1, taking $\boldsymbol{I}_H$ as the total spin vector of the hydrogen nuclei, $\boldsymbol{I}_N$ as the spin vector of a nitrogen nucleus, $\boldsymbol{F}_1 = \boldsymbol{I}_H + \boldsymbol{J}$; $\boldsymbol{F} = \boldsymbol{F}_1 + \boldsymbol{I}_N$. Not all the components of the hyperfine structure are of equal importance in the operation of a molecular oscillator. Oscillation occurs at a transition with $\Delta F = 0$ and $F_1 = 0$. The corresponding line is made up of twelve components with different values of the quantum numbers F and F_1 (Fig. XI.7). The distance between the components is too small for them to be observed separately, but the form of the resulting line depends upon the relative intensities of the individual components, which are determined in their turn by the state of the gas. For example, the central frequency ω_0 of the emission line from the molecular beam depends on the voltage across the focusing system since it controls the molecular state distribution function.

The dependence of the central frequency of the 3,3 ammonia line, and therefore of the oscillation frequency, on the mode of operation of the molecular oscillator is an unpleasant fact if we remember that such devices are often used as frequency standards. In order to eliminate this effect, or at least to reduce it we have to turn to different spectral lines. For example the 3,2 line of the same $N^{14}H_3$ or the lines of $N^{15}H_3$ have no electric quadrupole structure so they should not depend upon the field in the focusing system. Although the 3,2 line (f_0 = 22,384 MHz) is several times less intense than the 3,3 line its advantages justify its use in certain cases (Basov, Nikitin and Oraevskii, 1961).

The formaldehyde ($C^{12}H_2O^{16}$) molecule, which can also be used to obtain oscillation, is an elongated, slightly asymmetrical top (Krupnov and Skvortsov, 1963a and 1963b). The transition between the lowest rotational levels 1_{01} and 0_{00} has a frequency f_0 = 72,838 MHz, i.e. lies in the 4 mm waveband. The Stark effect in formaldehyde molecules is shown in Fig. XI.8.

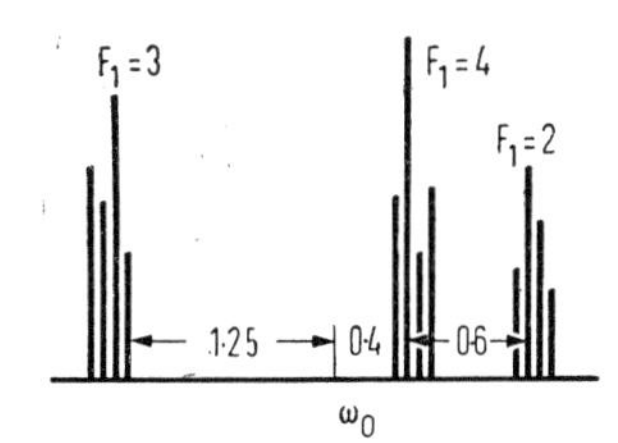

FIG. XI.7. Fine structure of the $J = 3$, $K = 3$ line of the inversion spectrum of $N^{14}H_3$. The frequencies are given in kilohertz.

Until the field strength exceeds a value corresponding to the maximum of the curve $J = 1$, $M = 0$ (150 kV/cm) there is focusing of the molecules in the $J = 1$, $M = 0$ state and defocusing of all the other molecules. A favourable point is the absence of any hyperfine structure in the spectral line.

The transition between the lower rotational levels of the linear molecule HCN (hydrocyanic acid) lies in the 3 mm waveband; to be more precise, it has a frequency of 88·63 GHz. The spectral line consists of three resolved components of the hyperfine quadrupole structure. The self-excitation conditions have been successfully reached in two of these components and oscillation observed (Marcuse, 1961).

The atomic hydrogen line due to the transition between the hyperfine levels $F=1$, $m=0$ and $F=0$; $m=0$ of the $^2S_{1/2}$ ground state has a simple

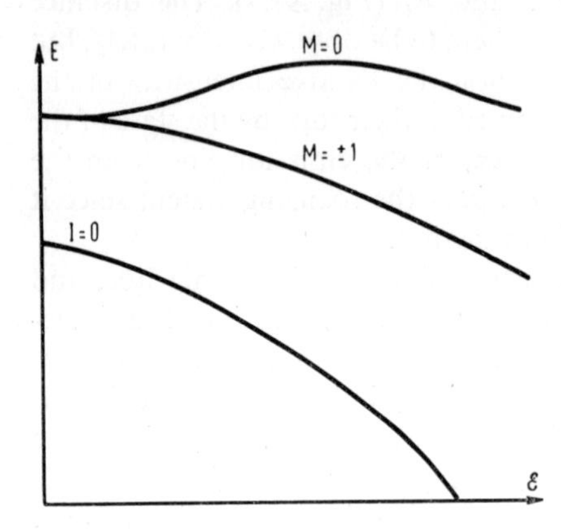

FIG. XI.8. Stark effect in the lower rotational levels of a formaldehyde (CH_2O) molecule.

structure; its frequency is 1420·405 MHz. The transition is magnetic dipole and so is far weaker than the inversion and rotational transitions of molecules discussed above. There are consequently particular difficulties in achieving oscillation in such a maser (Kleppner *et al.*, 1962). The principle of sorting the H atoms and focusing the excited atoms is clear from Fig. XI.9, which shows the Zeeman splitting of the ground state.

We have given above all the working substances which have so far been used in molecular oscillators. The list will, of course, be extended but we cannot help noticing that it is not being done very quickly at present. The search for new substances is stimulated by two practical problems: the requirement

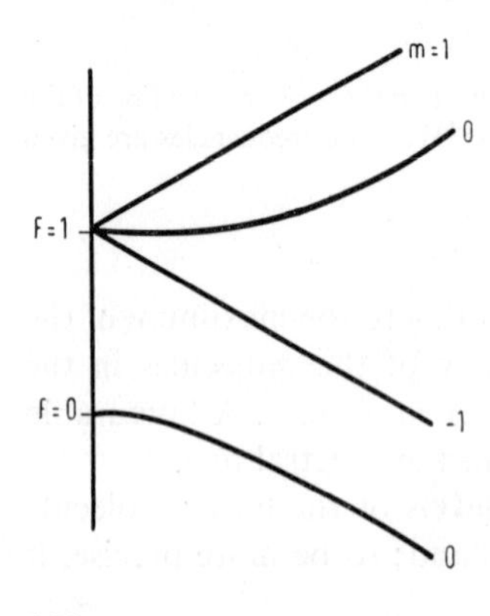

FIG. XI.9. Zeeman effect in the ground state of atomic hydrogen.

for very high stability in frequency standards, and the problem of penetrating into the sub-millimetric waveband. In both cases optical maser oscillators are serious competitors. A theoretical analysis permits us to select certain gases (Zhabotinskii and Zolin, 1959; Barnes, 1960) which are the most promising from the point of view of using them in a molecular oscillator. Hydrogen halides are a case in point.

52.4. The Design of a Molecular Oscillator

Figure XI.10 shows in the most general outline how a molecular oscillator is constructed. Its essential parts include: a beam source, a separation or focusing system, a resonator with a waveguide output, a system for maintaining a vacuum within the oscillator. Now that other gases (HCN, CH_2O and H) are being used in molecular oscillators as well as NH_3, and that the field of application of frequency standards is extremely extensive, it is impossible in a brief description to cover the whole range of designs of the device and its various parts. The hydrogen maser oscillator differs most from the others. The transition at $\lambda = 21$ cm is magnetic dipole, and the

TABLE XI.1. PARAMETERS OF QUADRUPOLE-TYPE SEPARATION SYSTEMS

Length of electrodes (cm)	Distance between electrodes (cm)	Working potential (kV)	References
56	0·2	15	Gordon *et al.*, 1954
20	4	30	Gordon *et al.*, 1954
10	0·6	30	Basov, 1956

spectral line connected with it is very weak. The separation of the hydrogen atoms is based on the Zeeman effect and is achieved with a non-uniform magnetic field. The other gases named have electric dipole transitions and the differences in their design are not of such a major nature. We shall therefore concentrate our attention on describing an ammonia beam oscillator and to conclude we shall deal briefly with the features of a hydrogen maser.

The source of the molecular beam is a chamber containing ammonia, generally at room temperature, at a pressure slightly greater than in the rest of the device. The gas molecules leave the chamber, forming a directional molecular beam, through the holes in a grid or a single orifice directed towards the separation system. In any case it is desirable to obtain as intense and directional a beam as possible; the cross-section required depends upon the aperture of the separation system. The output of such sources can be as much as 10^{18} molecules/sec. The source chamber is connected via a pressure reducing valve to a reservoir to act as a supply of gas. The separation system most often used is the quadrupole capacitor mentioned above. The parameters of some systems of this kind are given in Table XI.1.

Of the other designs we should mention the ring-type systems (Krupnov, 1959) (Fig. XI.6) which are a series of coaxial annular electrodes with alternating charge signs. The flat modification of this kind of system is particularly

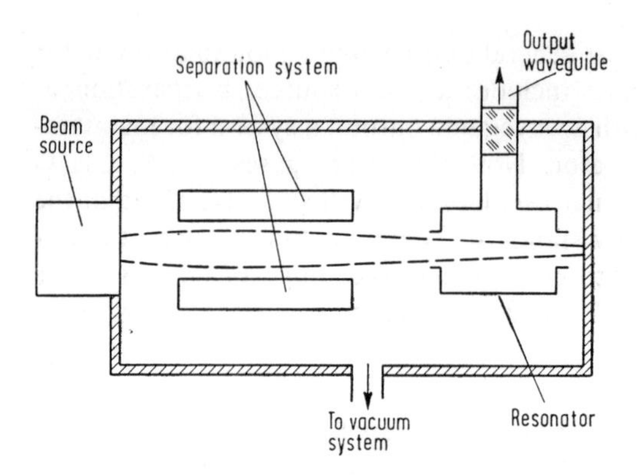

FIG. XI.10. Diagram of a molecular oscillator.

good when working with a Fabry–Perot resonator (Barchukov *et al.*, 1963; Krupnov and Skvortsov, 1965).

As the voltage across the electrodes of a sorting system is increased its efficiency rises for two reasons. Firstly, there is an increase in the effective angular aperture, which can be determined from equality of the Stark energy and the kinetic energy of radial motion. Secondly, the degree of separation is improved with large field gradients. However, if the dependence of the Stark energy of the molecule on the field strength changes sign, the voltage across the separation system electrodes must not be raised above a certain limit. This is the situation for the 1_{01} rotational level of a CH_2O molecule (Fig. XI.8).

The flux of molecules leaving the source is many times greater than the flux of the active molecules reaching the resonator. Losses occur for many reasons. Because the angular divergence of the beam is greater than the acceptance angle of the separation system only a certain number of the molecules reach the latter. Not all the molecules are in the necessary 3,3 rotational state; there is only about 6% of these molecules in the beam. Altogether not more than 10^{14} active molecules enter the resonator per second.

The requirements demanded of a resonator are dictated by the desire to obtain oscillations of the highest possible power, spectral purity and stability. The threshold intensity of the molecular flux is given by (52.22). By slightly altering this expression we can write the starting flux as

$$N_0 = \frac{\hbar \bar{v}^2 A_c}{4\pi\,|\mu_{12}|^2\, l Q_L} = \frac{\hbar v^2}{4\pi\,|\mu_{12}|^2 M}, \tag{52.37}$$

taking A_c as the cross-sectional area of the resonator. The quantity

$$M = \frac{l Q_L}{A_c}\left(\frac{8}{\pi^2}\right)^n \tag{52.38}$$

is sometimes called the figure of merit of the resonator. If the field along the resonator axis is uniform, then $n = 0$. We have discussed this case earlier; the starting flux (52.37) corresponds to it. For a mode with a single maximum on the axis $n = 1$. The figure of merit of the resonator therefore drops, and the starting flux rises, when the field configuration of the working mode is made more complex. As we know, when the oscillations have a uniform field distribution along the resonator axis the Doppler broadening of the spectral line is minimal; this configuration is therefore generally chosen for the operation of a molecular oscillator. It includes oscillations of the TH_{001} type in a cylindrical, and TE_{011} in a rectangular resonator.

The coefficient M rises when the length and Q-factor of the resonator are increased; these are obviously physical considerations. The inverse proportional dependence on A_c strictly speaking holds only when there is a narrow beam; for a broad beam there is no strong dependence of M on A_c. Optimum coupling of the resonator to the waveguide system, which will ensure maximum output power from the oscillator, is obtained when $Q_a = Q_e$, i.e. $Q_L = Q_a/2$. The loaded Q-factor Q_L is determined only by the internal losses of the resonator, i.e. by the material, the manufacturing accuracy and the temperature. Table XI.2 (Shimoda *et al.*, 1956) gives an indication of the properties of cylindrical resonators.

As an alternative to the conventional cavity resonator the Fabry–Perot type of resonator is coming into use in molecular oscillators (Marcuse, 1961; Barchukov *et al.*, 1963; Krupnov and Skvortsov, 1965); the reader will find general information about these in Appendix II. From the design standpoint resonators of the open type are very convenient to use in a molecular beam oscillator, particularly in the millimetric waveband. The design of the resonator also determines the design of the other component parts. There is a possibility, in principle, of increasing the number of beams. In a cylindrical cavity resonator we can work with a maximum of two opposed beams; this method is sometimes used to make the oscillator more symmetrical. It is not difficult to inject a large number of beams into a Fabry–Perot resonator, but they obviously end up by interfering with each other (Krupnov and Skvortsov, 1965).

From the time the beam molecules leave the source until they leave the resonator they should not undergo any collisions. The molecular oscillator must therefore be enclosed in a leak-free vessel, and measures must be taken to maintain an adequate vacuum inside it. For a mean free path of the order of 20 cm the permissible pressure in the system should not exceed 10^{-5} torr.

Taking into consideration that about 10^{18} molecules of ammonia enter the vessel per second, high-speed pumps (10^4 l/sec) are required to maintain the necessary vacuum. Another method, consisting of cooling the resonant cavity, is more practicable. At the temperature of liquid nitrogen ammonia is solid and has a low vapour pressure (10^{-6} torr) so even a small diffusion pump is capable of maintaining the required low pressure. In the oscillator made by Gordon, Zeiger and Townes the nitrogen filled the electrodes of the quadrupole capacitor. This also solved the problem of reducing the pressure at the source outlet, which rises because of reflection of some of the molecules from the separation system. Because of condensation of NH_3 on the electrodes the aperture gradually grows smaller and finally the oscillator has to be stopped and cleaned. The oscillator cavity can be cooled by surrounding it with a nitrogen jacket as Basov (1956) has done. In this case a cooled diaphragm must be placed between the source and the separation system.

Such in general outline is the construction of an ammonia molecular beam oscillator. In a hydrogen maser (Kleppner *et al.*, 1962) the main difficulty is the low intensity of the spectral line. Special measures therefore have to be taken to lengthen the lifetime of the hydrogen atoms in the resonator. A quartz flask covered with Teflon is put into the resonator and the excited hydrogen atoms enter it through a single hole. Collisions with theTeflon do not alter the state of the atom. The extremely narrow line width m akes it possible to satisfy the self-excitation conditions. It should be pointed out that there is hardly any Doppler line broadening since the atom is in an excited state for more than 1 sec and its velocity when averaged over this time is close to zero. The beam of hydrogen atoms is passed through the non-uniform field of a multipole magnet for sorting and focusing.

TABLE XI.2. PARAMETERS OF CYLINDRICAL RESONATORS 12 cm LONG FOR DIFFERENT MODES OF OSCILLATION, $\lambda = 1{\cdot}25$ cm, $e = 4{\cdot}27 \times 10^{-5}$ cm (THE SKIN DEPTH OF COPPER AT ROOM TEMPERATURE) (SHIMODA *et al.*, 1956)

Type of oscillation	Radius (cm)	Q_a	*Me* narrow beam	*Me* broad beam
TE_{111}	0·37	6,100	12·2	5·9
TH_{010}	0·48	10,800	28·4	7·7
TH_{011}	0·48	10,400	22·2	6·0
TE_{211}	0·61	8,100	0	2·9
TE_{011}	0·76	17,800	0	4·1

52.5. Remarks on the Frequency Stability of a Molecular Oscillator†

The accuracy with which we can determine the frequency of an oscillator is set by the natural width of its spectrum. The latter is determined by such factors as the thermal fluctuations, the shot effect which is due to fluctuations in the density of the stream of molecules in the resonator, and the noise component of the spontaneous emission. Thermal noise is the most important. The broadening of the oscillation spectrum of a maser oscillator due to thermal noise has been estimated by Gordon *et al.* (1954) and has been considered more closely by Troitskii (1958a and 1958b). In the case of $\Delta\omega_0 \ll \Delta\omega_c$, which is valid for molecular oscillators,

$$\Delta f = \frac{1}{\tau^2}\,\frac{kT}{P_0}\,. \tag{52.39}$$

The natural line width for $\tau = 10^{-4}$ sec, $P_0 = 10^{-10}$ W and $T = 300$°K is only 5×10^{-3} sec, i.e. the maximum relative stability of a molecular oscillator is $0{\cdot}2\times 10^{-12}$.

It is quite a complicated business to achieve such high stability since unavoidable subsidiary effects come into play to make the actual frequency stability worse. Amongst them we must first mention the drift connected with the variation of the resonator tuning. The change of the oscillator frequency when the resonator tuning is varied is given by (52.29) and amounts to

$$\Delta f = \Delta f_c\,\frac{Q_L}{Q_0} \ll \Delta f_c. \tag{52.40}$$

The resonator tuning may vary as a result of a change in its temperature. Even a resonator made of Invar changes its frequency by $10^{-9} f_c$ with a 1°C change in temperature. High stability is therefore unattainable without a thermostatically controlled resonator.

The other effects which reduce the stability of molecular oscillators are outside the framework of the rough mathematical model which we used above. Some of them are connected with the multiplet nature of the spectral line. Under ordinary conditions the hyperfine components cannot be resolved. However, the peak of the line as a whole is determined by the intensity distribution of these components, and, since the interaction of a molecule with the field of the separation system depends upon the quantum numbers F and F_1, the frequency of the molecular line is a function of the voltage across the separation system electrodes. In addition, the various spectral components differ in the value of their dipole moment matrix elements. Because of this the position of the maximum of the line depends on the strength of the field in the resonator, i.e. on the intensity of the molecular beam.

† Oraevskii (1963a and 1963b) discusses questions of stability in detail.

These effects are absent in lines which have no hyperfine structure. In this respect the 3,3 line of ammonia $N^{14}H_3$ is far from optimal. It has only one advantage, maximum power, which is also important from the point of view of stability. Its 3,2 line and the lines of the inversion spectrum of $N^{15}H_3$ have no quadrupole structure. Therefore the oscillation frequency of a 3,2 line oscillator is far less dependent on the voltages across the electrodes of the separation system and in the oscillating circuit, but this dependence cannot be entirely eliminated. The point is that the time-of-flight distribution of the molecules in the separated beam is not $(1/\tau)\, e^{-t/\tau}$; worse still, it depends on the voltage across the separation system and is not Lorentzian in form.

The representation of the resonator by an equivalent circuit with lumped parameters, which was used above, is not completely justified even for modes like TH_{010}. As the molecules pass through the resonator they do not radiate uniformly along its length. In addition a lack of symmetry is introduced by the coupling to the output waveguide. All this leads to the appearance in the resonator of a non-uniform energy distribution and a resultant shift in the emission frequency. Systems with opposed beams of equal intensity and energy take-off exactly at the centre of the resonator are used to reduce this effect.

The frequency stability of existing molecular oscillators reach 10^{-11} after a few hours of operation. Any further increase in the stability, and particularly the absolute stability,† depends on improving the factors contained in (52.43): narrowing the spectral line while preserving a high power. This result can be achieved by using beams of slow molecules.

53. Two-level Solid-state Quantum Oscillators

53.1. Survey of Experimental Results

In practically every case ARP (adiabatic rapid passage) is used for the excitation of a solid two-level working substance. The resonator containing the working substance is placed in a cryostat and held in the gap between the poles of a magnet. The change in the magnetic field necessary for ARP and the subsequent matching of the frequencies of the substance and the resonator is achieved by a modulating coil. A special unit supplies this coil with a current of a given form. This same unit controls the operation of a powerful pulsed oscillator (the pumping generator). As a rule sine-wave modulation of the magnetic field is used. Its frequency is a few times greater than the repetition frequency of the pumping pulses. A standard signal generator is connected in parallel with

† Oraevskii (1963 b) defines the absolute stability as the accuracy with which the frequencies of two independent oscillators can be compared with the frequency of the spectral line. The relative stability is defined as the accuracy with which the difference frequency of the two oscillators is maintained in time.

the pumping generator to determine the maser parameters. The rest of the installation for investigating a two-level maser does not differ from a paramagnetic radiofrequency spectroscope and is largely made from standard components. As a rule rectangular resonators are used in two-level masers, since they are simpler and more convenient; they are either cavity resonators or structures with a solid filling. In the latter case the resonator is the single paramagnetic crystal itself (Thorp *et al.*, 1961; Thorp, 1961).

The use of active substances with a high ε allows us to reduce the volume of the resonator, which is sometimes very desirable, particularly when we move into the millimetric and sub-millimetric wavebands.

The design and position of the modulating coils depend on the modulating field amplitude required. If the working frequency is the same as the pump frequency, or is close to it, no difficulties arise. The modulating field in this case is small (of the order of the line width) and is parallel to the constant magnetizing field. The coils are located outside the cryostat. Helmholtz coils are used when greater uniformity of the field is required. The position is far more difficult for masers in the short wavelength of the millimetric band. Since the magnetic fields required are tens of thousands of oersteds one has to use the techniques connected with powerful pulsed magnetic fields. In this case the pumping is at a frequency much lower than the working frequency of the maser. The modulating field is many times more than the magnetizing field and the mutual orientation of these two fields is no longer so significant. All that is important is to maintain a constant angle between the magnetic field and axis of the crystal. One possible design for a maser with a pulsed magnetic field is shown in Fig. XI.11. The pulse coil takes the form of a solenoid whose axis is parallel to the axis of the cryostat. When it is necessary to reduce the diameter of the coil it can be put inside the nitrogen dewar so that it can be cooled at the same time.

It is convenient to classify two-level paramagnetic masers into two groups according to closeness of the working and secondary frequencies. We shall call them masers with matching frequencies and masers with separated frequencies for the sake of the discussion.

The basic data relating to masers with matching frequencies are summarized in Table XI.3. The first who tried to make a paramagnetic two-level maser were Combrisson, *et al.* (1956). Their work was of fundamental importance, although they did not achieve the expected effect. By substituting the actual figures from Table XI.3 in the inequality (44.22) it can be checked that the self-excitation condition was not satisfied here. Combrisson, Honig and Townes used as the working substance a single crystal of silicon doped with phosphorus: the phosphorus atoms are a donor impurity, and are responsible for the paramagnetic properties of the substance. Its record spin–lattice relaxation time, which is tens of seconds at liquid-helium temperatures, distinguishes it from a number of other paramagnetics.

TABLE XI.3. EXPERIMENTAL DATA ON TWO-LEVEL PARAMAGNETIC MASERS

Maser parameters	Working substance					
	Natural Si with P impurity	Isotopically pure Si^{28} with P impurity	Quartz containing *F*-centres	MgO containing *F*-centres	Ruby	Diphenylpicrylhydrazyl
Method of excitation	—	ARP	ARP	ARP	ARP	45° pulse
f (GHz)	9	9	9	9	33·1	9·3
Q	10^4	2×10^4	6×10^3	6×10^3	$2·5 \times 10^3$	30
$T_c = Q/\omega$ (sec)	2×10^{-7}	4×10^{-7}	10^{-7}	10^{-7}		5×10^{-10}
V_c (cm³)	4·4	—	—	—	3×10^{-2}	—
V_s (cm³)	0·6	0·3	—	—	3×10^{-2}	—
N	6×10^{16}	$1·2 \times 10^{16}$	10^{18}	10^{17}	10^{17}–10^{18}	10^{20}
T (°K)	2	1·2	4·2	4·2	1·2–4·2	300
ΔH (oersted)	2·7	0·22	1	1	20	
T_1 (sec)	5–30	60	2×10^{-3}	$2·5 \times 10^{-3}$	6×10^{-2}	4×10^{-8}
T_2 (sec)	$4·2 \times 10^{-8}$	$5·2 \times 10^{-7}$	$1·1 \times 10^{-7}$	10^{-7}–10^{-8}	6×10^{-9}	4×10^{-8}
Magnetic moment	$6·5 \times 10^{-5}$	$2·2 \times 10^{-5}$	$5·1 \times 10^{-4}$	$5·1 \times 10^{-5}$		
Duration of oscillation (sec)		50×10^{-6}	5×10^{-6}		25×10^{-6}	4×10^{-8}
Oscillation power (W)		$2·5 \times 10^{-6}$	12×10^{-3}		70×10^{-6}	
$G^{1/2} f$ (MHz)		–	5	–	–	
G_{max} (db)			21	20	14	
Remarks	Oscillation conditions not achieved	Amplifier parameters not determined			*Q*-factor shown corresponds to oscillation threshold	Cannot work in the amplification mode
References	Combrisson *et al.* 1956	Feher *et al.*, 1958	Chester *et al.*, 1958	Chester *et al.*, 1958	Thorp *et al.*, 1961; Thorp, 1961	Kaplan and Browne, 1959

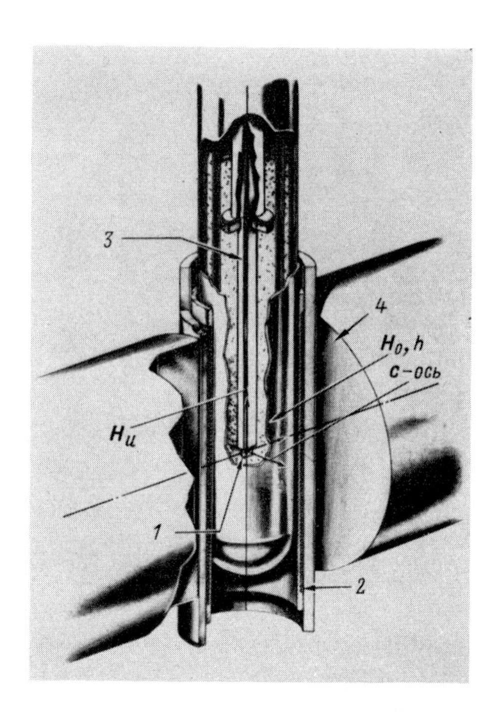

FIG. XI.11. Design of a maser with a pulsed magnetic field (Foner *et al.*, 1960): 1—crystal resonator; 2—solenoid for producing a pulsed field; 3—sapphire; waveguide section; 4—electromagnet.

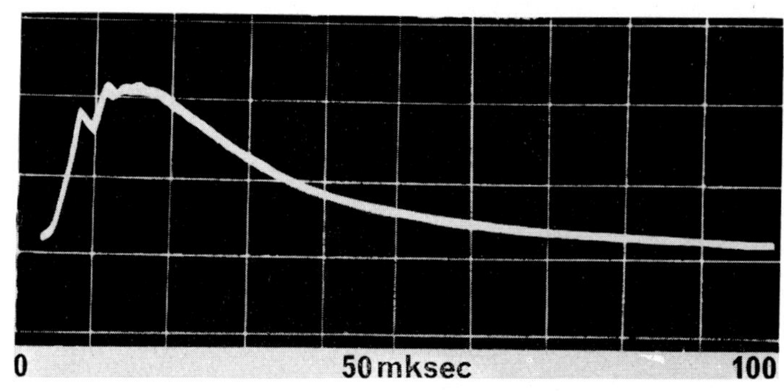

FIG. XI.12a. Oscillogram of the output of a phosphorus-doped silicon maser (Feher *et al.*, 1958); (1 mksec ≡ 1 μsec).

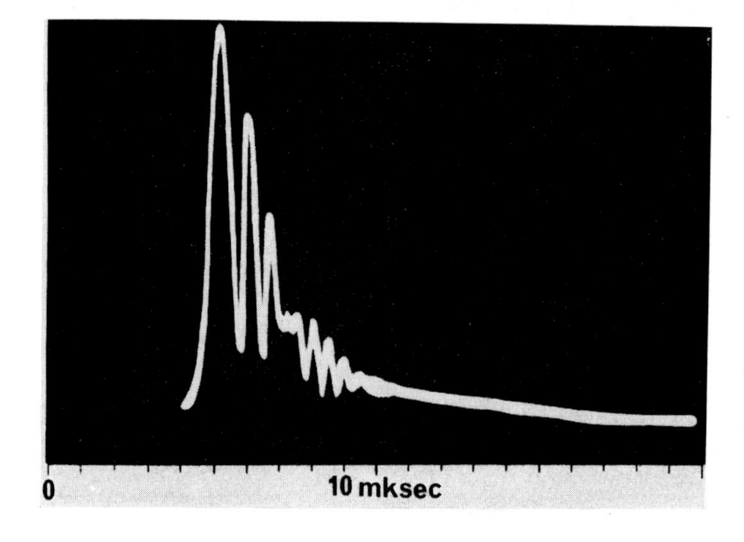

FIG. XI.12b. Oscillogram of the output of a maser using quartz irradiated with fast neutrons (Chester *et al.*, 1958); (1 mksec ≡ 1 μsec).

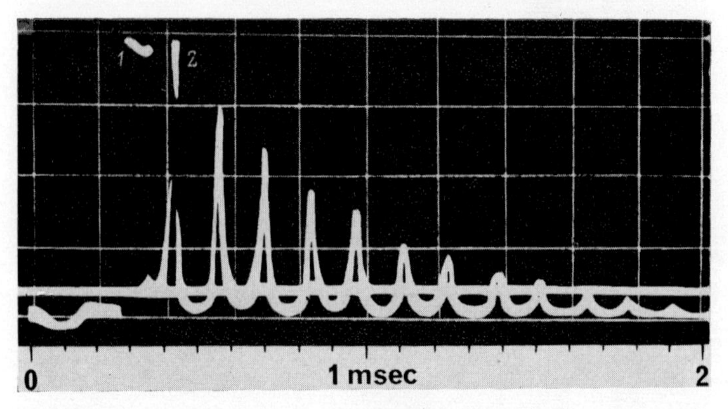

FIG. XI.13. Drop in amplification factor with time. The first peak on the oscillogram (1) is caused by the inverting signal, the second (2) by oscillation, the subsequent ones by amplification of the test signal during successive passes through resonance. The horizontal straight line is drawn at the $G = 1$ level (Chester *et al.*, 1958).

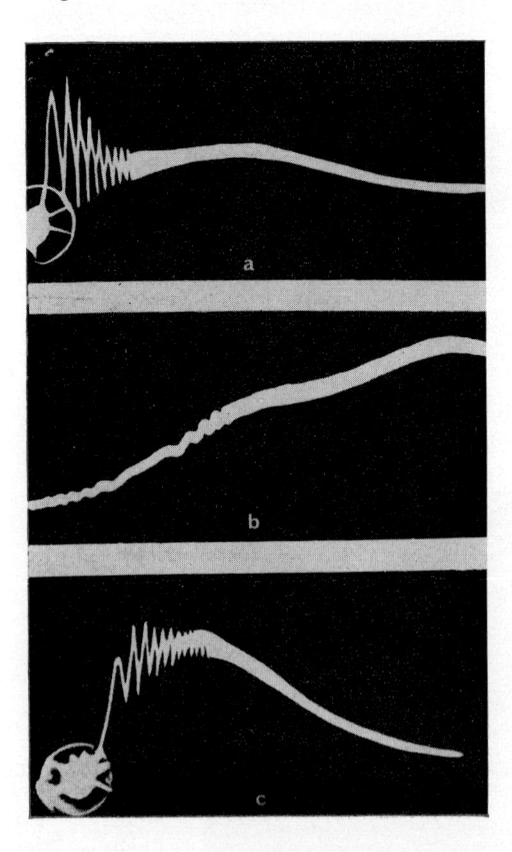

FIG. XI.17. Dependence of form of signal of a two-level paramagnetic oscillator on the initial conditions and rate of sweeping of the field ($K_1 = 1{\cdot}33$): (a) $n_0 = 2{\cdot}7$, $\alpha = 4{\cdot}5 \times 10^{-2}$, duration of sweep 200 μsec; (b) $n_0 = 10, \alpha = 9 \times 10^{-3}$, duration of sweep 200 μsec; (c) $n_0 = 10$, $\alpha = 0{\cdot}23$, duration of sweep 30 μsec, $\alpha = T_2^2(d\omega_0/dt)$.

In their experiment Feher *et al.* (1958) also used silicon doped with phosphorus. Unlike their predecessors, however, these researchers used a specimen of isotopically pure Si^{28}. In this way they eliminated the inhomogeneous broadening of the resonance line caused by the hyperfine interaction of the impurity electrons and the Si^{29} nuclei. The line was narrowed to 0·22 Oe, and thus the self-excitation condition was satisfied. An oscillogram of the emission pulse is reproduced in Fig. XI.12a.

Paramagnetism in a crystal can be achieved by defects of another type besides impurity atoms. Amongst these are *F*-centres, i.e. electrons replacing ions that have been knocked out of their sites in the lattice. One way of obtaining *F*-centres is irradiation of the crystal with fast neutrons. Crystals of quartz and magnesium oxide used in masers were processed in just this way (Chester *et al.*, 1958). An oscillogram of the oscillation of a quartz maser is shown in Fig. XI.12b. We shall now consider briefly the methods of measuring the amplification. A sine-wave modulated magnetic field ensures periodic passage through resonance. During the first pass the pumping generator is triggered and population inversion of the levels of the working substance occurs. If the degree of inversion is high enough, the second pass is accompanied by oscillation. In subsequent passes amplification is observed. In time the amplification factor drops as the number of molecules decreases because of longitudinal relaxation (Fig. XI.13). After the molecular system has reached a state of thermal equilibrium the pumping is started again and the cycle is repeated.

Ruby can also be used as the working substance for a two-level maser. The paramagnetism of ruby, as was pointed out above, is caused by the chromium ions which have a spin of 3/2, i.e. a ruby has four energy levels. Any pair can be used for the operation of a two-level maser although the corresponding transitions slightly in the magnitude of their matrix elements. It should be pointed out that ions with a high spin are of no use in two-level devices since some of them are in states which play no part in the maser action.

The experimental results given above allow us to distinguish the following characteristic features of a two-level maser.

(a) The amplitude of the emitted signal may vary during the pulse, i.e. the pulse may possess a fine structure.

(b) The duration of the emitted pulse may be large compared with the characteristic attenuation times T_2 and T_c.

(c) On completion of the oscillation process the working substance remains in an excited state and may be used for amplification.

An experiment by Kaplan and Browne (1959) with free radicals is rather an exception amongst those discussed. The relaxation parameters of the free radicals do not permit ARP and a 45° pulse is used for excitation. The pumping pulse must decay very rapidly because of the short duration of the emission process. Therefore the resonator's *Q*-factor must be correspondingly low. Under these conditions the resonator's influence on the motion of the magnetic moment is negligible and the latter can be looked upon as free precession. The theory agrees with experiment and predicts for this case an emitted pulse with an exponential form with a time constant T_2.

Masers with separated frequencies are dealt with in papers by Hoskins (1959a and 1959b), Hoskins and Birnbaum (1960), Foner *et al.* (1959) and Foner *et al.* (1960), who deal with design of a maser for the short wavelength part of the millimetric and sub-millimetric bands.

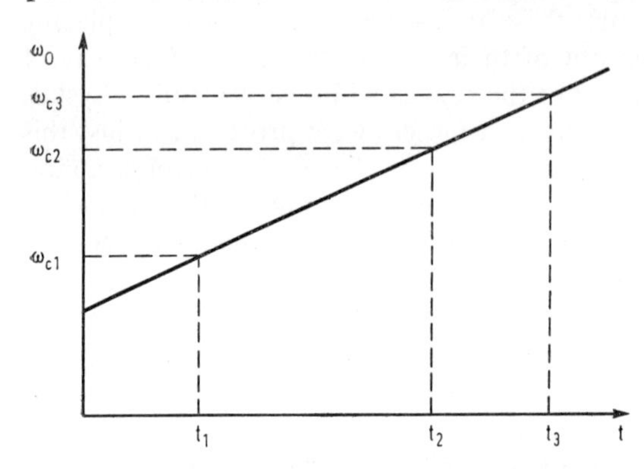

FIG. XI.14. Diagram to explain working principle of maser with variable transition frequency. The working substance is excited by secondary emission at the frequency of one of the lower modes ω_{c1}; the working frequencies used are those of the higher modes ω_{c2}, ω_{c3}, etc.

We shall use Fig. XI.14 to explain the working principle of a maser with separated frequencies. The magnetizing field H_0 is selected so that the initial frequency of the paramagnetic resonance is close to the fundamental frequency ω_{c1} of the resonator. The pump generator works at the frequency ω_{c1}. Once the paramagnetic is in an excited state the pulsed magnetic field changes its frequency ω_0 until it is the same as one of the higher eigenfrequencies ω_{cn} of the resonator. At these frequencies the paramagnetic emits.

Experiments of this kind described in published papers differ in practice only in their methods of excitation. Hoskins (1959a and 1959b; Hoskins and Birnbaum, 1960) uses three of the four levels of a ruby for excitation, with 70° orientation (Fig. XI.15). The magnetizing field is 6 kOe; the pulsed field is sinusoidal, of amplitude 4 kOe, and is parallel to the magnetizing field. The 2–3 and 1–2 transitions come in turn into resonance with the pumping generator. As these resonances are passed adiabatic inversion of the populations occurs first between the 2–3 levels and then the 1–2, after which the pumping pulse stops. As a result the number of active molecules $\sigma_{22} - \sigma_{11}$ is $\sigma_{11}^0 - \sigma_{33}^0$, i.e. more than $\sigma_{11}^0 - \sigma_{22}^0$ with single inversion. This method is called step excitation (Siegman and Morris, 1959; Wagner *et al.*, 1960). Oscillation at frequencies of 14, 18, 24 and 28 GHz is observed on the return of the pulsed field.

Foner (Foner *et al.*, 1959; Foner *et al.*, 1960) excited a ruby with a three-level arrangement. In the absence of a pulsed field the pump generator saturates the 1–3 transition, inverting the populations of levels 1 and 2. Pumping is at the frequency of 12·7, MHz the resonator's fundamental mode. The amplitude of the 30 kOe pulsed field corresponds to a paramagnetic frequency of about 70 GHz. The authors note that the form of the pulse depends on the rate of change of the magnetic field. In particular the nature of the beats and the power in the pulse change.

A frequency of 70 GHz ($\lambda \sim 4$ mm) is at present the maximum for a two-level maser. This is not the limit, and in principle the frequency can be increased indefinitely, but there are considerable practical difficulties in doing this. They are due to the requirements imposed on the magnetic field. In the first place the field must be strong enough: for example, a wavelength of 1 mm corresponds to $H \sim 100$ kOe. In the second place it should change rapidly enough for the excited state of the paramagnetic to be maintained during the change. For $T_1 = 10^{-1}$ sec the rise time of the pulse should be of the order of 10^{-2} sec. In the third place fairly rigid requirements are imposed on the uniformity of the field in the crystal. Non-uniformity of the field leads to a spread of the resonance frequencies of the paramagnetic centres located in different sections of the crystal, and reduces the number of active molecules taking part in the process. Quantitatively, the non-uniformity of the field should not exceed the width of the resonance line, i.e. a few oersteds.

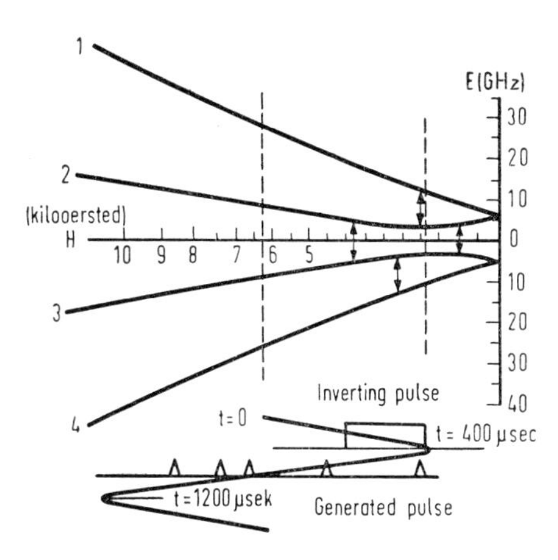

FIG. XI.15. Diagram of levels in ruby for $\theta = 70°$, and the form of the magnetic field pulse that permits adiabatic population inversion of levels 3 and 2 and then 2 and 1 (Hoskins, 1959 a).

The latter requirement is difficult to reconcile with the first two. The volume of the field should be increased in order to increase the uniformity, whilst an increase in the amplitude and the rate of rise of the pulse introduces the necessity for reducing the size of the coils. One of the ways of resolving the contradiction is to reduce the dimensions of the resonator and the paramagnetic crystal. Substances with a high ε such as ruby and rutile are particularly valuable here.

Among the disadvantages of a two-level maser is the variation of the magnetic field during the oscillation and amplification process. This has a particularly marked effect with powerful pulsed fields, when it is hard to find a flat enough section at the peak of the pulse. Changes in the field lead to variation of the output frequency; the idea has been proposed of working in a zero magnetic field in order to avoid this difficulty (Hoskins and Birnbaum, 1960).

Now a few words about pulsed oscillators for the centimetric and millimetric bands using ferromagnetic dielectrics (ferrites) as the working substance (Aleksandrov *et al.*, 1960; Stiglitz and Morgenthaler, 1960; Elliott *et al.*, 1960). We shall not deal with the properties of ferrites here, but refer those interested in this question to a book by Gurevich (1960). In ferrites the density of the spins is several orders higher than in paramagnetics. With the magnetic fields generally used the magnetization of ferrites reaches saturation and is weakly temperature-dependent, which is also a favourable property of these materials. Their basic defect as working substances, however, is their very short relaxation time. Yttrium iron garnets have the narrowest resonance lines but even for them $\Delta H \gtrsim 0.1$ Oe, which corresponds to relaxation times of $\tau \lesssim 10^{-6}$ sec.

In its operating principle a ferrite oscillator is similar to a maser with separated frequencies. However, the three or four orders of magnitude which distinguish the τ of ferrites and the T_1 of paramagnetics considerably complicate affairs since pulsed magnetic fields with very steep leading edges are required. Adiabatic excitation of a ferrite is also impossible because of the smallness of τ.

The second disadvantage of a ferrite in the capacity described is the presence of the so-called critical angle of excitation θ_c. As soon as the angle between $\boldsymbol{M}$ and $\boldsymbol{H}_0$ reaches θ_c uniform precession stops because of the pumping of energy into the higher non-uniform types of oscillation (spin waves). The connection of the uniform precession with the spin waves is non-linear. Since the critical angle of excitation θ_c is very small ($\theta_c \sim 10^{-2}$) a large amount of energy cannot be accumulated in a ferrite.

Solid-state two-level masers mark a definite stage in the development of quantum electronics. One of the basic directions in which the achievements of quantum electronics has, or may have, important practical significance is in entering into new bands of short electromagnetic waves. Amongst these are the sub-millimetric waves, i.e. wavelengths from 10 μ to 1 mm. For some time great hopes were held out of making sub-millimetric amplifiers and oscillators with solid-state two-level masers. There has not been much development in the field of two-level masers, however, which can be explained by the obvious disadvantages and design complexity of the device. For the present we can speak of a two-level maser as an interesting physical object but not as a device with any technical application.

53.2. The Theory of a Two-level Oscillator

A solid-state two-level maser differs from the devices discussed above by the absence of continuous pumping. Consequently, periodic processes cannot take place in it, the emission is limited to a certain finite time interval, and the system finally returns to a state of thermal equilibrium. This difference does not, however, affect the mathematical apparatus describing the emission process. We must, of course, distinguish two possible maser operating modes here: in the first the magnetic field H_0, and therefore the emission frequency, do not change in magnitude during the whole process; in the second the quantities H_0 and ω_0 do not remain constant during the working of the maser. This situation is unavoidable when using ARP for inverting the populations of the working levels, as is most often the case. Mathematically, the second case differs from the first in that it has equations with time-dependent coefficients.

Let us examine a fixed frequency oscillator. As always we shall consider that a point paramagnetic specimen is placed in a rectangular resonator and interacts with type TE_{01n} oscillations. This kind of device can be described by equations (41.5) whose solutions in the present case are of the form:

$$\begin{aligned} H_x &= h_s(t) \cos [\omega t + \psi_1(t)], \\ M_x &= m(t) \cos [\omega t + \psi_2(t)], \\ M_y &= m(t) \sin [\omega t + \psi_2(t)]. \end{aligned} \tag{53.1}$$

The change in the amplitudes $h_s(t)$ and $m(t)$ and also the phases $\varphi(t) = \psi_1 - \psi_2$ are assumed to be slow. The further transformations are already well known to us. Substitution of (53.1) in (41.5), averaging over the high-frequency period and ignoring the small terms leads to the following set of equations:

$$\begin{aligned} \dot{m} + \frac{1}{T_2} m &= \frac{1}{2} \gamma M_z h_s \sin \varphi, \\ \dot{M}_z + \frac{1}{T_1} M_z &= \frac{1}{T_1} M_0 - \frac{1}{2} \gamma h_s m \sin \varphi, \\ \dot{h}_s + \frac{1}{2T_c} h_s &= -\frac{1}{2} \omega_c a V_s m \sin \varphi, \\ \dot{\varphi} &= -(\omega_0 - \omega_c) + \left(\frac{1}{2} \frac{\gamma h_s M_z}{m} - \frac{1}{2} \omega_c a V_s \frac{m}{h_s} \right) \cos \varphi. \end{aligned} \tag{53.2}$$

Changing to the dimensionless form of the equations we use the relations:

$$\begin{aligned} &t_{\text{new}} = t_{\text{old}}/T_2, \quad v = m/\overline{M}, \quad n = M_z/\overline{M}, \\ &x = \tfrac{1}{2}\gamma T_2 h_s, \quad K_2 = T_2/T_1, \quad K_1 = T_2/T_c, \\ &\Delta = (\omega_0 - \omega_c)\, T_2. \end{aligned} \tag{53.3}$$

A convenient transformation coefficient to use is

$$\bar{M} = -\frac{2}{\gamma T_2 T_c \omega_c a V_s}. \tag{53.4}$$

Equations (53.2) in the notation introduced in (53.3) and (53.4) can be written as:

$$\begin{aligned} \dot{v} + v &= nx \sin \varphi, \\ \dot{n} + K_2 n &= K_2 n_0 - vx \sin \varphi, \\ \dot{x} + \tfrac{1}{2}K_1 x &= \tfrac{1}{2}K_1 v \sin \varphi, \\ \dot{\varphi} &= -\Delta + \left(\frac{x}{v} n + \frac{1}{2} K_1 \frac{v}{x}\right) \cos \varphi. \end{aligned} \tag{53.5}$$

The physical object described by the equations (53.5) generally speaking has two equilibrium positions:

$$n_1 = n_0, \quad v_1 = 0, \quad x_1 = 0, \tag{53.6a}$$

$$n_2 = 1/\sin^2 \bar{\varphi}, \quad v_2^2 = (n_0 - n_2) K_2/\sin^2 \bar{\varphi},$$

$$x_2^2 = K_2(n_0 - n_2), \quad \sin^2 \bar{\varphi} = \frac{(K_1 + 2)^2}{4\Delta^2 + (K_1 + 2)^2}. \tag{53.6b}$$

The singular point (53.6b) has a real physical meaning and defines the equilibrium position only if

$$(n_0 - n_2) \geqq 0. \tag{53.7}$$

The working phase of a two-level maser occurs after the pumping generator is switched off, when n_0 corresponds to thermal equilibrium: $n_0 = M_0/\bar{M} < 0$, and there is no stationary self-oscillation mode or modulated oscillation mode. When there is any deviation of the system from the only equilibrium position it will try to return to it. But the nature of the processes occurring during this depends essentially on the initial conditions.

If the initial inverse magnetization is less than a certain threshold value, i.e. $n(0) < n_{\text{thresh}}$, the transition to equilibrium proceeds without emission by longitudinal relaxation. The characteristic time of this process is of the order of T_1. In the opposite case $[n(0) > n_{\text{thresh}}]$ radiative transitions become important and part of the energy is transformed into coherent emission. It is clear that the emission process should be completed in a time of the order of T_2 or T_c. Since $T_2 \ll T_1$ and $T_c \ll T_1$ the longitudinal relaxation can have no effect on the process.

We need only limit ourselves to discussing the process in the short initial period when the change in n is negligibly small, in order to find n_{thresh}.

In this case the present system does not differ from a maser with continuous pumping, and we can take it that $n_0 = n(0)$. Now from (53.6b) the self-excitation condition can be written in the form

$$n(0) > n_2 = n_{\text{thresh}} = 1/\sin^2 \bar{\varphi} \tag{53.8}$$

in the same way as in § 52.3. The radiation instability condition depends on the magnitude of the frequency difference and in its expanded form can be written as

$$\begin{aligned} &M_z < 0, \\ &|M_z| > \frac{2}{\gamma T_2 T_c \omega_c a V_s} \cdot \frac{4\Delta^2 + (K_1 + 2)^2}{(K_1 + 2)^2}. \end{aligned} \tag{53.9}$$

The threshold is most easily achieved with zero frequency difference, when it will be sufficient if

$$|M_z| > \frac{2}{\gamma T_2 T_c \omega_c a V_s}. \tag{53.10}$$

The linear theory gives no information on how the emission process takes place. The system of non-linear equations (53.5) must be solved to determine the shape and duration of the emitted signal and the change in the components of the magnetic moment. These equations can be solved by electronic computers. A few examples corresponding to different values of the system parameters are given in Fig. XI.16. When the threshold is slightly exceeded at the start the emitted signal takes the form of a simple pulse (Fig. XI.16a). An increase in $|M_z(0)|$ is accompanied by a more complicated form of the pulse; 100% modulation of the pulse appears, which can be explained by the successive transfer of the energy flux from the paramagnetic to the resonator field and back again.

A comparison of the results of calculation and experiment (Fig. XI.12) shows a certain qualitative similarity. The serious quantitative discrepancy, however, leaves no hope for the validity of the theory as applied to these experiments. The theory gives too short a duration with an anomalously high signal amplitude; the calculated modulation depth is also too great. A theory which took into consideration the fact that the resonance frequency of the paramagnetic sweeps through the pass band of the resonator would be more realistic.

In almost all the experiments listed in § 53.1 adiabatic rapid passage was used for the excitation. The excited system is then returned to resonance conditions by changing the strength of the magnetizing field H_0. Since the conditions in a maser are continuously changing the emission process can be imagined as follows. At the start the frequency difference Δ is so great that the state of the system is stable. The difference then decreases and at some point in time exceeds the threshold value, which is approximately the same as (53.8). If we fix the value of Δ reached the quantum system emits a definite amount of energy and remains in an excited state with $n < n(0)$. We then change the field H_0 by an amount ΔH, which is sufficient for the maser to be above the threshold again. Once again a pulse will be emitted but with a

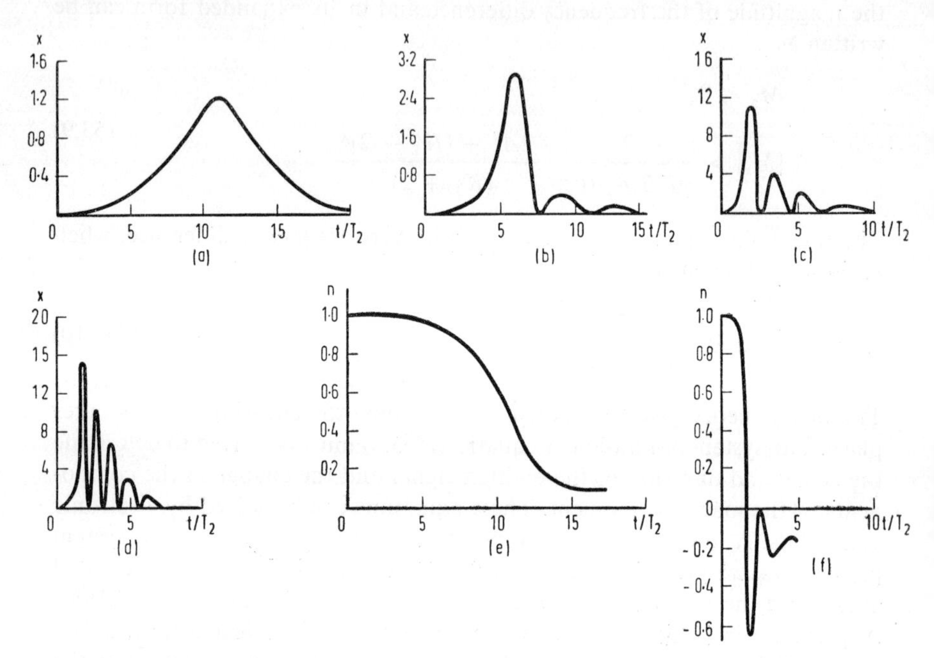

FIG. XI.16. Solutions of equation (53.5) obtained on an electronic computer for $K_1 = 1{\cdot}33$; $K_2 = \infty$; $T_2 = 2 \times 10^{-7}$ sec. In Figs. (a)–(d) the curves of $x(t)$ are given for the following values: (a) $n_0 = 2$; (b) $n_0 = 6{\cdot}6$; (c) $n_0 = 20$; (d) $n_0 = 100$. In Figs. (e), (f) the curves of $n(T)$ are given for (d) $n_0 = 2$; (e) $n_0 = 10$.

slightly different frequency. By continuing the process of reducing the frequency difference in steps we obtain a series of pulses which finish when n reaches the value $n = 1$. Modulation of the signal can therefore be brought about by purely kinematic effects. In reality the frequency difference is altered continuously; this however, does not change the essentials of the matter.

Kemp (1961) has drawn attention to the fact that the change in the frequency difference may have a significant effect on the oscillation process. He made a series of experiments with a two-level oscillator in which the rate of passage through resonance in the emission process was varied in a controlled manner. Kemp's results, reproduced in Fig. XI.17, confirm the assumption that was made.

In the mathematical description of a tunable maser we shall take equation (42.5) as the basis as before. All that is necessary to do is to remember that now $\omega_0 = \omega_0(t)$ and some of the coefficients of the equations become variable. If the change in the frequency difference $\omega_0(t) - \omega_c$ proceeds sufficiently slowly, the solutions of the set (42.5) can be found in the form of the functions:

$$H_x = h_s(t) \cos \left[\int_{t_0}^{t} \omega(t)\, dt + \psi_1(t) \right].$$

$$M_x = m(t) \cos \left[\int_{t_0}^{t} \omega(t)\, dt + \psi_2(t) \right], \tag{53.11}$$

$$M_y = m(t) \sin \left[\int_{t_0}^{t} \omega(t)\, dt + \psi_2(t) \right].$$

We change to the dimensionless abbreviated equations in just the same way as in the problem discussed. The form of the shortened equations does not differ from (53.5) if the inequality $d\omega/dt \ll \omega^2/Q_L$ holds. Examples of solutions of the equations (53.5) for linear sweeping at a rate α are given in (Khanin, 1966). The numerical values of the parameters in the calculations are taken from the paper by Kemp. Nevertheless the solution is given in the form of a sequence of isolated bursts of emission which differs significantly from Fig. XI.17. It should be noted, however, that this type of emission was found in an experiment (Firth and Bijl, 1961), although the interpretation given by Firth and Bijl (1961; Firth, 1963) can hardly be considered satisfactory. The non-uniform broadening of the paramagnetic absorption line must be allowed for, as well as the sweeping, in order to explain the results of Kemp (1961) and in many other experimental papers (Thorp *et al.*, 1961; Thorp, 1961; Feher *et al.*, 1958; Chester *et al.*, 1958; Shamfarov and Smirnova, 1963). In this case, as shown by Khanin (1966), spins with different resonance frequencies take part in the oscillation and the emission process is very similar to the process for the establishment of a stationary amplitude in continuous quantum oscillators, the part of the longitudinal relaxation time being played by a parameter that is inversely proportional to the sweep rate. The quantitative comparison of the theory developed by Khanin (1966) with experimental work is made difficult by the absence in the latter of the necessary numerical data. In the cases when this comparison can be made the agreement is quite satisfactory.

CHAPTER XII

Lasers

ONE OF the most recent and most important achievements in quantum electronics has been the practical advance of its methods into the optical and infrared wavebands. In 1960 Maiman (1960b) designed the first working optical oscillator based on the quantum electronic principles. This device was called a laser. For the first time a source of coherent monochromatic oscillations was available in the field of optics. The significance of this event is comparable with the invention of the vacuum tube, which produced a revolution in radio technology. The rapid tempo of the development of research in the field of lasers was consequently quite natural; considerable progress has been achieved in a short time.

The extension of quantum methods from the radio band into optics was not unexpected. At the beginning of the development of quantum electronics Fabrikant and his collaborators put forward the idea of molecular amplification of light (Fabrikant *et al.*, 1951). A more concrete discussion of the problem of optical maser oscillators and amplifiers, first undertaken by Townes and Schawlow (Schawlow and Townes, 1958), was based on practical achievements in the field of radiofrequency maser devices on the one hand, and of optical excitation of atoms on the other. There has also been a clarification of the question of optical frequency resonators.

In their operating principle lasers with optical excitation of the working substance resemble three-level masers most closely. A powerful light from a secondary source stimulates transitions between the ground level of the system and a certain excited level. Between them there is a third, metastable level into which the excited electrons rapidly pass. When the population of this level sufficiently exceeds the population of the lower level, amplification and generation of electromagnetic waves are possible at the frequency of the corresponding transition.

In the design of a laser the most important points are the selection of the working substance, and the method of exciting it, and the selection of the electrodynamic system which will ensure effective interaction of the field with the working substance. The above method of excitation with noncoherent light is applicable to any state of the substance. Another mechanism, excitation of the working molecules by collisions with particles of a different kind, can be used only in a gaseous medium. The medium must, of course, be removed from its state of thermal equilibrium by means, for example, of a beam of electrons or by exciting a gas discharge in the medium itself.

The particular wavelength region involved has a considerable effect on the design of the electrodynamic system of a laser. Optical resonators operating in low order modes are unsuitable. In the resonators used in practice the working frequency corresponds to a very high order mode, and consequently a large number of the resonator's eigenfrequencies fall within the width of the spectral line of the working medium.

The effect of working in the optical band is also shown up when the various relaxation mechanisms are considered. The reason for this is, first, satisfaction of the inequality $\hbar\omega_{kl} \gg kT$ (ω_{kl} is the quantum transition frequency, T is the temperature of the substance) and, secondly, the essential part played by the process of spontaneous emission. We recall that in the radiofrequency region the opposite is the case. This feature excludes the possibility of direct application of the results of maser theory to the optical waveband. The second, more serious difficulty in building up the theory of a laser is caused by the fact that the numerous modes of the resonator are coupled by the working substance and by the walls of the resonator. It is not possible to neglect this coupling when finding the amplitude and spectral distribution of the output.

54. Methods of obtaining Negative Temperatures

In this section we expound the principles of the method of obtaining a population inversion of levels in solid paramagnetic materials and gases. Idealized situations are discussed, from which general laws can be derived. The features inherent in real working substances which make it necessary to correct the general scheme will be discussed in the following sections, which are devoted to practical devices.

54.1. Population Inversion by Optical Excitation of a Substance

We can use the example of molecules with three energy levels to explain the essentials of the method of optical excitation (Fig. XII.1). Transitions between the levels lie in the optical region ($\hbar\omega_{kl} \gg kT$) so when there are no external perturbations practically all the molecules are in the ground state.

The secondary pumping field stimulates transitions between levels 1 and 3, thus altering the distribution of populations among the levels. When there are no other fields the new distribution can be found from the balance equations:

$$\dot{N}_3 = -\frac{N_3 - N_2\,\mathrm{e}^{-\hbar\omega_{32}/kT}}{T_{32}^*} - \frac{N_3 - N_1\,\mathrm{e}^{-\hbar\omega_{31}/kT}}{T_{31}^*} - \frac{N_3}{\tau_{32}} - \frac{N_3}{\tau_{31}}$$

$$- W_{31}(N_3 - N_1), \tag{54.1}$$

$$\dot{N}_2 = \frac{N_3 - N_2\,\mathrm{e}^{-\hbar\omega_{32}/kT}}{T_{32}^*} - \frac{N_2 - N_1\,\mathrm{e}^{-\hbar\omega_{21}/kT}}{T_{21}^*} - \frac{N_2}{\tau_{21}} + \frac{N_3}{\tau_{32}},$$

$$N_1 + N_2 + N_3 = N.$$

Here N_k is the density of particles in the k level; τ_{kl} and T_{kl}^* are the spontaneous emission and non-radiative relaxation times respectively.

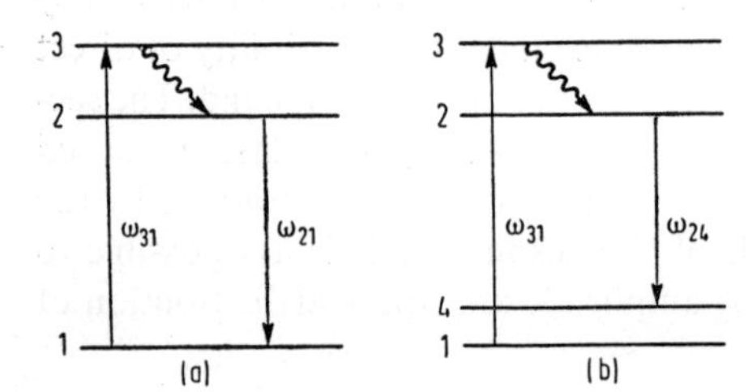

FIG. XII.1. Possible arrangements of levels in a working substance.

Equations (54.1) can be used, generally speaking, to describe an optically thin, evenly illuminated volume of a substance, when we need not take into consideration the dependence of the pumping energy density on the coordinates.

Assuming a constant pumping rate we find that the stationary solutions of (54.1) are

$$N_1 = N\frac{W_{31}T_{31}T_{32} + T_{31} + T_{32}}{W_{31}(T_{21} + 2T_{32})\,T_{31} + T_{31} + T_{32}},$$

$$N_2 = N\frac{W_{31}T_{31}T_{21}}{W_{31}(T_{21} + 2T_{32})\,T_{31} + T_{31} + T_{32}}, \tag{54.2}$$

$$N_3 = N\frac{W_{31}T_{31}T_{32}}{W_{31}(T_{21} + 2T_{32})\,T_{31} + T_{31} + T_{32}},$$

where we have used the notation $T_{kl}^{-1} = T_{kl}^{*\,-1} + \tau_{kl}^{-1}$. A maser is capable of oscillation if the inverted population difference exceeds a certain threshold value, let us say $N_2 - N_1 = N_{21}$. Using (54.2) we can estimate the conditions under which this threshold can be reached, what energy has to be expended in creating the required population difference, and what power is required to maintain it.

The pump power absorbed by the substance is

$$P_{31} = W_{31}(N_1 - N_3)\,\hbar\omega_{31}V_s. \tag{54.3}$$

The transition probability W_{31} is determined by the energy distribution function of the molecules and we can define it as a function of the quantity N_{21}:

$$W_{31} = \frac{(N + N_{21})(T_{31} + T_{32})}{NT_{31}(T_{21} - T_{32}) - N_{21}T_{31}(T_{21} + 2T_{32})}. \tag{54.4a}$$

Generally the threshold values of $N_{21} \ll N$, and the fact that N_{21} is finite has little effect on the pump power required. Therefore to simplify the calculations we can put $N_{21} = 0$, obtaining

$$W_{31} = \frac{T_{31} + T_{32}}{T_{31}(T_{21} - T_{32})}. \tag{54.5a}$$

It is clear that in three-level systems a state with a negative temperature can be achieved only if $T_{21} > T_{32}$. The most favourable relation between the lifetimes, which leads to minimum pump power, is

$$T_{32} \ll T_{21};\quad T_{31}. \tag{54.6}$$

If (54.6) is satisfied we can use, instead of (54.4a) and (54.5a), the simpler expressions

$$W_{31} = \frac{N + N_{21}}{N - N_{21}}\,T_{21}^{-1} \tag{54.4b}$$

and

$$W_{31} = T_{21}^{-1}. \tag{54.5b}$$

In addition, if (54.6) is satisfied, $N_3 \ll N_1$ and the power absorbed by the substance to maintain $N_2 = N_1$ is

$$P_{31} = \tfrac{1}{2}N\hbar\omega_{21}T_{21}^{-1}V_s. \tag{54.7}$$

As an example let us take some typical parameters: $N = 10^{19}$ cm^{-3}, $T_{31} = 25 \times 10^{-8}$ sec, $T_{32} = 5 \times 10^{-8}$ sec, $T_{21} = 5 \times 10^{-3}$ sec, $f_{31} = 6{\cdot}6 \times 10^{14}$ Hz.† Inequalities (54.6) are satisfied, and calculation using (54.7) leads to $P_{31} = 450$ W/cm^3. We have calculated the power absorbed by a specimen in which half the lasing molecules are kept continually in the

† We are taking an idealized three-level system as the example. It differs from pink ruby by the absence of level splitting and degeneracy.

metastable level. If we allow for the losses in the pumping system, the power consumed by the power sources must be far greater. There are many reasons for the energy losses: in the first place not all the electric energy used is converted into light. Next, only part of the output of the lamp reaches the working substance. Lastly, emission is lost uselessly in the part of the lamp spectrum which does not lie within the absorption band of the substance.

With pulsed excitation it is convenient to consider the pump system in terms of the energy it uses in one flash. The specimen should absorb an energy of not less than $N\hbar\omega_{31}V_s/2$ in a time $t < T_{21}$ in order to achieve a state with equal population of the ground and metastable levels. In the example given this energy is 2·2 J/cm³. In practice a ruby laser uses from 100 to several thousand joules in a flash, depending on the efficiency of the pump system and the quality of the ruby.

The requirement for high pumping powers is characteristic of substances operating on the three-level scheme, where it is necessary to transfer more than half the particles to the upper level. Systems with four energy levels arranged as shown in Fig. XII.1b have considerable advantages in this respect. The additional level 4 is between 2 and 1, and laser action takes place in the $2 \rightarrow 4$ transition. When $E_4 - E_1 \gg kT$, level 4 is hardly populated at all, and a far smaller number of molecules have to be excited to create the necessary difference $N_{24} = N_2 - N_4$ than in the three-level version. If the lifetimes of the different transitions satisfy the conditions

$$T_{32} \ll T_{31}; \quad T_{24}; \quad T_{21}, \qquad (54.8)$$

$$T_{41} \ll T_{24},$$

then calculation leads to an expression similar to (54.5):

$$W'_{31} \simeq \frac{N_{24}}{N} T_{24}^{-1}. \qquad (54.9)$$

When we compare these expressions we can see that when the lifetimes of the metastable state are the same

$$W'_{31}/W_{31} \simeq N_{24}/N \ll 1, \qquad (54.10)$$

i.e. a four-level system is N_{24}/N times more economical than a three-level one.

We can use a different approach to the calculation of the absorbed pump power for a given population N_2 (Kaiser *et al.*, 1961). A unit volume of a substance located at a distance z from the surface absorbs a power

$$P = \int_\lambda [I(\lambda, z)\, e^{-K(\lambda, z)z} - I(\lambda, z+1)\, e^{-K(\lambda, z+1)(z+1)}]\, d\lambda.$$

Here $I(\lambda)$ is the energy flux per unit wavelength interval.

Assuming that the absorption is sufficiently weak that $I(\lambda, z) \simeq I(\lambda, 0)$, it is not difficult to find for the total absorbed power

$$P = V_s \int_{\lambda} I(\lambda) K(\lambda) \, d\lambda. \tag{54.11}$$

The population of level 2 is determined by the absorbed pump power, the fluorescence efficiency and the lifetime of the atom in this state. The fluorescence efficiency η is equal to the ratio of excited atoms brought into the required state to the number of absorbed photons. Multiplying the integrand in (54.11) by $\eta(\lambda)/\hbar\omega$ we can write

$$\int_{\lambda} \frac{I(\lambda) K(\lambda) \eta(\lambda)}{\hbar\omega} \, d\lambda = \frac{N_2}{T_{21}}. \tag{54.12}$$

The same equation when expressed in terms of quantities averaged over the absorption spectrum becomes

$$\frac{\bar{I}\bar{K}\bar{\eta} \, \Delta\lambda}{\hbar \omega_{31}} = \frac{N_2}{T_{21}}. \tag{54.13}$$

Noticing that $\bar{I}\bar{K} \, \Delta\lambda = P_{13}/V_s$ we have

$$P_{31} = N_2 \hbar\omega_{31} V_s / T_{21} \eta. \tag{54.14}$$

This becomes (54.7) if $N_2 = \frac{1}{2}N$ and $\eta = 1$. This method of calculation is useful since it allows us to take into consideration any shape of pump spectrum or of the substance's absorption spectrum.

The light sources generally used in lasers, pulsed gas-discharge lamps, emit over the whole optical spectrum. Substances with wide absorption bands characteristic of solids are necessary for efficient use of such pumping. The absorption lines of gases are far narrower.

The possibility of using gaseous working substances in lasers with optical excitation has been discussed by a number of authors (Schawlow and Townes, 1958; Bergmann, 1960; Cummins *et al.*, 1961). The use of gases at low pressures has a number of advantages connected with the weak dependence of the properties of the medium on physical conditions, which may vary in an uncontrollable manner (e.g. on the temperature). There are, of course, no broad absorption bands in gas spectra, which complicates the problem of obtaining a large enough population difference. Nevertheless, this problem is not hopeless, as shown by the successful operation of a caesium vapour laser excited by a helium discharge lamp (Rabinowitz *et al.*, 1962). A fortunate circumstance is the coincidence of the frequencies of one of the emission lines of the helium lamp and the absorption line of caesium.

There are obviously optical excitation mechanisms applicable to gases that are different from those discussed. Rautian and Sobel'man (1961) suggest the use of the phenomenon of photodissociation. One of the decay products may be in an excited state. The advantage of the suggested mechanism is the possibility of using non-monochromatic light sources.

54.2. Pumping Systems

As we shall see from what follows the majority of working substances are pumped in the visible and near infrared regions of the spectrum. Pulsed xenon lamps (Marshak, 1962) possess good characteristics in this range and

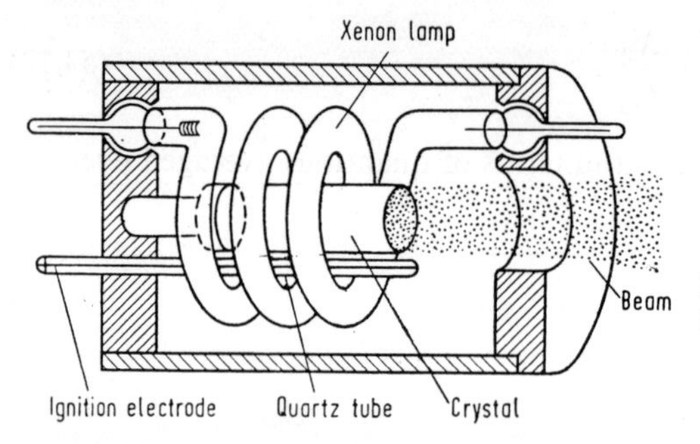

FIG. XII.2. Pumping of a crystal by a spiral xenon lamp.

are the most frequently used laser pump sources. Since the working substance is as a rule in cylindrical form spiral and rod lamps are the most suitable. A design for a laser with a spiral lamp is shown in Fig. XII.2. The crystal is mounted on the axis of the spiral, and most of the illumination comes from the surface of the lamp facing the axis. Only a small part of the rest of the light is reflected back to the crystal by a cylindrical reflector surrounding the lamp.

A spiral lamp ensures a more uniform excitation of a crystal than a straight lamp. One reason for the remaining non-uniformity is the focusing properties of the crystal itself (Devlin *et al.*, 1962). Let us consider a cylindrical crystal with a polished side wall (Fig. XII.3). The material of the specimen has a refractive index of $n > 1$. The light rays, which arrive from every possible direction at some point on its surface, are refracted into the specimen into a conical beam with an apex angle of

$$\Theta = 2 \arcsin (1/n). \tag{54.15}$$

The outside rays of the beam pass at a distance R/n from the axis of the cylinder. Therefore inside the working substance there is a region of enhanced pumping in the form of a cylinder with a radius R/n, whilst the periphery of the specimen is excited more weakly. It is therefore better to use composite rods whose central part contains the active medium, whilst the outer shell is made up of a non-absorbent material with the same refractive index. For example an inner ruby rod ($Al_2O_3:Cr^{3+}$) can be surrounded by a sapphire shell (Al_2O_3). Among the advantages of such a system is the good dissipation of the heat (thanks to the high thermal conductivity of the sapphire) generated in the central rod by the non-radiative 3–2 transitions.

The main disadvantage of a spiral lamp is the large loss of light energy; designs with straight lamps are far more economical. In the simplest case one or several lamps are placed near the crystal and parallel to it, and the whole system is surrounded by a reflector. The most successful design is a reflector in the form of an elliptical cylinder with the lamp and the crystal at its foci. In this case almost all the light from the lamp is focused into the crystal, and the pumping efficiency is considerably higher than that of a spiral lamp. At the same time, however, there is a decrease in the uniformity of the pumping. Direct experiments (Li and Sims, 1962) confirm that the energy from the lamp is concentrated into a small part of the crystal, thus reducing the useful volume of the working substance.

Considerable cooling of the crystal is required in a number of cases. For example, it is necessary to cool materials operating in a four-level arrangement, unless the condition $E_4 - E_1 \gg kT$ is satisfied at room temperature. In these cases the crystal is put in a transparent cryostat filled with a refrigerant. In all other respects the pumping arrangement is the same. One design of a laser operating at a low temperature is shown in Fig. XII.4. To prevent

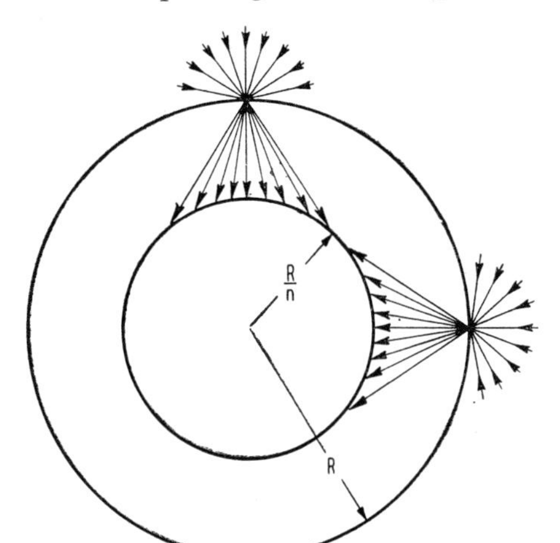

FIG. XII.3. Focusing of light inside a cylinder with $n > 1$.

unnecessary heating of the specimen the unused part of the lamp spectrum is removed by a selective absorbent filter, which surrounds the cryostat. This design is suitable for working with liquid oxygen, which is cooled below its boiling point to avoid boiling when the lamp is switched on. Part of the energy absorbed by the substance in the pumping process is released in the transition to the metastable level (3–2 transition). As a rule there is no radiation from this transition and the energy evolved goes to heating the substance. In many cases the extremely high pumping powers and the severe thermal conditions make it impossible to operate the laser continuously.

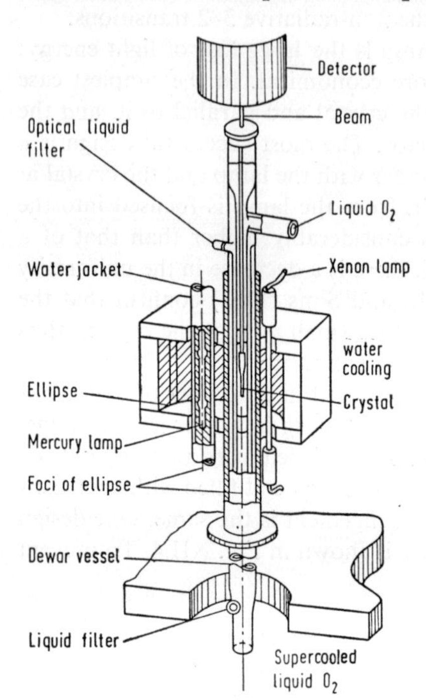

FIG. XII.4. Design of a laser with liquid-oxygen cooling of the working substance.

The last of the laser designs discussed above is suitable for operating in both pulsed and continuous modes. Two lamps are used for this purpose: near the crystal, at one of the foci of the ellipse, is placed a pulsed xenon lamp, and at the other focus there is a continuously operating mercury lamp. Four-level materials are the most suitable working substances for continuous lasers, since a lower power is required to excite them. Particularly suitable materials for this purpose are $CaWO_4:Nd^{3+}$, and $CaF_2:Dy^{2+}$, and we shall discuss these a little later. It is difficult to produce a continuously operating ruby laser. Apart from cooling, special measures have to be taken to use the light from the lamp as efficiently as possible. One way is shown in Fig. XII.5 (Nelson and Boyle, 1962). The output of a mercury lamp is focused by a system of mirrors onto the front face of a specimen consisting of two parts: a ruby rod and a sapphire horn. Because of internal reflection a ray entering the face cannot leave through the side and is almost entirely absorbed by the ruby. The development of high-efficiency systems of this kind is also of importance for solving the problem of a laser excited by sunlight.

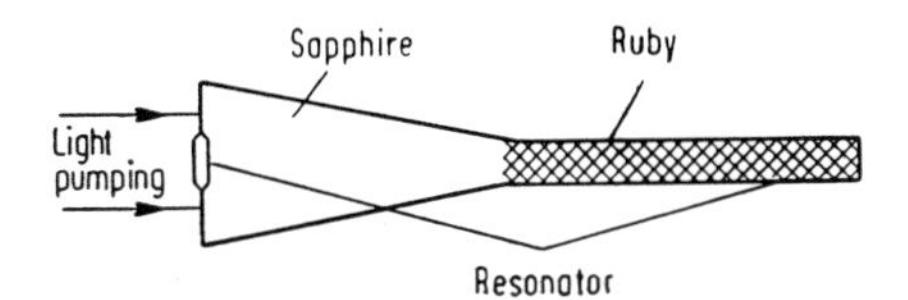

FIG. XII.5. Ruby crystal pumped through a sapphire horn. Used in a continuous laser (Nelson and Boyle, 1962).

54.3. Excitation of Gas Molecules during Collisions with Electrons

If the working substance is a gas, optical excitation is not the only possible way of obtaining negative temperatures. Under suitable conditions the same effect can be achieved by collisions of the gas molecules with electrons or with molecules of another gas. For a start let us examine the simplest case of a single-component gas (Javan, 1959; Sanders, 1959; Basov *et al.*, 1960). Excitation occurs when there is a collision between a molecule and an electron.† The necessary number of electrons having sufficient energy can either be introduced into the medium from outside or formed in a gas discharge in the medium itself. We shall consider that the electron energies have a Maxwellian distribution, their temperature being T_e and density N_e.

Let the energy level diagram of the molecules be similar to that shown in Fig. XII.1a;‡ we shall consider the levels to be non-degenerate. While there are no electrons present only level 1 is populated, since $E_i - E_1 \gg kT_0$ (T_0 is the temperature of the gas). When there are electrons present transitions between the levels occur for two reasons: collisions and spontaneous emission. Using σ_{ik} to denote the cross-section of the collision process causing a transition between levels i and k, and v to denote the relative velocity of the colliding particles we can express the probability of a transition Θ_{ik}^{-1} in terms of these quantities:

$$\Theta_{ik}^{-1} = N\overline{\sigma_{ik}v}. \tag{54.16}$$

The bar denotes averaging over all velocities. The principle of detailed balancing connects the probability of forward and reverse transitions by the relation

$$\Theta_{ik}/\Theta_{ki} = e^{(E_k - E_i)/kT_e}. \tag{54.17}$$

We can determine the populations of the excited levels of the gas from the

† Ions are present in a gas-discharge plasma as well as electrons. Their mean energy is low, however, so the process of molecule excitation in collisions with ions can be excluded from the discussion.

‡ The energy level spectrum of the gas molecules is ,of course, more complicated and we are dealing with a few selected levels. Ignoring the rest of the levels is justified if their presence has no significant effect on the process under consideration.

balance equations

$$\dot{N}_3 = \frac{N_1}{\Theta_{13}} + \frac{N_2}{\Theta_{23}} - N_3\left(\frac{1}{\Theta_3} + \frac{1}{\tau_3}\right),$$
$$\dot{N}_2 = \frac{N_1}{\Theta_{12}} - N_2\left(\frac{1}{\Theta_2} + \frac{1}{\tau_2}\right) + N_3\left(\frac{1}{\Theta_{32}} + \frac{1}{\tau_{32}}\right) \tag{54.18}$$

for the stationary state of the system. The quantities τ_i which appears in (54.18) are the lifetimes of the excited states against spontaneous emission into all the rest of the states. The lifetimes Θ_i connected with collisions are to be interpreted in a similar way.

A situation is possible in which transitions between excited states are not the main factor in determining the state of the system. From (54.18) we obtain

$$\frac{N_i}{N_1} = \frac{1/\Theta_{1i}}{1/\Theta_{i1} + 1/\tau_i}. \tag{54.19}$$

And at the limit, when $1/\Theta_{i1} \gg 1/\tau_i$, it turns out that

$$\frac{N_i}{N_1} = \frac{\Theta_{i1}}{\Theta_{1i}} = e^{-(E_i-E_1)/kT_e},$$
$$\frac{N_3}{N_2} = e^{-(E_3-E_2)/kT_e}, \tag{54.20}$$

i.e. there is a Boltzmann distribution with a positive temperature T_e. Some hope of achieving a negative temperature exists in the opposite case $1/\Theta_{i1} \ll 1/\tau_i$, when spontaneous emission plays a major part in depopulating the excited levels:

$$\frac{N_i}{N_1} = (\tau_i/\Theta_{i1})\, e^{-(E_i-E_1)/kT_e},$$
$$\frac{N_3}{N_2} = (\tau_3\Theta_{21}/\tau_2\Theta_{31})\, e^{-(E_3-E_2)/kT_e}. \tag{54.21}$$

When there are favourable relations between the transition probabilities $N_3/N_2 > 1$. An example in the case of helium confirms the feasibility of such a proposal: for the excited levels 3^1D and 2^1P it turns out that $\tau_3/\tau_2 = 35$, $\Theta_{31}/\Theta_{21} = 15$.

Two circumstances prevent reliable theoretical estimates of how good actual gases are, and for a solution of the problem one must turn to experiments. Firstly, the cross-sections σ_{ik} are unknown and therefore the probabilities $1/\Theta_{ik}$ also; all that can be stated as a rule is that allowed transitions correspond to large cross-sections. Secondly, the parameters τ_i in (54.21) are not generally speaking the spontaneous emission lifetimes of an isolated molecule. An excited molecule is surrounded by a number of unexcited ones which are able to reabsorb an emitted photon.† This effect, the trapping of resonance radiation, is particularly strong when the gas is sufficiently dense and the matrix element of the i–1 transition is large (an allowed transition). Under these conditions the effective life can rise so much that the level i becomes quasi-metastable. It is therefore advantageous if the upper of the excited levels in question (level 3) is connected optically with the ground state. On the same argument the 2–1 transition should be forbidden by the selection rules. But under these conditions τ_2 can be small only if there is another level 4 between 1 and 2, connected with 2 by an allowed transition. The above calculations are, of course, valid only when there is weak excitation of the gas, and the population of level 4 is negligibly small. In the example given a transition from both the helium levels (3^1D and 2^1P) to the ground state is allowed. Thanks to this oscillation cannot be achieved at the 3^1D–2^1P transition. In the spectra of inert gases there are favourable transitions which can be used to obtain oscillation. In helium such a transition is 7^3D–4^3P (Patel, Bennett *et al.*, 1962).

54.4. Excitation in a Gas Discharge when there is an Impurity present

The problem of attaining the necessary negative temperature can be simplified if we use a discharge in a two-component mixture of gases. In many transitions oscillation has been obtained simply by a successful choice of mixture, such as He–Ne, He–Xe, Ne–O_2, A–O_2 (Bennett, 1962a). There are several mechanisms involving the direct and indirect action of the impurity gas B on the basic gas A which affect the populations of the levels of the latter.

1. The process of resonance transfer of energy from an excited molecule of B to an unexcited molecule of A when they collide produces selective population of certain levels of gas A. Excitation of the molecules of gas B is achieved by electron collisions (Javan *et al.*, 1961; Javan, 1961; Bennett, 1961; Basov and Krokhin, 1960; Basov and Krokhin, 1962) or by optical excitation.

2. The basic process for the excitation of the molecules of gas A is collisions with electrons. The presence of the secondary gas B acts on the electrons in the discharge, altering the density of the electrons and their velocity distribution, which affects the state of the basic gas (Bennett, 1962a; Aisenberg, 1963; Powers and Harned, 1963; Patel, Faust and McFarlane, 1962).

† A similar effect is discussed in Chapter VII for systems which are small compared with λ.

3. As a result of the triple collision of atoms of A and B and an electron a molecular ion is formed which dissociates to give an atom of A in a definite excited state (Aisenberg, 1963).

4. The excited state into which the molecule A goes in an inelastic collision with B is unstable. The molecule dissociates, giving off one of the decay products in an excited state (Bennett, 1962a; Bennett, Faust, and McFarlane, 1962).

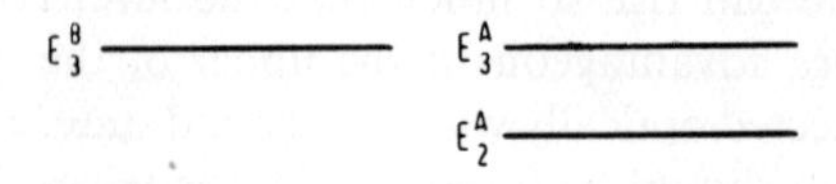

FIG. XII.6. Diagram of arrangement of levels of basic (A) and impurity (B) gases.

The cross-sections of these processes are essentially determined by features inherent in the actual gas mixture. Investigations into their relative efficiency are still far from complete. We shall limit ourselves to discussing the kinetics of the process, following the approach of Basov and Krokhin (Basov and Krokhin, 1960 and 1962).

The structure of the levels of the basic and secondary gases is shown in Fig. XII.6. We are interested in the distribution function of the molecules of gas A, which we can find from the following set of equations:

$$\dot{N}_3^A = N_1^A\left(\frac{1}{\Theta_{13}} + \frac{1}{t_{BA}}\right) - N_3^A\left(\frac{1}{\Theta_3} + \frac{1}{t_{AB}} + \frac{1}{\tau_3}\right) + \frac{N_2^A}{\Theta_{23}},$$
$$\dot{N}_2^A = N_1^A\frac{1}{\Theta_{12}} + N_3^A\left(\frac{1}{\Theta_{32}} + \frac{1}{\tau_{32}}\right) - N_2^A\left(\frac{1}{\Theta_{21}} + \frac{1}{\tau_{21}} + \frac{1}{\Theta_{23}}\right). \tag{54.22}$$

It differs from (54.18) in the terms which allow for the effect of the secondary gas. Here $1/t_{BA} = N_3^B\overline{\sigma_{BA}v}$ is the probability of the transfer of excitation from molecule B to A, $1/t_{AB} = N_1^B\overline{\sigma_{AB}v}$ is the probability of the reverse process; the meaning of the rest of the suffixes is self-evident. The process of transfer of energy in collisions between molecules is of a resonant nature. Its cross-section is greatest when there is no frequency difference $\Delta E_{BA} = |E_3^B - E_3^A|$ and small when $\Delta E_{BA} > kT$ (there is a decrease according to the law $e^{-\Delta E_{BA}/kT}$). For the secondary gas to have an effect on the state of the basic one their energy levels should coincide with an accuracy of

$$|E_3^B - E_3^A| \lesssim kT. \tag{54.23}$$

Assuming that (54.23) is satisfied with something to spare we can consider that $\sigma_{AB} \approx \sigma_{BA}$ and $t_{BA}/t_{AB} = N_1^B/N_3^B$. As previously, we neglect direct transitions between excited levels of gas A. Then the stationary state can be characterized by the following population ratio:

$$\frac{N_3^A}{N_2^A} = \frac{\left(\dfrac{1}{\Theta_{21}} + \dfrac{1}{\tau_2}\right)\left(\dfrac{1}{\Theta_{13}} + \dfrac{1}{t_{BA}}\right)}{\dfrac{1}{\Theta_{12}}\left(\dfrac{1}{\Theta_{31}} + \dfrac{1}{\tau_3} + \dfrac{1}{t_{AB}}\right)} . \tag{54.24}$$

Assuming $N_3^A/N_2^A > 1$, we can find from (54.24) that this is possible when

$$F < \Theta_{13}/t_{BA} \text{ when } 1 + \Theta_{21}/\tau_2 > (N_1^B/N_3^B)\, e^{-(E_2-E_1)/kT_e}, \tag{54.25a}$$

$$F > \Theta_{13}/t_{BA} \text{ when } 1 + \Theta_{21}/\tau_2 < (N_1^B/N_3^B)\, e^{-(E_2-E_1)/kT_e}, \tag{54.25b}$$

where

$$F = \frac{\left(\dfrac{\Theta_{31}}{\tau_{31}} + 1\right) e^{(E_3-E_2)/kT_e} - 1 - \dfrac{\Theta_{21}}{\tau_2}}{1 - \dfrac{N_1^B}{N_3^B} e^{-(E_2-E_1)/kT_e} + \dfrac{\Theta_{21}}{\tau_2}} .$$

The condition that the denominator in the left-hand side of the inequalities (54.25) is zero divides the $(\Theta_{31}/\tau_3;\ \Theta_{21}/\tau_2)$ plane into two regions with different physical conditions. If N_1^B/N_3^B is a fixed parameter of the system,

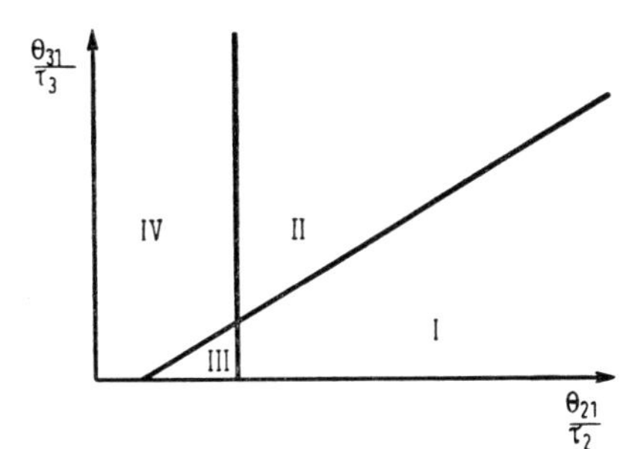

FIG. XII.7. The $(\Theta_{31}/\tau_3;\ \Theta_{21}/\tau_2)$ plane.

which is possible when gas A has no noticeable effect on the state of gas B (in particular when $N^A \ll N^B$), then the boundary is a vertical straight line (Fig. XII.7):

$$\Theta_{21}/\tau_2 = (N_1^B/N_3^B)\, e^{-(E_2-E_1)/kT_e} - 1. \tag{54.26}$$

To the right of it a negative temperature is attained if inequality (54.25a) is satisfied, i.e. when $1/t_{BA}$ is sufficiently large. Hence the requirement for a large density N_3^B and high temperature of the impurity gas.

The conclusion can also be drawn that in the right-hand half-plane the gas B helps the excitation of gas A. In the left-hand half-plane the situation is the direct opposite and gas B plays a negative part.

The numerator of (54.25) becomes zero on the straight line

$$\Theta_{31}/\tau_3 = \frac{\Theta_{21}/\tau_2 + 1 - e^{(E_3-E_2)/kT_e}}{e^{(E_3-E_2)/kT_e}}. \tag{54.27}$$

The regions into which this straight line divides the plane differ in the diametrically opposed part played by the process of direct excitation of gas A by electrons. In the region below the straight line (54.27) collisions of molecules of A with the electrons make it easier to obtain a negative temperature, above it they prevent it. Therefore the $(\Theta_{31}/\tau_3; \Theta_{21}/\tau_2)$ plane is divided into four regions. In region I attainment of a state with a negative temperature is assisted by collisions of molecules of A with molecules of B and with electrons. In region II states of this kind can be attained only by collisions with molecules of the impurity gas, and electron impact plays a negative part. In region III, on the other hand, the effect is attained by collisions with electrons, despite the presence of an impurity gas. Region IV in principle permits only positive temperatures. It is obvious that region I offers the most favourable possibilities for exciting the gas.

The above analysis leads to a number of remarks regarding the selection of the working conditions. When a two-component mixture is used the working point on the $(\Theta_{31}/\tau_3, \Theta_{21}/\tau_2)$ plane is located to the right of the boundary (54.26). The position of the boundary is determined by the parameter N_1^B/N_3^B. While it is large, i.e. the secondary gas is weakly excited by electrons, pumping of energy proceeds from gas A to gas B and the latter is a quenching agent.† On increasing the excitation there is an increase in the density of the excited molecules and a decrease in the density of the unexcited molecules, the direction of the transfer of excitation changes and gas B becomes the exciter. In order that collisions with electrons should not destroy the inversion in the populations N_3^A and N_2^A the working point should be in region I. The condition for this, together with

$$\Theta_{21}/\tau_2 > (N_1^B/N_3^B)\, e^{-(E_2-E_1)/kT_e}, \tag{54.28}$$

is a second inequality that follows from (54.27):

$$\frac{(\Theta_{21}/\tau_2) + 1}{(\Theta_{31}/\tau_3) + 1}\, e^{-(E_3-E_2)/kT_e} > 1. \tag{54.29}$$

Under the conditions of an actual experiment $(E_3 - E_2) < kT_e$ and (54.29) can be satisfied only when $\Theta_{21} \gg \tau_2$ and $\Theta_{31} \gg \tau_3$. Then (54.29), of course, is the same as the criterion (54.21), which we already know for a single-component gas.

† At this point it is essential to stress again that N_1^B/N_3^B is considered to be a fixed parameter.

The significance of the theory that has been expounded should not be overestimated. Several simplifying assumptions were made in the statement of the problem. This made it easier to elucidate the physical picture of the processes taking place, and a number of qualitative indications were obtained as a result. This scheme is, however, far too idealized to be used for calculations in practical cases. As an illustration let us take the example of the popular gas mixture of neon and helium. The structure of the energy levels of neon (Fig. XII.38) differs quite considerably from the system of three levels of a gas *A* which we have been discussing. In the first place there is another group (1*s*), which, moreover, have long lifetimes, between the 2*p* levels and the ground state (i.e. between 2 and 1). In the second place all the excited levels are split, and we must know the distribution of the sub-levels. It is clear that when all these additional details are taken into consideration the calculation becomes extremely laborious and not very reliable. Other excitation mechanisms may also play a significant part. Experiment is therefore the decisive factor in selecting the optimum conditions for the excitation of gases. Methods of experimental investigation and the results for a neon–helium mixture are given in Javan (1961) and Bennett (1961).

In conclusion we should mention that the mechanism of resonance transfer of excitation in an inelastic collision can be used not only for populating the required level but also for depleting a level that is interfering. For example, for the depopulation of the 1*s* levels of neon Javan has suggested the use of an argon impurity whose ionization potential is lower than the excitation potential of the 1*s* state of Ne.

55. The Elements of Laser Theory

From the point of view of their working principle there is no difference between microwave and optical maser oscillators. But the specific feature of the optical region, the large ratio of the dimensions of the device to the wavelength, has a considerable effect on their operating characteristics and considerably complicates the development of an adequate theory. In the analysis of microwave maser devices the non-uniformity of the radiation field in the working substance can be allowed for comparatively simply by the filling factor; a laser, however, is essentially a distributed non-linear system which excludes the possibility of similar simplifications.

Because of the high value of L/λ a large number of the eigenfrequencies of the unperturbed resonator fall within the spectral line of the working substance. The susceptibility of the substance at adjacent resonator frequencies is approximately the same, which leads to the possibility of exciting a laser in a whole spectrum of "modes". The suppression of one kind of oscillation and the excitation of another is possible in the oscillation process, accompanied by a change in the spatial structure of the field inside the resonator. Similar arguments supported by experimental data lead to the conclusion that we should be very careful when allowing for spatial effects. Unfortunately, only a small number of problems have as yet been successfully solved by a rigid theory. In the majority of cases it is necessary to limit oneself to treating extremely rough models, e.g. a single-mode model, which provide very approximate information on the processes actually taking place in the laser. We shall not touch on the single-mode theory again since it was discussed in sufficient detail in Chapter XI.

55.1. Linear Approximation

The linear approximation is limited by the condition of a small field which allows us to consider unperturbed resonator modes and neglect coupling between them. It is obvious that it is suitable for finding the self-excitation conditions and the limits of the spectrum within which self-excitation is possible† (Fain and Khanin, 1961; Genkin and Khanin, 1962).

Considering for the sake of argument that the working transition is electric dipole we shall start with the set of equations for the generalized spin (19.42)–(19.45). The equations describe a system of two-level molecules when there are no inhomogeneous line broadening mechanisms. The latter circumstance limits their sphere of application to certain solid working substances. Although a large number of levels (three or four) takes part in the operation of the laser the two-level idealization is justified. The important point is that levels 3 and 4 (Fig. XII.1) are short lived.

The vector potential of the electromagnetic field is expanded into the series (3.59) in the eigenfunctions of the unperturbed resonator. We introduce the notation

$$(\boldsymbol{A}_\lambda \cdot \boldsymbol{e}_1)/\hbar = \alpha_{1\lambda}, \; (\boldsymbol{A}_\lambda \cdot \boldsymbol{e}_2)/\hbar = \alpha_{2\lambda}, \; \alpha_{2\lambda} - i\alpha_{1\lambda} = \alpha_\lambda \tag{55.1}$$

and change to the variables

$$s_1 \pm i s_2 = s_\pm,$$

after which the equations become:

$$\dot{s}_+ + (T_2^{-1} - i\omega_0)\, s_+ = -\sum_\lambda \alpha_\lambda q_\lambda s_3, \tag{55.2a}$$

$$\dot{s}_- + (T_2^{-1} + i\omega_0)\, s_- = -\sum_\lambda \alpha_\lambda^* q_\lambda s_3, \tag{55.2b}$$

$$\dot{s}_3 = \frac{1}{T_1}(s_3^0 - s_3) + \frac{1}{2}\sum_\lambda (s_+\alpha_\lambda^* + s_-\alpha_\lambda)\, q_\lambda, \tag{55.2c}$$

$$\ddot{q}_\lambda + \frac{\omega_{c\lambda}}{Q_\lambda}\dot{q}_\lambda + \omega_{c\lambda}^2 q_\lambda = -\frac{i\hbar}{2}\int_{V_c} (s_+\alpha_\lambda^* - s_-\alpha_\lambda)\, dV. \tag{55.2d}$$

† The method suggested by Schawlow and Townes (1958) provides no information about the spectrum.

We shall assume further that the working substance fills the whole volume of the resonator uniformly. In this case it is convenient to expand $s_+(\boldsymbol{r}, t)$ in a system of functions $\alpha_\lambda(\boldsymbol{r})$, whose completeness and orthogonality follow from the properties of the system $\boldsymbol{A}_\lambda$ provided that the angles between the vectors $\boldsymbol{A}_\lambda$, $\boldsymbol{e}_1$ and $\boldsymbol{e}_2$ are constant. The latter condition is satisfied by a crystal placed in a field of plane waves. Let there be no oscillations in the system initially and let $s_3 = s_3^0$. The substitution of s_3 by $s_3^0 = \text{const.}$ in (55.2a) and (55.2c) linearizes them, allowing us to separate the equations relating to the different suffixes λ. We substitute in (55.2) the results of the expansion

$$\begin{aligned} s_+(\boldsymbol{r}, t) &= \sum_\lambda \alpha_\lambda(\boldsymbol{r})\, s_{+\lambda}(t), \\ s_-(\boldsymbol{r}, t) &= \sum_\lambda \alpha_\lambda^*(\boldsymbol{r})\, s_{-\lambda}(t), \end{aligned} \tag{55.3}$$

then multiply (55.2a) by $\alpha_{\lambda'}^*$, equation (55.2b) by $\alpha_{\lambda'}$ and integrate them over the volume of the resonator. After integration equations (55.2) reduce to

$$\begin{aligned} &\dot{s}_{+\lambda} + (T_2^{-1} - i\omega_0)\, s_{+\lambda} + s_3^0 q_\lambda = 0, \\ &\dot{s}_{-\lambda} + (T_2^{-1} + i\omega_0)\, s_{-\lambda} + s_3^0 q_\lambda = 0, \\ &\ddot{q}_\lambda + \frac{\omega_{c\lambda}}{Q_\lambda}\dot{q}_\lambda + \omega_{c\lambda}^2 q_\lambda = \frac{i\hbar a}{2}(s_{-\lambda} - s_{+\lambda}). \end{aligned} \tag{55.4}$$

We now investigate the set (55.4) for stability with respect to a small perturbation of form $e^{i\xi_\lambda t}$, where $\xi_\lambda = \omega_\lambda - i\delta_\lambda$. The complex characteristic equation

$$\xi_\lambda^4 - i\xi_\lambda^3(\omega_{c\lambda}/Q_\lambda + 2/T_2) - \xi_\lambda^2(\omega_{c\lambda}^2 + \omega_0^2 + T_2^{-2} + 2\omega_{c\lambda}/Q_\lambda T_2) + i\xi_\lambda\omega_{c\lambda}$$

$$\times\,[2\omega_{c\lambda}/T_2 + (\omega_2^0 + T_2^{-2})/Q_\lambda] + \omega_{c\lambda}^2(\omega_0^2 + T_2^{-2}) + \hbar a\omega_0 s_3^0 = 0 \tag{55.5}$$

makes it possible to determine ω_λ and δ_λ as a function of the parameters of the system. If the density of the active molecules exceeds a certain critical value, the oscillation rise ($\delta_\lambda > 0$) at frequencies close to ω_0. Outside this region, whose width is determined by s_3^0, the oscillations are damped ($\delta_\lambda < 0$). To find the boundaries of the region we substitute the value $\delta_\lambda = 0$ in (55.5) and obtain

$$\omega_\lambda^2 = \frac{\omega_{c\lambda}(\omega_0^2 T_2 + T_2^{-1} + 2Q_\lambda\omega_{c\lambda})}{\omega_{c\lambda}T_2 + 2Q_\lambda}, \tag{55.6}$$

$$s_3^0 = \frac{\omega_{c\lambda}[(\omega_0^2 - \omega_\lambda^2 + T_2^{-2})^2 + 4T_2^{-2}\omega_\lambda^2]}{2Q_\lambda T_2^{-1}\omega_0\hbar a}. \tag{55.7}$$

The connection between s_3^0 and the boundary frequencies of the self-excitation region is illustrated qualitatively in Fig. XII.8. For instability to occur in the limiting case at just one frequency we have to satisfy

$$s_3^0 > s_{3\text{cr}}^0 \simeq 2\omega_0^2/\hbar a Q_\lambda T_2 . \tag{55.8}$$

Expression (55.8) is obtained from (55.7) by putting $\omega_0 = \omega_{c\lambda} = \omega_\lambda$ and $T_2^{-2} \ll \omega_0^2$. The coefficient a is defined by the expression

$$a = \frac{1}{\hbar^2} \int_{Vc} [(\boldsymbol{A}_\lambda \cdot \boldsymbol{e}_1)^2 + (\boldsymbol{A}_\lambda \cdot \boldsymbol{e}_2)^2 \, dV$$

and in the case of $\boldsymbol{e}_1 \parallel \boldsymbol{e}_2 \parallel \boldsymbol{A}_\lambda$ it is equal to $16\pi\omega_0^2 \, |d|^2/\hbar^2$. From this follows the self-excitation condition

$$N_0 = 2s_3^0 > \frac{\hbar}{4\pi \, |d|^2 \, QT_2} . \tag{55.9}$$

We must stress that the conclusions about the self-excitation region are valid only within the framework of the linear approximation, and this breaks down a certain time after the appearance of the oscillating process. Expression (55.6) gives the frequencies of the possible stationary self-oscillations in a single mode with a small oscillation amplitude. This expression can be used

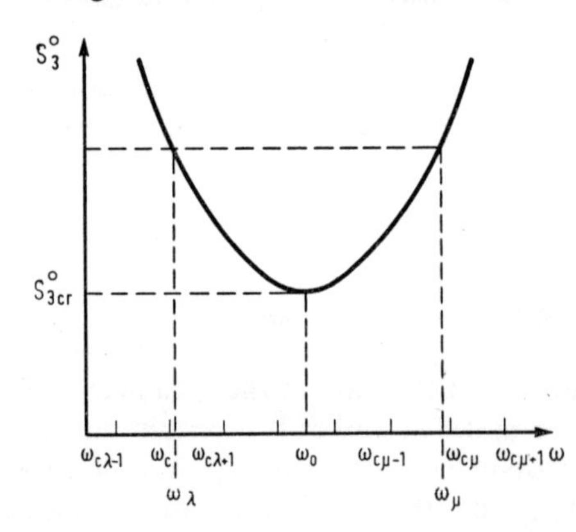

FIG. XII.8. Diagram giving a qualitative explanation of the dependence of the threshold density of active molecules on the oscillation frequency.

conveniently if we rewrite it in a slightly different form, remembering that $\omega_\lambda/\omega_{c\lambda} \simeq 1$, $\omega_0/\omega_{c\lambda} \simeq 1$, $T_2^{-2} \ll \omega_0^2$:

$$\omega_\lambda = \frac{\dfrac{\omega_0^2}{Q_\lambda} + \dfrac{2\omega_{c\lambda}}{T_2}}{\dfrac{\omega_{c\lambda}}{Q_\lambda} + \dfrac{2}{T_2}} = \frac{\omega_0 \, \Delta\omega_{c\lambda} + \omega_{c\lambda} \, \Delta\omega_0}{\Delta\omega_0 + \Delta\omega_{c\lambda}} . \tag{55.10}$$

These results essentially exhaust the information that the linear approximation can provide. The solution of equations (55.2) in the non-linear case is very difficult. It should be pointed out that it is hardly convenient to use an expansion of the field in the eigenwaves of the unperturbed resonator to find the non-linear operating conditions of a laser. The field distribution that is established in the non-linear condition may differ considerably from the configuration of the linear mode.

55.2. The Steady-state Oscillations of an Active Layer without Volume Losses

The expansion of the field in terms of the resonator modes in the initial equations is one of the methods of solving problems concerned with the excitation of oscillations in a resonator. This method often simplifies the treatment, but it is not the only one and, applied to lasers, may not be the best. A number of simple problems have been solved without recourse to expansion of the field in the "linear" modes (Ostrovskii and Yakubovich, 1964; Ostrovskii and Yakubovich, 1965). Amongst them is the problem of the stationary oscillations of an active plane layer.

The initial set of equations is well known. If we assume that the molecules have magnetic dipole transitions we can eliminate the electric field from Maxwell's equations and write them in the form

$$\text{curl curl}\, \boldsymbol{H} + \frac{\varepsilon}{c^2}\frac{\partial^2 \boldsymbol{H}}{dt^2} + \frac{4\pi\varepsilon}{c^2}\frac{\partial^2 \boldsymbol{M}}{dt^2} = 0\,, \tag{55.11a}$$

$$\text{div}\,(\boldsymbol{H} + 4\pi \boldsymbol{M}) = 0. \tag{55.11b}$$

Here ε is the permeability of an isotropic dielectric; $\boldsymbol{M}(\boldsymbol{r}, t)$ is the magnetic moment of unit volume of the substance.

The latter can be expressed in terms of the dipole moment matrix and the density matrix

$$\boldsymbol{M} = \text{Tr}\,(\hat{\boldsymbol{\mu}}\hat{\varrho}) = \boldsymbol{\mu}_{12}\varrho_{21} + \boldsymbol{\mu}_{21}\varrho_{12}. \tag{55.12}$$

For the system to become closed it should be complemented by the equations for the components of the density matrix†

$$\frac{\partial \varrho_{12}}{\partial t} - i\omega_0\varrho_{12} + \frac{1}{T_2}\varrho_{12} = \frac{i}{\hbar}(\boldsymbol{H}\cdot\boldsymbol{\mu}_{21})\,N, \tag{55.11c}$$

$$\frac{\partial N}{\partial t} + \frac{1}{T_1}(N - N_0) = \frac{2i}{\hbar}\,[(\boldsymbol{H}\cdot\boldsymbol{\mu}_{21})\,\varrho_{12} - (\boldsymbol{H}^*\cdot\boldsymbol{\mu}_{12})\,\varrho_{21}], \tag{55.11d}$$

where $N = \varrho_{22} - \varrho_{11}$.

† Equations (55.11) are equivalent to the equations for the generalized spin (55.2).

The method of finding the approximate solutions of the system (55.11) suggested by Ostrovskii and Yakubovich (1964) consists of successive applications of the method of averaging. Since the relaxation and non-linear terms $(T_1^{-1}, T_2^{-2}, |(\boldsymbol{\mu} \cdot \boldsymbol{H})|/\hbar)$ are small compared with the transition frequency ω_0 the oscillatory process is almost a sine wave and the solution can be found in the form

$$\varrho_{12} = \sigma(\bar{\boldsymbol{r}}, t)\, e^{i\omega t}, \quad \boldsymbol{H} = \boldsymbol{h}(\boldsymbol{r}, t)\, e^{i\omega t} + \boldsymbol{h}^*(\boldsymbol{r}, t)\, e^{-i\omega t}. \tag{55.13}$$

The amplitudes of σ and h, generally speaking, are slowly varying functions of time. Substitution of (55.13) in (55.11) and averaging over the period $2\pi/\omega$ leads to a set of abbreviated equations:

$$\text{curl curl}\, \boldsymbol{h} - \frac{\varepsilon\omega^2}{c^2}\boldsymbol{h} + \frac{2i\varepsilon\omega}{c^2}\frac{\partial \boldsymbol{h}}{\partial t} = \frac{4\pi\varepsilon\omega^2}{c^2}\sigma\boldsymbol{\mu}_{21},$$

$$\text{div}\,(\boldsymbol{h} + 4\pi\sigma\boldsymbol{\mu}_{21}) = 0,$$

$$\frac{\partial N}{\partial t} + \frac{1}{T_1}(N - N_0) = \frac{2i}{\hbar}\,[\sigma^*(\boldsymbol{\mu}_{12} \cdot \boldsymbol{h}) - \sigma(\boldsymbol{\mu}_{21} \cdot \boldsymbol{h}^*)], \tag{55.14}$$

$$\frac{\partial \sigma}{\partial t} + (i\Delta\omega + T_2^{-1})\,\sigma = \frac{i}{\hbar}\,(\boldsymbol{\mu}_{12}\boldsymbol{h})N,$$

which with an accuracy up to second-order quantities are equivalent to the original set (55.11) but are of a lower order. $\Delta\omega$ denotes the quantity $(\omega - \omega_0)$.

To examine the steady-state oscillation mode we put all the time derivatives equal to zero. After this it is not difficult to eliminate the variables σ, N and obtain the equation for the amplitude of the magnetic field:

$$\text{curl curl}\, \boldsymbol{h} - k^2\boldsymbol{h} = \frac{i\alpha k^2 \dfrac{\boldsymbol{\mu}^*}{\mu^*}\left(\dfrac{\boldsymbol{\mu}}{\mu} \cdot \boldsymbol{h}\right)}{1 + \beta^2\left|\left(\dfrac{\boldsymbol{\mu}}{\mu} \cdot \boldsymbol{h}\right)\right|^2}. \tag{55.15}$$

The following notation is used in (55.15):

$$k^2 = \frac{\varepsilon\omega^2}{c^2},$$

$$\alpha = \frac{4\pi T_2 N_0\, |\mu|^2(1 - iT_2\Delta\omega)}{\hbar[1 + T_2^2(\Delta\omega)^2]}, \tag{55.16}$$

$$\beta^2 = \frac{4T_1T_2\, |\mu|^2}{\hbar^2[1 - T_2^2(\Delta\omega)^2]}.$$

In the one-dimensional case, when $\boldsymbol{h} \| \boldsymbol{\mu}$† and the variables depend only on the coordinate z, equation (55.15) becomes

$$\frac{d^2F}{dz^2} + k^2F = -i\alpha k^2 \frac{F}{1 + FF^*}, \tag{55.17}$$

where $F = \beta h$ is a dimensionless variable.

The next stage of the solution consists of averaging (55.17) over the spatial period of the wave $2\pi/k$. This can be done if the right-hand side of (55.17) is small, and therefore the function $F(z)$ is almost a sine wave. The characteristic values for a solid of $T_2 \sim 10^{-11}$ sec, $|\mu| \sim 10^{-20}$, $N_0 \sim 10^{19}$ cm^{-3} lead to $\alpha \sim 10^{-4}$. In this case (55.17) can be written in the form

$$F = C_1(z)\,e^{-ikz} + C_2(z)\,e^{ikz}, \tag{55.18}$$

where $C_1(z)$ and $C_2(z)$ are the complex amplitudes of the wave travelling to the right and the left and are assumed to be slowly varying functions of the coordinate z.

The differential equations for the amplitudes

$$\frac{dC_{1,2}}{dz} = \pm\frac{\alpha k}{2} \frac{C_{1,2} + C_{2,1}\,e^{\pm 2ikz}}{1 + |C_1|^2 + |C_2|^2 + C_1^* C_2\,e^{2ikz} + C_1 C_2^*\,e^{-2ikz}}, \tag{55.19}$$

that follow from (55.17) and (55.18), must then be averaged over the spatial period. The right-hand side is a fairly complex function of z and the procedure for averaging is not as simple as before. Oscillating terms proportional to $e^{\pm 2ikz}$ make a considerable contribution to the result of the averaging.

Having obtained as a result of averaging

$$\frac{dC_{1,2}}{dz} = \pm\frac{\alpha k}{4} C_{1,2} \times \frac{\sqrt{1 + 2(|C_1|^2 + |C_2|^2) + (|C_1|^2 - |C_2|^2)^2} \pm |C_2|^2 \mp |C_1|^2 - 1}{|C_{1,2}|^2 \sqrt{1 + 2(|C_1|^2 + |C_2|^2) + (|C_1|^2 - |C_2|^2)^2}} \tag{55.20}$$

we then change to the real variables $m_{1,2}$ and $\varphi_{1,2}$, introducing them by the substitutions

$$C_{1,2} = |C_{1,2}|\,e^{i\varphi_{1,2}}, \quad m_{1,2} = |C_{1,2}|^2. \tag{55.21}$$

This allows us to pick out two equations that associate the intensities of two waves travelling towards each other:

$$\pm\frac{dm_{1,2}}{dz} = \frac{k\,\mathrm{Re}\,\alpha}{2} \frac{\sqrt{1 + 2(m_1 + m_2) + (m_1 - m_2)^2} \pm m_1 \mp m_2 - 1}{\sqrt{1 + 2(m_1 + m_2) + (m_1 - m_2)^2}}. \tag{55.22}$$

† This is possible if $\boldsymbol{\mu}$ is real.

The remaining equations

$$\frac{d\varphi_{1,2}}{dz} = \frac{1}{2} \frac{\operatorname{Im} \alpha}{\operatorname{Re} \alpha} \frac{d}{dz} (\ln m_{1,2}) \tag{55.23}$$

establish the relation between the amplitude and phase of each wave. The latter can be easily integrated. Since $\operatorname{Im} \alpha / \operatorname{Re} \alpha = -T_2 \Delta\omega$, as can be seen from (55.16), the result of integrating (55.23) will be

$$\varphi_{1,2} = -\tfrac{1}{2} T_2 \Delta\omega \ln m_{1,2} + \psi_{1,2}. \tag{55.24}$$

This result contains two arbitrary constants ψ_1 and ψ_2.

Let us now return to equations (55.22). Introducing the variables

$$u = m_2 - m_1,$$

$$v = \sqrt{1 + 2(m_1 + m_2) + (m_1 - m_2)^2} - 1$$

reduces them to

$$\frac{du}{dz} = -k \operatorname{Re} \alpha \frac{v}{v+1}, \quad \frac{dv}{dz} = -k \operatorname{Re} \alpha \frac{u}{v+1}. \tag{55.25}$$

The integrals of (55.25) are, as can easily be checked,

$$v^2 - u^2 = 4A,$$

$$u + \ln |u + \sqrt{u^2 + 4A}| = -(\operatorname{Re} \alpha) kz + B$$

or, returning to the variables $m_{1,2}$,

$$(m_1 - A)(m_2 - A) = A,$$

$$m_{1,2} - A - \frac{A}{m_{1,2} - A} + \ln (m_{1,2} - A) - \frac{1}{2} \ln A = \pm (\operatorname{Re} \alpha) kz. \tag{55.26}$$

Expressions (55.26) do not contain the integration constant B since the origin is at the point $(1/k \operatorname{Re} \alpha)(B + \ln 2\sqrt{A})$ at which

$$m_1 = m_2 = A + \sqrt{A} \tag{55.27}$$

and the mean energy of the field, which is proportional to $m_1 + m_2$, has its minimum value.

The integration constant A is determined by the boundary conditions. If the device in question is a layer of active material bounded by the planes

$z = -L_1$ and $z = L_2$ (Fig. XII.9), whose emission is the only source of the electromagnetic field, then the intensities of the travelling waves at the boundaries are connected with the reflection coefficients by the relations

$$\frac{m_1}{m_2} = r_1 \quad \text{when} \quad z = -L_1,$$
$$\frac{m_2}{m_1} = r_2 \quad \text{when} \quad z = L_2. \tag{55.28}$$

The quantity A is uniquely determined by the values of the reflection coefficients and the thickness of the layer.

With given boundary conditions equations (55.26) permit the formulation

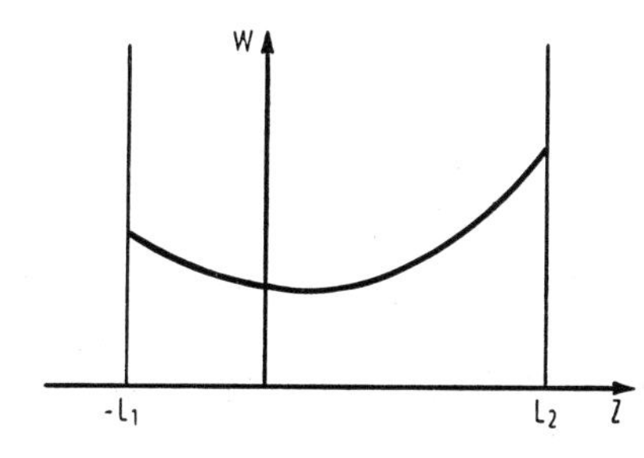

FIG. XII.9. Distribution of field energy density in an oscillating layer without volume losses.

of the necessary criterion for the existence of stationary oscillations in the layer. We transform (55.26) to

$$m_1 - m_2 + \frac{1}{2} \ln \frac{m_1 - A}{m_2 - A} = (\operatorname{Re} \alpha)\, kz, \tag{55.29}$$

from which follows

$$(m_1 - m_2)_{z=L_2} - (m_1 - m_2)_{z=-L_1} + \frac{1}{2} \ln \left(\frac{m_1 - A}{m_2 - A}\right)_{z=L_2} - \frac{1}{2} \ln \left(\frac{m_1 - A}{m_2 - A}\right)_{z=-L_1} = (\operatorname{Re} \alpha)\, kL. \tag{55.30}$$

Here $L = L_1 + L_2$. Expressions (55.26) lead to positive $m_{1,2}$ only if $A > 0$, where $m_{1,2} > A$. Since m_1 is an increasing, and m_2 a decreasing function of z the left-hand side of (55.30) decreases as A drops, reaching the extreme value of $\ln (1/\sqrt{r_1 r_2})$ for $A = 0$. Therefore stationary oscillations with a finite amplitude are possible only if

$$L > L_{cr} = \frac{\ln (1/\sqrt{r_1 r_2})}{k \operatorname{Re} \alpha}. \tag{55.31}$$

Substitution of the value of Re α from (55.16) in (55.30) allows us to transform this condition to the form

$$N_0 > \frac{\hbar[1 + T_2^2(\Delta\omega)^2]}{4\pi T_2 |\mu|^2} \frac{C \ln (1/\sqrt{r_1 r_2})}{\sqrt{\varepsilon}\omega L}. \tag{55.32}$$

This is the same as the condition for self-excitation given in (55.9) if one puts $\Delta\omega = 0$ and uses the fact that the Q of a resonator consisting of plane parallel reflectors, and having loss only at the reflecting surfaces, is

$$Q = \frac{\sqrt{\varepsilon}\omega L}{c \ln (1/\sqrt{r_1 r_2})}. \tag{55.33}$$

With $r_1 = r_2 = r \approx 1$ the Q-factor can be found from

$$Q = \frac{\sqrt{\varepsilon}\omega L}{c(1 - r)}.$$

The results obtained make it possible to determine the frequencies at which excitation of continuous oscillations is possible. The resonance condition is found from the fact that the total change of phase of a wave which has passed through the layer in both directions and returned to the original point must be a multiple of 2π, i.e.

$$2kL + (\varphi_1 - \varphi_2)_{z=-L_1} + (\varphi_2 - \varphi_1)_{z=L_2} = 2\pi p, \tag{55.34}$$

where $p = 0, \pm 1, \pm 2, \ldots$ Substituting (55.24) in this and using the boundary conditions we obtain the expression for the possible oscillation frequencies†

$$\omega_p = \omega_0 \frac{2\pi p + \omega_0 T_2 \ln (1/\sqrt{r_1 r_2})}{2k_0 L + \omega_0 T_2 \ln (1/\sqrt{r_1 r_2})}. \tag{55.35}$$

We have used k_0 to denote the quantity $\sqrt{\varepsilon}\omega_0/c$. Expression (55.35) describes the equidistant spectral lines of the steady-state oscillations of a laser. By virtue of the non-linearity of the system superposition of the oscillations with frequencies described by (55.35) is not a solution. The spectrum of the "non-linear" modes is not the same as the spectrum of the resonator eigenfrequencies

$$\omega_{c,p} = \omega_0 \frac{\pi p}{k_0 L}, \tag{55.36}$$

† If the reflection coefficients are complex ($r_1 = |r_1| e^{i\gamma_1}$, $r_2 = |r_2| e^{i\gamma_2}$), then in (55.35) the place of $r_1 r_2$ is occupied by $|r_1 r_2|$ and in addition $-(\gamma_1 + \gamma_2)$ must be added to the numerator.

which can easily be checked:

$$\omega_{c,p+1} - \omega_{c,p} = \frac{\pi\omega_0}{k_0 L} > \frac{2\pi\omega_0}{2k_0 L + \omega_0 T_2 \ln(1/\sqrt{r_1 r_2})} = \omega_{p+1} - \omega_p. \quad (55.37)$$

We notice that the (55.35) gives the same frequencies as (55.10), which was obtained in the linear approximation, also for the case of stationary oscillations.

Not all the modes in the spectrum (55.35) correspond to large oscillation amplitudes. The number of dominant modes depends on how small the interval $(\omega_{p+1} - \omega_p)$ is compared with the width T_2^{-1} of the spectral line. If $(\omega_{p+1} - \omega_p)T_2 \ll 1$, there are many intense modes. In the case of $(\omega_{p+1} - \omega_p)T_2 \gtrsim 1$ they are small in number, or there is only one mode. As an estimate let us take actual figures for a solid, $r = 1$, $k_0 L = 10^5$ and $\omega_0 T_2 = 3 \times 10^3$ which lead to $(\omega_{p+1} - \omega_p)T_2 \sim 0{\cdot}1$. In a similar case only one non-linear mode may exist.

Let us move on to the question of the spatial distribution of the field inside the oscillating layer. As has already been pointed out, the mean energy density in the period $2\pi/k$ reaches its minimum at the point where $m_1 = m_2$. In a special case when one of the boundaries is an ideal reflector ($r = 1$) the position of the minimum coincides with this boundary. If both the boundaries are ideal reflectors, then $W(z) = \text{const}$. When $r_1 \neq r_2$ it is obvious that the minimum is closer to the less reflecting boundary and the energy density reaches its maximum value on the more reflecting boundary. The form of the energy distribution inside the resonator is illustrated qualitatively in Fig. XII.9.

The output parameter of a laser is the emitted power. Let us assume that the left-hand boundary of the layer is perfectly reflecting ($r_1 = 1$) and find the energy flux (in non-dimensional units) emitted through the right-hand boundary

$$P = m_1(1 - r). \quad (55.38)$$

This can be done by putting $z = L_2$ in (55.26) and then eliminating the constant A. As a result the connection between P and r can be expressed by the equation

$$r = \frac{P + 1 - e^{-2[(\text{Re}\,\alpha)kL - P]}}{P - 1 + e^{2[(\text{Re}\,\alpha)kL - P]}}. \quad (55.39)$$

The emitted power rises monotonically from $P = 0$ for $r = r_{cr}$, when the oscillation threshold is reached, to $P = (\text{Re}\,\alpha)\,kL$ for $r = 1$. The function $P(r)$, for several values of the parameter $L' = (\text{Re}\,\alpha)\,kL$, is shown in Fig. XII.10.

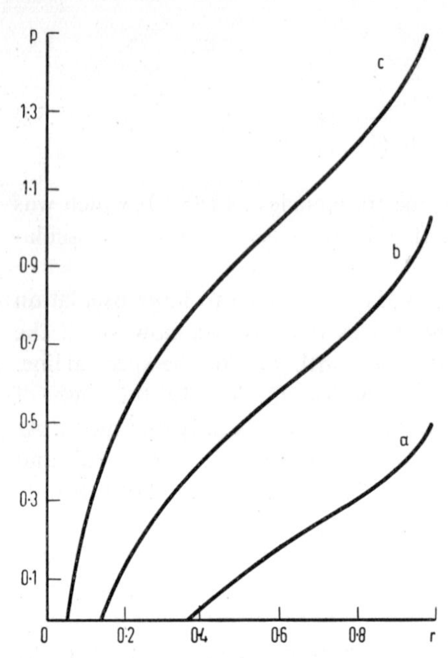

FIG. XII.10. Dependence of the power emitted through the boundary of the layer on the magnitude of the reflection coefficient. The curves are plotted for the following values of the parameter (Re α) kL: (a) 0·1; (b) 0·5; (c) 1.

The fact that the output power rises monotonically when $r \to 1$ is due to the fact that all sources of losses have been neglected, except transmission through the boundary.

55.3. The Steady-state Oscillations of an Active Layer with Volume Losses

In the majority of cases the assumption that there are no volume losses within the resonator of the laser is not justified. Solid working substances contain a large number of scattering inhomogeneities and the internal losses are comparable with the losses at the boundaries. The volume losses can be allowed for by an additional term in (55.17), which now becomes

$$\frac{d^2h}{dz^2} + k^2h = -ik^2\alpha \frac{h}{1 + \beta^2 hh^*} + ik^2gh. \tag{55.40}$$

The small attenuation coefficient g generally depends on ω. Proceeding as in (55.20) we change to the equations for the intensities of the travelling waves

$$\pm \frac{dm_{1,2}}{dz} = \frac{k \operatorname{Re} \alpha}{2} \times \frac{\sqrt{1 + 2(m_1 + m_2) + (m_1 - m_2)^2} \mp m_2 \pm m_1 - 1}{\sqrt{1 + 2(m_1 + m_2) + (m_1 - m_2)^2}} - gkm_{1,2} \tag{55.41}$$

and to the relations between the intensities and the phases:

$$\varphi_{1,2} = -\frac{T_2 \Delta\omega}{2} (\ln m_{1,2} \pm gkz) + \psi_{1,2}. \tag{55.42}$$

The last equations, if the boundary conditions are specified, give the spectrum of the possible frequencies of monochromatic oscillations

$$\omega_p = \omega_0 \frac{2\pi p + \omega_0 T_2[\ln (1/\sqrt{r_1 r_2}) + gk_0 L]}{2k_0 L + \omega_0 T_2[\ln (1/\sqrt{r_1 r_2}) + gk_0 L]}. \tag{55.43}$$

If g does not depend on ω, (55.43) shows that the oscillation frequencies are uniformly spaced. In the other case $[g = g(\omega)]$, (55.43) should be treated as the equation from which ω_p can be determined, and the spacing is not uniform. Comparison of (55.43) and (55.35) shows that introducing the volume losses does not alter the qualitative form of the spectrum of the steady-state oscillations of the laser.

Integration of equations (55.41) is carried out in the same way as for equations (55.22). First we change the variables

$$u = m_1 + m_2 + 1 - (\mathrm{Re}\,\alpha)/g,$$

$$v = \sqrt{1 + 2(m_1 + m_2) + (m_1 - m_2)^2} - (\mathrm{Re}\,\alpha)/g,$$

and then the integrals of (55.41) are found in the form

$$v^2 - u^2 = \text{const.} \tag{55.44a}$$

$$z - z_0 = -g \int \frac{[v + (\mathrm{Re}\,\alpha)/g]dv}{\sqrt{v^2 - B^2}\sqrt{[v + (\mathrm{Re}\,\alpha)/g]^2 - 2\sqrt{v^2 - c^2} + 1 - 2(\mathrm{Re}\,\alpha)/g}}, \tag{55.44b}$$

where z_0 and B are constants. Equation (55.44a) can be rewritten in the $m_{1,2}$ notation as

$$(m_1 - A)(m_2 - A) = A + \frac{g}{\mathrm{Re}\,\alpha} m_1 m_2 \cdot \left[1 + m_1 + m_2 - 2A - \frac{g}{\mathrm{Re}\,\alpha} m_1 m_2\right]. \tag{55.45}$$

A considerable amount of information about the possible solutions of equations (55.41) can be obtained from an analysis of the structure of their phase plane, i.e. the (m_1, m_2) plane. First of all we find the singularities by making the derivatives equal to zero. If $(\mathrm{Re}\,\alpha)/g < 1$, the only singularity has the coordinates $m_1 = m_2 = 0$. In this case the amplification of the medium cannot compensate for the losses and the solutions are damped waves. There is more interest in the opposite case $(\mathrm{Re}\,\alpha)/g > 1$, which gives the set of integral curves shown in Fig. XII.11. It contains four singularities:

$$(0) \quad m_1^{(1)} = 0, \quad m_2^{(1)} = 0 \text{ -- saddle point,}$$

$$(A) \quad m_1^{(2)} = \frac{\mathrm{Re}\,\alpha}{g} - 1, \quad m_2^{(3)} = 0 \text{ -- saddle point,}$$

$$(B) \quad m_1^{(3)} = 0, \quad m_2^{(3)} = \frac{\mathrm{Re}\,\alpha}{g} - 1 \text{ -- saddle point,} \qquad (55.46)$$

$$((M \quad m_1^{(4)} = m_2^{(4)} = m$$

$$\frac{1}{2}\left[\frac{\mathrm{Re}\,\alpha}{g} - \frac{1}{4} - \sqrt{\frac{1}{2}\left(\frac{\mathrm{Re}\,\alpha}{g} + \frac{1}{8}\right)}\right]$$

Each singularity is a solution of (55.41) with an amplitude that is constant in space; two of them describe travelling waves and one a standing wave.

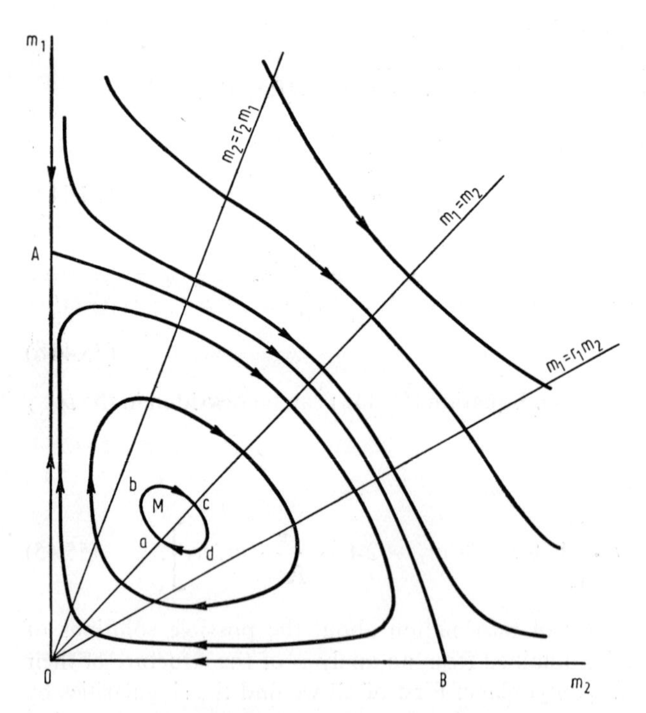

FIG. XII.11. Phase plane of the variables m_1, m_2 (Ostrovskii and Yakubovich, 1965).

From the integral curves of equations (55.41) we must isolate the separatrix curves that connect the singularities. The curves OA and OB correspond to travelling waves that have an amplitude dependent on z. These waves are excited in a semi-infinite active medium when an external wave is incident on the boundary. The curve AB corresponds to the interaction of a pair of waves propagated in opposite directions from infinity. This transition curve is the boundary of regions in which the behaviour of the phase trajectories is qualitatively different. The solutions lying outside OAB are damped waves; m_1 decreases and m_2 rises with a rise in z. This can be explained physically by the fact that the active molecules in the medium cannot maintain such intense oscillations. Inside OAB lies the region of solutions that are periodic in space. The appearance of these solutions in this case is due to the inclusion of the distributed losses.

Finding the spatial period (Λ) of the process depends on the evaluation of the integral (55.44b). It is larger for the outer trajectories than the inner, and near the boundary $\Lambda \to \infty$ (OA and OB are purely travelling waves). We can readily find the minimum "period of rotation" on trajectories of small radius. To do this we must linearize equations (55.41) in the vicinity of the singularity M and look for solutions of the form

$$m_{1,2} = m + \eta_{1,2}\, e^{i\varkappa kz},$$

which leads to

$$\varkappa^2 = g^2 \left(1 - \frac{2m}{\sqrt{1 + 4m} - 1}\right)\left[\frac{2m}{(1 + 4m)(\sqrt{1 + 4m} - 1)} - 1\right]. \quad (55.47)$$

The periodic solutions hold only when the attenuation coefficient lies in the range $0 < g < \operatorname{Re}\alpha$. When $g \to 0$ the singularities A, B and M approach infinity and the phase trajectories, as follows from (55.45), are a family of hyperbolae [see (55.26)]

$$(m_1 - A)(m_2 - A) = A.$$

When $g \to \operatorname{Re}\alpha$ all the singularities move towards the origin.

Let us now discuss processes in a layer of finite thickness. The presence of volume losses is included in the self-excitation condition, which is easily obtained from the linear approximation

$$L > \frac{\ln (1/\sqrt{r_1 r_2})}{k(\operatorname{Re}\alpha - g)}. \quad (55.48)$$

Plotting on the phase plane the straight lines

$$m_1 = r_1 m_2, \quad m_2 = r_2 m_1, \quad (55.49)$$

we notice that the mapping point must move from the first to the second. This confines the possible phase trajectories to the region OAB. Of all the trajectories the solution will be that for which the change in z between the limits (55.49) is the thickness L of the layer. This latter condition determines the choice of the constants z_0 and B in (55.44b).

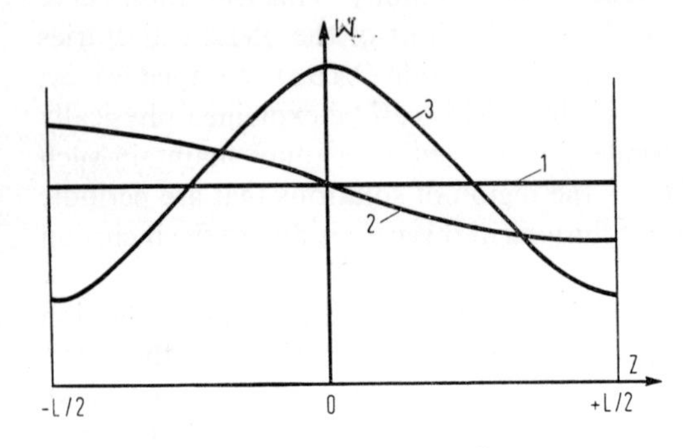

FIG. XII.12. Some possible distributions of the energy density of the field in an oscillating layer with volume losses. Curve 1 corresponds to a system at M; curves 2 and 3 motion along *cda* and *abcda* (see Fig. XII.11).

Under certain conditions the solution for a finite active layer is not unambiguous. This can easily be seen in the case of a layer with $r_1 = r_2 = 1$, when the limiting straight lines (55.48) merge and pass through the singularity M. The obvious solution is the singularity itself. But if $L > \pi/\varkappa$, where $\varkappa$ is given by (55.47), another pair of non-symmetrical solutions appears. On the phase plane they are similar to curves of the type *abc* and *cda*. Increasing L to $2\pi/\varkappa$ results in fresh solutions (*abcda* and *adcba*), etc.

Examples of the spatial distributions of $W(z)$ are given in Fig. XII.12.

55.4. The Applicability of the Balance Equations

The complexity of a rigorous and argued approach to the problem of an optical maser oscillator leads us to search for simplified methods suitable for the solution of special problems. The probability method or the use of balance equations are of obvious application (Stepanov and Gribkovskii, 1964; Stepanov *et al.*, 1962; Ivanov *et al.*, 1962). Balance equations are successfully used in the analysis of certain models of maser oscillators with lumped parameters (see § 45.4). But a laser is a more complex system, a system with distributed parameters, which considerably changes the situation. The probability of a transition, which is proportional to the energy density of the field in the active medium, is an unknown function of the coordinates. It is hard to find the true spatial structure of the field in the framework of the probability method for the following reason. In the case of a plane active layer the field is formed by waves travelling towards each other and interacting non-linearly. The phase relations between the waves must be taken into consideration to obtain the correct result; the balance equations by their very nature cannot allow for this. To change to them from equations of the type (55.19) we must neglect oscillating terms of the form $e^{\pm 2ikz}$. This neglect is justified only in the linear approximation when the amplitude of the oscillatory function is small compared with unity. The region where the probability method can be applied is also clearly limited to the linear approximation and only in the case of a travelling wave.

Now a few words about the terminology used in works on lasers. We very often speak of the modes of a laser, having in mind the frequencies of a discrete spectrum at which oscillation occurs, and the field distributions in the resonator corresponding to them. In the linear theory we take the modes to be the eigenfunctions of the system, their superposition also being an eigenfunction. In non-linear systems there are no modes of this kind and the use of this word is taken to mean an oscillation with a frequency close to one of the resonator frequencies. We speak in a similar way of a mode when we wish to make use of the similarity of the spatial distribution of the radiation to the structure of the corresponding mode of the resonator.

56. Solid-state Lasers

56.1. The Requirements imposed on Working Substances

Active substances used in optical maser oscillators must satisfy definite requirements. One of the most important of these is the desirability of reaching the oscillation threshold with the lowest possible excitation energy.

(a) The substance should have a suitable system of energy levels. At least one of them must have a long life, i.e. be metastable. The population of this level must be made large enough; this in itself is insufficient since the population of the initial level is not as important as the population difference between it and the final level. It is therefore desirable that the latter should be well above the ground state ($E_4 - E_1 \gg kT$) and have as short as possible a life. All other things being equal, substances operating on a four-level scheme are better than three-level ones.

(b) Since the threshold density of active molecules is proportional to the width of the spectral line the latter should be as narrow as possible.

(c) The substance should have wide absorption bands or a large number of absorption lines through which the initial metastable level can be excited. The xenon and mercury lamps most widely used for pumping emit over a wide frequency spectrum and the efficiency of a laser depends on the absorption band of the working substance. The use of monochromatic sources of light such as, for example, semiconductor lasers eases the requirements imposed on the width of the absorption spectra.

(d) The most important characteristic of the working substance is the magnitude of the fluorescence efficiency, i.e. the ratio of the number of spontaneously emitted photons of a given frequency to the number of absorbed photons of the exciting light. The fluorescence efficiency depends on two factors. First the atom reaches the required metastable state through one or more intermediate levels; if the atom can return from these levels to the ground state by some way which misses out the metastable level, only a certain fraction of the excited atoms will pass through it. Secondly, a transition from the metastable level to lower ones may occur either with or without radiation. The higher the probabilities of these competing processes, the lower the magnitude of the fluorescence efficiency of the substance and the less valuable it is for lasers.

(e) The absorption in the substance at the frequency of the working transition should be minimal. We must make particular note of a loss mechanism which is confined to strongly excited substances and is that due to transitions between excited levels; absorption of this kind cannot be observed in ordinary spectroscopic measurements. Under the conditions which obtain in laser it is often very considerable if the working frequency is in an absorption band.

The requirements as formulated are satisfied in a large number of single crystals and glasses, some of which are listed in Table XII.1, p. 184.

Transitions used for pumping and oscillation are those between levels of impurity ions introduced in small amounts into the host lattice of the crystal. Certain elements of the transition and rare-earth groups have been found to be successful activators. The hosts used are Al_2O_3 (corundum), CaF_2 (fluorite), $CaWO_4$ (scheelite) and also BaF_2, SrF_2, etc., into which the individual activators may readily be introduced.

Amongst the crystals used in lasers a special place is occupied by ruby ($Al_2O_3{:}Cr^{3+}$). Because of their good mechanical, refractory and optical properties, corundums have been widely used in a number of branches of technology. The large demand for corundums resulted in the development of methods of industrial production, and the continuous improvement in manufacture and processing. Therefore the requirements of quantum electronics for a suitable material (of which ruby is a very good example) were readily satisfied as a result of work which had already been done. This circumstance distinguishes ruby from other possible materials.

56.2. Ruby as a Working Substance for Lasers

In many ways the homogeneity of a crystal determines its suitability for use in a laser. Existing techniques for growing ruby crystals cannot ensure an ideally uniform distribution of the chromium ions, a fixed direction of the optic axis and complete absence of internal stresses. All these factors lead to optical inhomogeneity of the crystal and may make the specimen useless if they are present to any marked extent.

A crystal intended for a laser is made in the form of a rod with a round or rectangular cross-section. The exciting light is introduced into the crystal through the side. The surface does not need to be particularly well finished to fulfil this function successfully. Moreover, the side surface of the crystal is sometimes made matt on purpose in order to prevent excitation of the laser in "whispering" modes, i.e. waves propagated round the perimeter of the

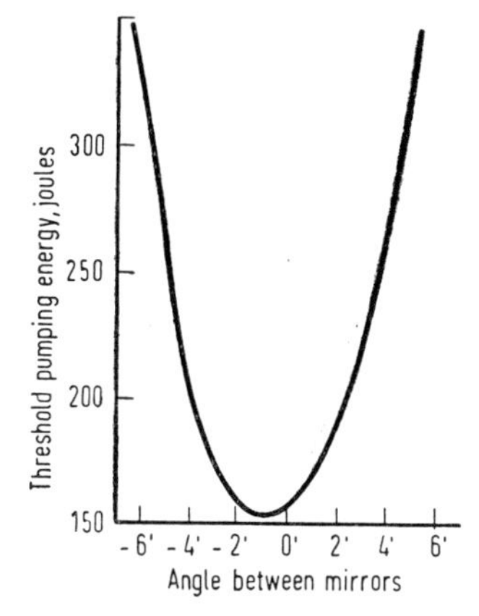

FIG. XII.13. Dependence of oscillation threshold on angle between mirrors (Ready and Hardwick, 1962).

crystal by total internal reflection. The finish of the end surfaces, on the other hand, is very important: they are therefore carefully worked in order to give them as ideal a form as possible (flat or spherical). The resonator of a laser is formed either by the actual faces of the crystal or by external mirrors. Silver films vacuum-evaporated on to the resonator surface are sometimes used as reflective coatings. Although simple to make, they do not provide a coefficient of reflection better than 94–96%, are very absorbent and therefore burn out rapidly during operation. Multi-layer dielectric interference coatings are of higher quality and may have a coefficient of reflection of up to 99%, with very low absorption.

Any deviation of the end surfaces of the crystal from the required form provides an additional source of loss in a laser resonator. They can be tolerated only as long as the contribution of this additional loss to the total amount is small. In a ruby of the highest quality the variations in the optical length caused by internal inhomogeneities are $\lambda/10$ (Stickley, 1963) when the crystal is a few centimetres long. Variations of the order of $\lambda/4$ are typical for good crystals. The surfaces should be worked to an accuracy up to these figures. Their requirements for parallelism are not very stringent.

Figure XII.13 shows the dependence of the oscillation threshold of a laser with external mirrors on the angle between them. It is obvious that there is no sense in trying to achieve parallelism with greater accuracy than 2′. The dimensions and shape of the ruby element are chosen according to the type of lamp, the optical pumping system and the method of cooling.

Some information on the spectrum of a ruby which is essential from the point of view of using this material in laser oscillators and amplifiers is given in Appendix III. There also, in Fig. A.III.4, are shown the energy levels of chromium ions in pink ruby that take part in the operation of a laser. In the fluorescence spectrum of a ruby there are two characteristic narrow intense lines which are caused by transitions from the metastable ^2E state to the ground state: they are called the R-lines. At a temperature of 300 °K the maximum of the R_1-line occurs at a wavelength of $\lambda_{R_1} = 6943$ Å, of the R_2-line at $\lambda_{R_2} = 6927$ Å. When the crystal is cooled the R-lines shift to a shorter wavelength, reaching values of $\lambda_{R_1} = 6934$ Å and $\lambda_{R_2} = 6920$ Å at 77 °K.

Excitation of the metastable levels is achieved via the broad bands $^4F_1$ and $^4F_2$ connected with ^2E by non-radiative transitions. The absorption band of ruby corresponding to a transition from the ground state $^4A_2$ to the $^4F_2$ band is in the green part of the spectrum ($\lambda \sim 5000$–6000 Å). The second absorption band is in the blue–violet region ($\lambda \sim 3600$–4400 Å). The absorption spectrum of ruby is shown in Fig. A.III.5. Excitation via the shorter wavelength band is energetically less profitable and, in addition, the non-radiative transitions $^4F_1$–^2E cause considerable heating of the crystal. The path $^4A_2 \rightarrow {^4F_2} \rightarrow {^2E}$ is better. After a time $T_{32} = 5 \times 10^{-8}$ sec ions from the $^4F_2$ band pass to the ^2E levels or after a time $T_{31} = 2 \times 10^{-7}$ sec emit spontaneously, returning to the ground state. These figures show the high lasing quantum efficiency, which is close to unity according to experiment. The life of the ^2E state is $T_{21} = 3$–5 msec.

The ratio of the densities of ions in the E and $2A$ states is established in accordance with Boltzmann's law. Since the distance between the levels is 29 cm^{-1} the difference in their populations is as much as 15%. The relaxation time between $\bar{E}$ and $2\bar{A}$ is extremely small; according to McClung and Hellwarth (1963) it is not greater than 10^{-7} sec. For this reason stimulated emission cannot be achieved under ordinary conditions on the R_2-line.

Let us examine the structure of the R_1-line in slightly greater detail. For pink ruby the width of this line at a temperature of 300 °K is 10 cm^{-1} (3×10^{11} Hz). When the crystal is cooled the frequency of the line shifts towards the violet and narrows at the same time (Fig. XII.14). As is well known, the crystal field of corundum splits the ground state of Cr^{3+} into two components 0·38 cm^{-1} apart. This leads to splitting of the R_1-line, which is already noticeable at 77 °K, when the width of the individual component is 0·3 cm^{-1}. With further cooling the rate of variation of the parameters of the lasing lines with temperature becomes less. The width of the line reaches a value of 0·07 cm^{-1} at $T = 4{\cdot}2$ °K. The shape of the R_1-line at the temperature of liquid helium is shown in Fig. XII.15.

From the form of the function $\Delta\nu(T)$ we can conclude that thermal lattice vibrations make the major contribution to the line width in the high temperature range. At liquid-helium temperatures $\Delta\nu$ varies quite considerably

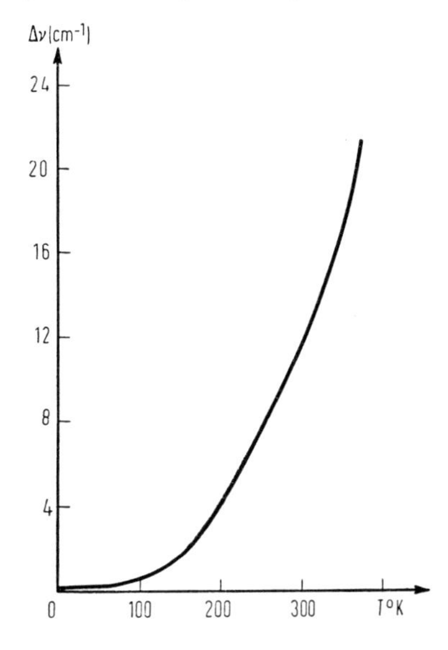

FIG. XII.14. Dependence of width of R_1-line of a ruby on temperature (Schawlow, 1961).

from specimen to specimen and the line width is evidently determined by the internal random stresses in the crystals. We should note that the data given refer to a crystal with little stress. Broader lasing lines are characteristic of lower quality crystals. Collins *et al.* (1960) quote an example of a stressed ruby with $\Delta\nu \sim 1\ \text{cm}^{-1}$ for $T = 77°\text{K}$.

Two different lines N_1 and N_2 appear in the spectrum of red ruby (0·5–0·7% Cr^{3+}) (Schawlow *et al.*, 1959; Wieder and Sarles, 1961; Schawlow and Devlin, 1961). The corresponding wavelengths at a temperature of 77 °K are 7041 Å and 7009 Å. The nature of the N-lines is connected with

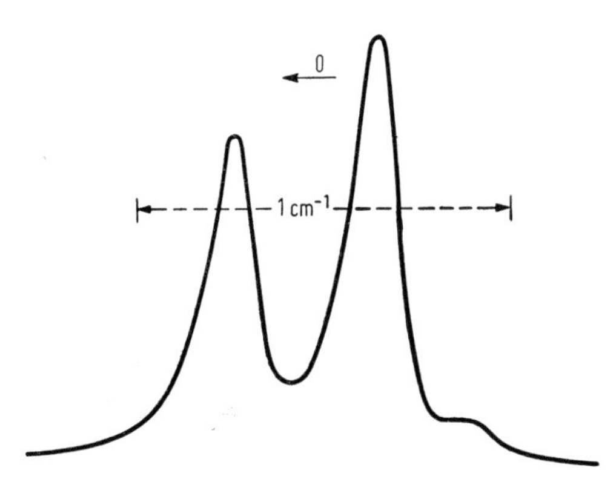

FIG. XII.15. Form of R_1-line at temperature of 4·2°K (Schawlow, 1961).

exchange interaction between the chromium ions, so their intensity is strongly dependent upon the concentration. The fact that two lines appear can be explained by the formation of non-equivalent chromium pairs. The laser output on the *N*-lines is only 0·1 % or three orders less than on the *R*-lines. The oscillation threshold, nevertheless, is reached fairly easily at liquid-nitrogen temperatures, since the lower levels of the transitions are 100 cm^{-1} higher than the ground level. Oscillation can be obtained simultaneously on both lines since they are due to different ion complexes.

We shall be discussing below chiefly pulsed lasers. The basic design of lasers has been described in § 54 and we shall not dwell further on this topic. The type of emitted radiation is determined by the population difference of the chromium levels achieved by the pumping. When it is small, and far from the critical value corresponding to the oscillation threshold, the ruby fluoresces in the ordinary way, i.e. the individual chromium ions emit spontaneously and independently of each other. The radiation is of a completely different nature when the threshold is passed. The major part is now played by the process of stimulated emission, which can be maintained for at least limited periods of time. The crossing of the threshold is accompanied by a sharp decrease in the line width of the emission and an increase in the spectral density of the energy, the appearance of spatial coherence and directional emission.

In the fluorescence spectrum of ruby the ratio of the intensities of the *R*-lines is close to unity and does not depend on the direction of observation. For emission through the side surface this is the case at any pumping level. Emission from the ends behaves differently. The onset of stimulated emission is accompanied by a sharp increase in the intensity of the R_1-line only. Self-excitation, as we have already said, does not occur at the R_2 transition. When oscillation starts the relaxation process that maintains equilibrium between the sublevels causes transitions from $2\bar{A}$ to $\bar{E}$. Therefore the sublevel $2\bar{A}$ may be considered to act as an additional pumping source.

Of course the *Q*-factor of the resonator can be reduced at the frequency of the R_1-line on purpose, thus creating the conditions for the laser to work on the R_2-line. McClung *et al.* (1962) discuss how this may be done. The multi-layer dielectric coatings of the mirrors forming the resonator may be made so that the coefficient of reflection drops sharply in the range of wavelengths between 6927 Å and 6943 Å. Therefore despite the greater population of the $\bar{E}$-level the self-excitation conditions can be reached earlier on the R_2-line, and the R_1 transition which is dominant under ordinary conditions can be completely suppressed.

Figure XII.16 shows the energy emitted by the laser as a function of the energy in one pumping pulse in a typical case. The initial flat slope of the curve is the fluorescent output before the threshold. The maximum energy output that can be achieved by a laser is determined by such factors as the dimensions

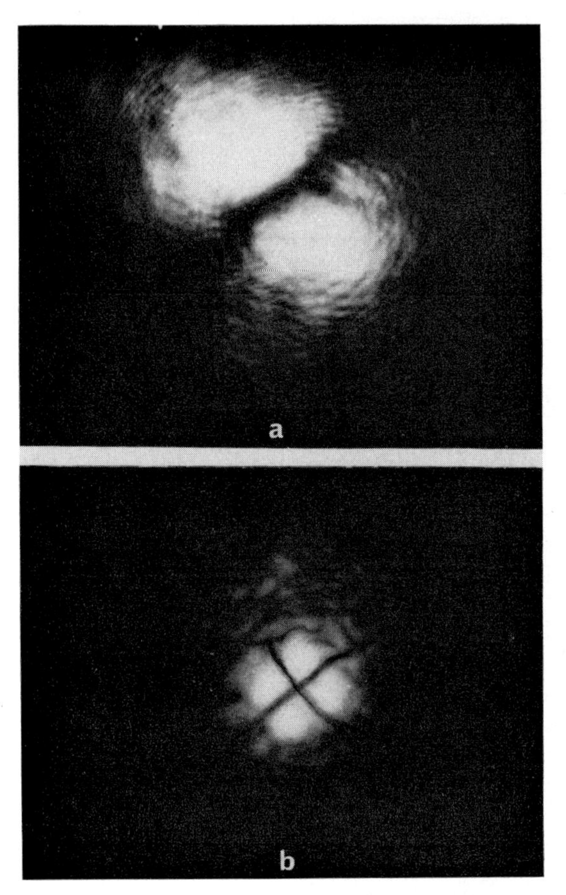

FIG. XII.17. Simple transverse structures of the near field of a ruby laser corresponding to (a) TEM_{01} mode, and (b) TEM_{11} mode.

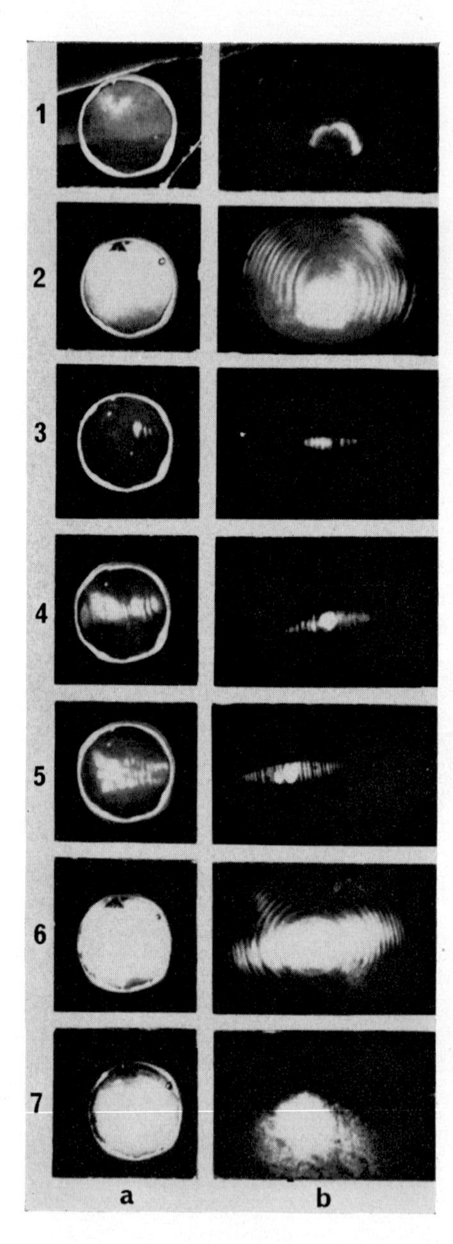

FIG. XII.18. Dependence of beam structure in near (a) and distant (b) fields on internal stresses in a ruby crystal (Lipsett and Strandberg, 1962): 1—no load, pump pulse energy 207 J; 2—no load, pump pulse energy 250 J; 3—load 22 g, energy 185 J; 4—load 32 g, energy 185 J; 5—load 43 g, energy 185 J; 6—load 43 g, energy 250 J; 7—load 150 g, energy 285 J.

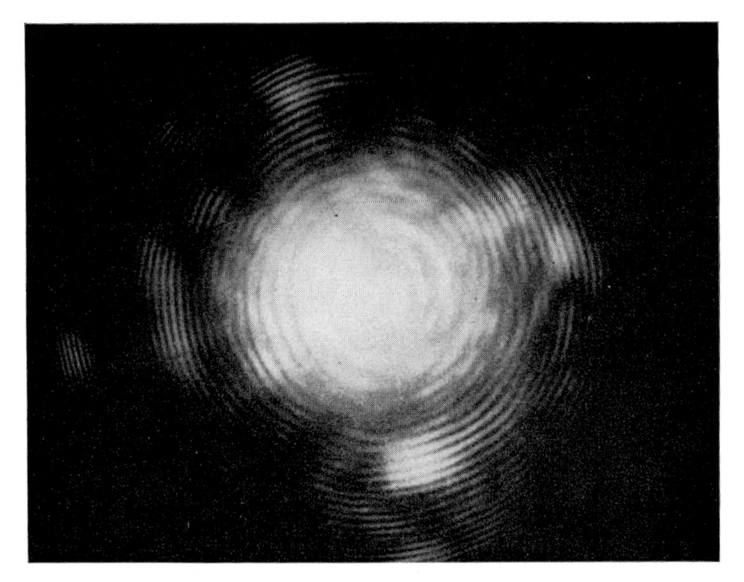

FIG. XII.20. Rings appearing as the result of interference of the direct and scattered waves.

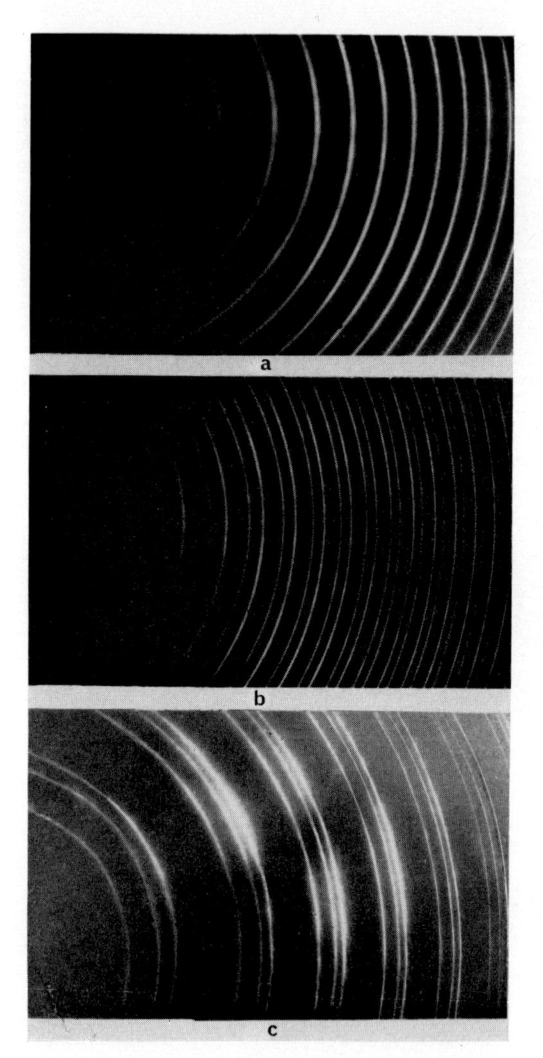

FIG. XII.21. The output of a laser passed through a Fabry–Perot etalon. In the spectrum there are one (a), two (b) and three (c) frequencies corresponding to different axial modes of the resonator.

and quality of the crystal, the care taken in processing it, the capabilities of the pumping system, etc. There are now lasers which will generate energies of the order of hundreds thousends of joules in a flash. The pulse power

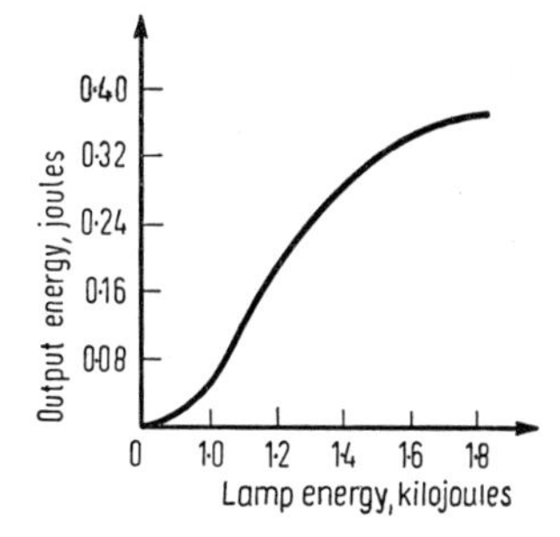

FIG. XII.16. Dependence of the output energy of a pulsed laser on the pumping energy.

reaches tens of kilowatts and even megawatts in devices with Q-switching; latter will the be discussed a little later.

The energy generated by a laser is emitted within a small solid angle which depends on the parameters of the crystal, the form of the resonator and the pumping level. For lasers with plane mirrors the beam spread varies between 0·1 and 1°, rising to 5–6° for spherical mirrors.

Oscillation on the N-lines of red ruby can be observed when it is cooled with liquid nitrogen (Schawlow and Devlin, 1961). The threshold pumping levels for the N_1-line and the N_2-line are not the same and the relation between them is not the same for all crystals, although, as a rule, it is easier to excite the N_1 transition (λ = 7041 Å). This can be explained only by the independence of the N_1- and N_2-lines. The threshold energy is not inherently large since the ruby is in this case working on a four-level scheme. There are clearly no particular practical advantages in using the N-lines, since the concentration of the ion complexes responsible for their appearance is low, the quantum efficiency negligible and the power of the laser correspondingly small. We shall not refer again to stimulated emission from the N-lines.

56.3. *The Spatial Distribution of the Output from a Laser*

No methods are known at present for measuring directly the field inside the crystal of a working laser; all conclusions about it must therefore be made by analysing the output. The only exception is the distribution of intensity on the surface of the mirrors forming the resonator. In order to obtain an output from the resonator, the mirrors used have a small transmission factor, usually a few percent. The negligible thickness of the reflecting coating justifies the assumption that the intensity distributions on its internal and external surfaces are identical; the latter can be determined by photographing the end of the crystal at the time of the flash. In the simplest case of the excitation of one mode of the resonator this distribution should correspond to that given by resonator theory (see Appendix II). The difficulty of comparing the theoretical and experimental results is that even when there is an ideally homogeneous working substance excitation of a large number of modes is possible during the exposure time. The picture is further complicated by the presence in actual crystals of numerous internal inomogeneities which break the crystal up into smaller regions and destroy the regularity of the field distribution. Only in exceptional cases can simple structures be observed corresponding to excitation of a single mode or a small number of modes. Examples of this kind of distribution are shown in Fig. XII.17.

The examples given of the simple structure of the field near the output end are exceptional for a ruby laser. As a rule we find far more complex, often quite irregular, pictures of the type shown in Fig. XII.18a. Here we see a characteristic mosaic structure with a grain size of the order of 0·05 cm which was found in the very first investigations (Collins *et al.*, 1960). The grains are combined in large groups which may occupy a considerable area of the end of the crystal. It should be pointed out that sometimes, particularly when the threshold is considerably exceeded, the mosaic structure disappears and the illumination becomes uniform.

One of the main reasons for the non-uniformity of the optical properties of a crystal is internal stresses. The part played by this factor has been clearly demonstrated by Lipsett and Strandberg (1962) who altered the stress in a ruby artificially. To do this one end of the rod was rigidly fixed in a holder and a small weight was hung on the free end. The degree of elastic bending of the rod can be indicated by the fact that a weight of 100 g altered the angle between the ends by 3′. Bending is accompanied by the appearance of tension above the axis and compression below the axis of the rod. If the internal stresses that were in the crystal before this are distributed in the reverse order, the action of the weight compensates for them and the operating conditions of the laser change. In Fig. XII.18a we can see how the beam structure changes across the end plane when the loading is varied. It may be noted that the oscillation region follows the movements of the compression region. The threshold pumping energy does not remain constant either: the minimum threshold corresponds to a certain optimum loading depending on the internal stressing of the crystal.

The location of the oscillation regions in the cross-section of the crystal depends to a great extent on the distribution of the excited ions through its volume. The latter is determined by the geometry of the pumping lamps and the reflectors and also by the shape and finish of the side surface of the crystal. For example, a polished cylindrical crystal evenly illuminated from the sides focuses the pumping into the axial zone of a radius R/n, from which intense emission can generally be observed. A crystal with a matt surface emits more uniformly from the end surface.

The structure of the beam in the far Fraunhofer zone of the laser provides a considerable amount of information, besides that which can be obtained from the energy distribution in the near zone. The structure of the beam in the distant zone is indicative of the angular distribution (the polar diagram) of the laser. A laser provides a weakly diverging beam and in this sense is a distant source. If a lens is placed in the path of the beam, each bundle of parallel rays emitted by the laser in a definite direction is focused to a point in its focal plane. Photography of the resultant picture in the focal plane of the lens allows us to deduce the angular distribution of the laser.

What factors can have an effect on the structure of the beam in the distant zone? First, there is diffraction at the aperture, over which the intensity distribution is the same as in the oscillating region. Secondly, there is the set of resonator modes excited by the operation of the laser. If there are many of these modes the angular distribution is a system of concentric rings whose radius r_p depends on the ring number p, as $r_p \sim p^{1/2}$. This result can be obtained by considering the interference of the rays in a Fabry–Perot etalon (see, for example, Landsberg, 1957). Thirdly, and lastly, non-uniformities in the crystal break up the regularity of the picture. Figure XII.18b shows distributions in the distant zone corresponding to the pictures in Fig. XII.18a in the near zone. On comparing them the conclusion may be drawn that the annular form in the distant zone occurs when there is no mosaic structure in the near one. A comparison of the results obtained with different crystals shows that the thickness of the ring is directly dependent on the imperfections of the crystal (Stickley, 1963). Thin rings are observed in the emission from high-quality rubies. With variations of the optical length of the order of λ there is generally no annular structure in the distant zone. When the laws governing the radii of the rings are directly checked the above relations are confirmed. The fact that the lasing region is not a perfect circle makes the hypothesis that the rings are caused by diffraction untenable.

A diffraction pattern can be obtained if we cover the end with an opaque screen with a hole not larger than the region of uniform illumination (Masters and Parrent, 1962). The pattern corresponds to the case of diffraction of a plane coherent wave. This method, however, can be used to check the spatial coherence of the emission for only a small part of the aperture ($\sim 100\,\lambda$).

A more suitable method of investigating the coherence is observation of the interference of rays leaving different points of the end of the crystal. Two slits are cut in the screen for this purpose and a photographic film is placed at a distance that ensures overlapping of the beams emitted by each slit (Masters and Parrent, 1962; Nelson and Collins, 1961). This method has been used to check the coherence of the oscillations right up to diametrically opposite points on the aperture ($\sim 3000\,\lambda$). We must point out that the system of interference fringes that indicates the presence of spatial coherence can be observed only when the oscillation threshold is exceeded.

Therefore the laser is an example, at the moment the only one, of a coherent light source. The coherence of the oscillations emitted from opposite ends has also been proved by observing interference (Kisliuk and Walsh, 1962).

An interesting laser design with two ruby rods is described by Masters (1962). Oscillations in the rods are coupled by a prism using total internal reflection, as shown in Fig. XII.19. The degree of coupling is adjusted by

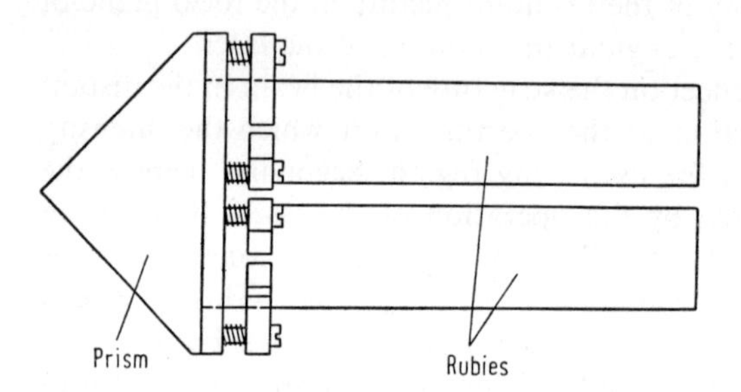

FIG. XII.19. Diagram of a laser with two coupled rods (Masters, 1962).

altering the transparency of the coatings on the ends of the rods facing the prism. The opposite ends have, as usual, coatings with a high coefficient of reflection; both rods are pumped. If the coupling between them is small, they oscillate independently of each other. When the coupling is fairly strong the system starts to operate as a unit and emits coherent waves from the ends of both rods to give a clear interference picture. No interference is observed when the coupling is small. It should be pointed out that a system of two rods coupled through a prism is not equivalent to one rod of double the length. The distinction is that the prism introduces differences into the condition for the propagation of waves travelling along the axis and at an angle to the axis, suppressing the latter. Therefore the spectrum of the modes excited in this kind of combined oscillator should be simpler than in the case of a single rod because of the suppression of the non-axial types of oscillation.

Annular structure in the output from a laser can be observed at places other than in the Fraunhofer zone. Very often it can be seen at moderate distances from the laser without the use of additional lenses (Stickley, 1963; Stoicheff and Szabo, 1963). The nature of the rings in this case has nothing to do with the Fabry–Perot resonator spectrum excited by the oscillation process. The reason for their appearance is scattering centres in the path of the ray. In the Fraunhofer zone ($d \gg a^2/\lambda$, where d is the distance to the scattering object, a is its size) the scattered radiation is a spherical wave with a phase centre coinciding approximately with the position of the object. The original beam is also a spherical wave, but with a different phase centre and a different polar diagram. Because of the spatial coherence of the radiation these two beams interfere to give an annular pattern (Fig. XII.20).

The polarization of the output from a laser is determined by the anisotropy of the crystal, and its nature depends on the orientation of the optic axis relative to the direction of propagation, i.e. relative to the geometrical axis of the rod (Cook, 1961). With 90° orientation all the emission is plane polarized. The electric field vector of the wave is orientated at right angles to the optic axis. In the case of 0° orientation there is no polarization. With arbitrary orientation, in particular with the 60° orientation commonly used, partial polarization is observed.

56.4. The Emission Spectrum of a Ruby Laser

Until pumping provides the threshold density of excited ions the working substance of the laser fluoresces in the usual way. But in the transition to conditions of stimulated emission the spectrum undergoes considerable changes. The intensity of the line rises considerably and its width decreases.

The width of the emission spectrum from a ruby laser is so small that instruments with extremely high resolution have to be used to measure it. Generally, Fabry–Perot etalons which have a resolution up to 10^{-2} cm^{-1} are used for this purpose. The laser beam passed through the etalon forms an annular interference picture whose structure provides information on the spectrum of the radiation being analysed. The radius of the interference ring depends on the wavelength of the radiation and its width on the breadth of the spectral line. Figure XII.21 a shows an example of a simple spectrum; the various rings correspond to different orders of interference. The spectra shown in Fig. XII.21 b, c are of a different nature: there is splitting into several components within a single order. The multiplet nature is caused by the excitation of several different longitudinal modes of oscillation. The distance between the components is approximately the same as the difference of the wave numbers of adjacent longitudinal modes of the ruby rod.

It must be pointed out that the structure of the spectrum also depends on the temperature of the ruby at the time of oscillation. The results shown in Fig. XII.21 relate to room temperature. By referring to Korobkin and Leontovich (1963) we can find data for the overall width of the emission spectrum. With a fluorescent line width of 10 cm^{-1} for an uncooled crystal and of 0·6 cm^{-1} for a temperature of about 100°K laser action gives a width of 1 cm^{-1} and 0·25 cm^{-1} respectively.

56.5. Lasers based on Solid Materials doped with Rare Earths

The most extensive class of working substances suitable for making lasers contains ions of rare-earth elements as the basic component. The starting point in the development of these materials was the work of Galkin and

Feofilov (1957 and 1959) on a study of the luminescence spectra of crystals activated with rare earths. Some information on the spectra of rare-earth ions is given in Appendix III.

The advantage of crystals with rare earths as working substances is that the great majority of them operate on a four-level scheme. Both the working

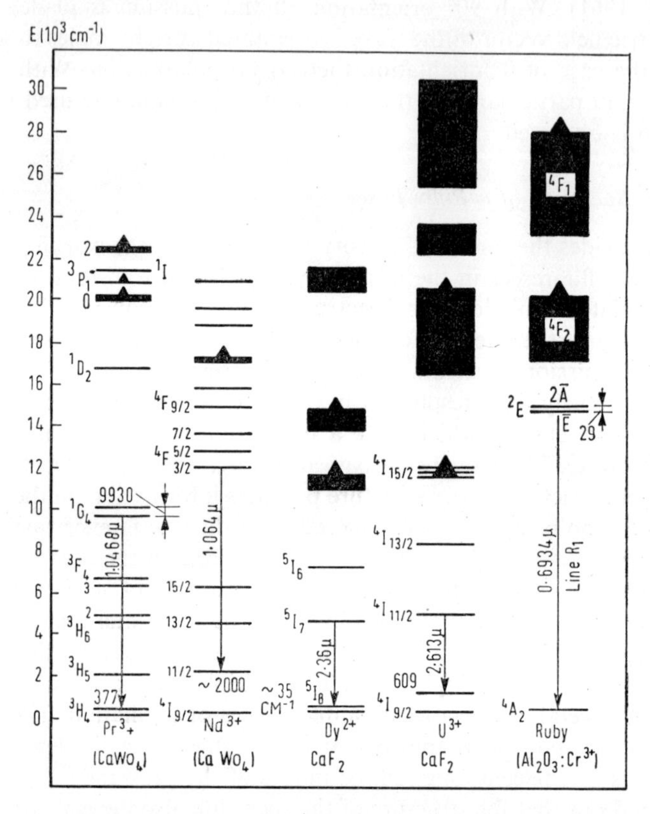

FIG. XII.22. Diagram of energy levels of some crystals used in lasers. The levels that play a part in the pumping process are marked with triangles.

and secondary transitions of the trivalent ions are of the $4f$–$4f$ type. These transitions are forbidden and correspond to narrow lines. This situation is not very favourable for pumping, but it is saved by the large fluorescence efficiency. In the case of divalent ions the $4f$–$5d$ transition, which provides a wide absorption band, is used for pumping. However, it is more difficult to obtain crystals with divalent ions and, in addition, there is a tendency for this state to change its valency.

Some brief information on individual materials with rare-earth impurities is given in Table XII.1. With the exception of $CaF_2{:}Sm^{2+}$ they all lase in the infrared range. Among the crystals doped with rare earths particular note should be taken of $CaF_2{:}Dy^{2+}$, $CaF_2{:}U^{3+}$ and $CaWO_4{:}Nd^{3+}$. The structure of the energy levels of some of these materials is shown in Fig. XII.22. The crystals noted above have the lowest values for the threshold pumping energy. For example, lasing action was achieved with $CaWO_4{:}Nd^{3+}$ with 5 J pumping (300°K) and 3 J pumping (77°K). An energy of 4 J suffices to excite laser action in $CaF_2{:}\,U^{3+}$ at 77°K and 1 J for $CaF_2{:}Dy^{2+}$. Owing to the comparatively low pumping powers these three crystals are suitable for continuous operation.

Glass doped with Nd^{3+} is widely used as a working substance. In its energy parameters it is similar to ruby and is used in cases when an infrared source of high power is required. Glass has the advantage that it is simpler to make than crystals, and it can be made into any shape.

The following papers refer to rare-earth materials: Johnson, 1962a, 1962b; Johnson and Nassau, 1961; Johnson and Soden, 1962; Johnson, Boyd and Nassau, 1962a, 1962b; Johnson, Boyd, Nassau and Soden, 1962; Sorokin and Stevenson, 1960, 1961; Boyd *et al.*, 1962; Porto and Yariv, 1962; Kiss and Duncan, 1962a, 1962b, 1962c; Yariv, 1962; Garrett *et al.*, 1961; Snitzer, 1961; Etzel *et al.*, 1962.

56.6. The Laser as a Pumping Source for Microwave Masers

In Chapter X we mentioned that the unsolved problem of high-frequency pumping sources is one of the basic obstacles to the operation of masers in the sub-millimetre band. Although paramagnetic maser amplifiers have been developed intensively there have been no such sources. But now lasers are available and are devices that in every way satisfy the requirements demanded of pumping oscillators. Moreover, successful operation of a microwave ruby paramagnetic amplifier pumped by a ruby laser has been achieved (Devor *et al.*, 1962).

The population inversion conditions in a three-level quantum system with optical pumping can be found from the balance equations (54.1). The conditions appropriate to this case are that $\hbar\omega_{31} \gg kT$, whilst $\hbar\omega_{21} \ll kT$.

Assuming that the pumping intensity is constant we obtain from (54.1) the set of algebraic equations

$$\begin{aligned} &w_{21}\left(1 - \frac{\hbar\omega_{21}}{kT}\right) N_1 - w_{21}N_2 + w_{32}N_3 = 0; \\ &W_{31}N_1 - (w_{32} + w_{31} + W_{31})\, N_3 = 0; \\ &N_1 + N_2 + N_3 = N. \end{aligned} \tag{56.1}$$

Solving these equations we find the ratio of the populations of the lower

TABLE XII.1. SOLID-STATE LASERS

Working substance	Operating wavelength (μ)	Transitions	Position of final level (cm^{-1})	Working temperature (°K)	Lasing conditions	Absorption spectrum (μ)	Remarks	References
$Al_2O_3:Cr^{3+}$	0·6943	$\bar{E}(^2E) \to {}^4A_2$	0	300	pulsed	0·5–0·6; 0·32–0·44		Maiman, 1960b; Collins *et al.*, 1960; Maiman *et al.*, 1961
	0·6934	$\bar{E}(^2E) \to {}^4A_2$	0	77	continuous			Nelson and Boyle, 1962
	0·6929	$2\bar{A}(^2E) \to {}^4A_2$	0	300	pulsed			McClung *et al.*, 1962
	0·701		100	77	pulsed		Transitions between ion pair spectrum	Schawlow, Wood and Clogston, 1959; Wieder and Sarles, 1961; Schawlow, 1961
	0·704		100	77	pulsed			
$CaWO_4:Nd^{3+}$	1.065	${}^4F_{3/2} \to {}^4I_{11/2}$	2000	300	pulsed	0·57–0·6		Johnson and Nassau, 1961
			2000	77	continuous			Johnson, Boyd, Nassau and Soden, 1962
glass: Nd^{3+}	1·06	${}^4F_{3/2} \to {}^4I_{11/2}$	2000	300	pulsed			Snitzer, 1961
$CaWO_4:Er^{3+}$	1·612	${}^4I_{13/2} \to {}^4I_{11/2}$	375	77	pulsed	0·38		Kiss and Duncan, 1962a

$CaWO_4:Tm^{3+}$	1·911	$^3H_4 \rightarrow {}^3H_6$	325	77	pulsed	0·46–0·48 1·7–1·8		Johnson, Boyd and Nassau, 1962a
$CaWO_4:Ho^{3+}$	2·046	$^5I_7 \rightarrow {}^5I_8$	230	77	pulsed	0·44–0·46		Johnson, Boyd and Nassau, 1962b
$CaWO_4:Pr^{3+}$	1·047	$^1G_4 \rightarrow {}^3H_4$	377	77	pulsed	0·45–0·5		Yariv, *et al.*, 1962
$CaF_2:U^{3+}$	2·613 2·613		609	77 77	pulsed continuous	0·9		Boyd *et al.*, 1962
$SrF_2:U^{3+}$	2·407			77	pulsed	1–1·3		Porto and Yariv, 1962
$BaF_2:U^{3+}$	2·556			20	pulsed	1·1–1·5		Porto and Yariv, 1962
$CaF_2:Nd^{3+}$	1·046			77	pulsed			Johnson, 1962a
$CaF_2:Dy^{2+}$	2·36	$^5I_7 \rightarrow {}^5I_8$	35	77	continuous	0·8–1·0		Kiss and Duncan, 1962b; Johnson, 1962b; Yariv, 1962
$CaF_2:Sm^{2+}$	0·708	$^5D_0 \rightarrow {}^7F_1$	263	20	pulsed	0·425–0·5 0·59–0·65		Kaiser *et al.*, 1961
glass: Yb^{3+}	1·015	$^2F_{5/2} \rightarrow {}^2F_{7/2}$		77	pulsed	0·91–0·98		Etzel *et al.*, 1962
glass: Gd^{3+}	0·3125	$^6P_{7/2} \rightarrow {}^8S_{1/2}$		77	pulsed	0·275		Gandy and Ginther, 1962

levels (Hsu and Tittel, 1963)

$$\frac{N_2}{N_1} = \frac{w_{32} \cdot W_{31}}{w_{21}(w_{32} + w_{31} + W_{31})} - \frac{\hbar\omega_{21}}{kT} + 1. \tag{56.2}$$

Expression (56.2) contains the criterion for the attainability of an inverted population. We assume that the secondary transition is completely saturated, i.e. $W_{31} \gg w_{31}, w_{32}$, and we require that $N_2/N_1 > 1$. The latter is equivalent to

$$\frac{w_{32}}{w_{21}} > \frac{\hbar\omega_{21}}{kT}. \tag{56.3}$$

If we compare the inversion condition (56.3) for optical pumping with the analogous equation (45.4) for the case of microwave pumping, we can see their similarity. Criterion (56.3) can be obtained from (45.4) if we replace w_{32} by $\hbar/kT$. We can estimate the way in which (56.3) can be satisfied by assuming that $f_{21} = 10^{10}$ Hz, i.e.

$$\frac{\hbar\omega_{21}}{kT} = \begin{cases} 1{\cdot}5 \times 10^{-3} & \text{when} \quad T = 300\,°\text{K}, \\ 10^{-1} & \text{when} \quad T = 4{\cdot}2\,°\text{K}. \end{cases}$$

For an estimate of the quantity w_{32}/w_{21} we turn to ruby. The lifetime of the metastable ^2E levels is of the order of 10^{-3} sec and is weakly temperature-dependent. The spin–lattice relaxation time of the sublevels of the ground state varies considerably with temperature from 10^{-8} sec at 300°K to 10^{-2}–10^{-3} sec at liquid-helium temperatures. Therefore

$$\frac{w_{32}}{w_{21}} \sim \begin{cases} 10^{-5} & \text{when} \quad T = 300\,°\text{K}, \\ 1 & \text{when} \quad T = 4{\cdot}2\,°\text{K}, \end{cases}$$

and inversion can be achieved only with considerable cooling of the paramagnetic crystal. Inequality (56.3) can also be treated as the condition for determining the maximum working frequency.

Let us estimate the power which the crystal must absorb to maintain equal populations $N_2 = N_1$. In the calculation we start from (54.3) which is applicable to the case under discussion; if $w_{32}/w_{21} \gg \hbar\omega_{21}/kT$ it becomes

$$P_{31} = N\hbar\omega_{31}w_{21}\frac{\hbar\omega_{21}}{kT} V_s. \tag{56.4}$$

According to (56.4) pink ruby cooled with helium consumes a power $P_{31} \sim 10^2$–10^3 W/cm³. A pulsed ruby laser can develop a considerably greater power.

Experimental confirmation has been obtained (Hsu and Tittel, 1963) that a microwave amplifier and oscillator will work when pumped by coherent light. Since population inversion of the lower levels of a ruby can be achieved only at very low temperatures the amplifying crystal is cooled with liquid helium. The same must be done with the laser crystal in order to prevent a temperature displacement of the frequencies of one crystal relative to the other. The distance between the energy levels depends non-monotonically on the strength of the magnetic field (Fig. XII.23). Because of this, resonance

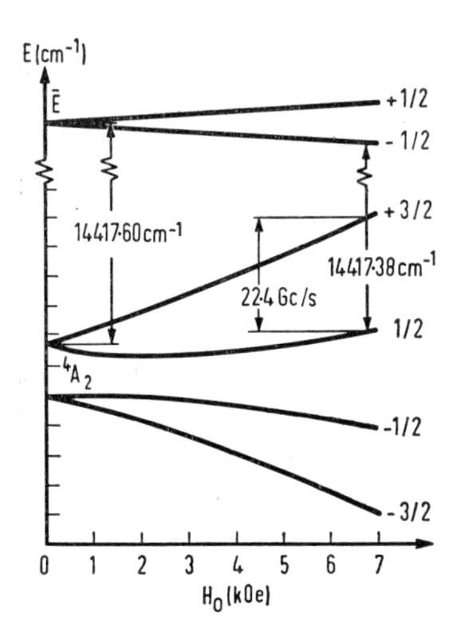

FIG. XII.23. The energy level distribution for ruby in a magnetic field, to show how a microwave maser can be pumped by a laser.

with the pumping radiation can be observed not only in zero field but also with a certain finite field (in the present example $H_0 \sim 6500$ Oe). The maser described by Hsu and Tittel (1963) amplified and oscillated at a frequency of $22 \cdot 4 \times 10^3$ MHz. With crystal dimensions of $1 \cdot 3 \times 3 \cdot 3 \times 2$ mm and a 0·05% chromium concentration a pumping energy of a few thousand joules is sufficient to obtain population inversion. In practice, of the 2·5 J supplied by the laser about 1 J reached the crystal.

From what has been said we can clearly see the difficulties standing in the way of developing masers with optical pumping. The basic difficulty is a result of the condition that the laser emission frequency should be the same as the amplifying crystal secondary frequency ω_{31}. This problem can be solved comparatively simply in the case of identical crystals, but it is very difficult to choose a pair of different materials. Since the pumping oscillator cannot in practice be tuned, the amplifying crystal must be suited to the pumping, which seriously limits the possible applications of the amplifier. The other major difficulty is the absence of powerful enough lasers which will operate continuously.

56.7. The Optical Maser Oscillator based on Stimulated Raman Emission†

Work on producing very high power lasers has led to an interesting discovery. When experimenting with a Q-switched ruby laser using an optical shutter (Kerr cell) Woodbury and Ng (1962) unexpectedly found that the spectrum of the laser output contained additional components whose origin was at first unexplained. Later, specially designed experiments found infrared emission shifted by 1344 cm^{-1} from the ruby laser line. Exactly the same displacement was known in the Raman spectrum of nitrobenzene (nitrobenzene was used in the optical shutter).

In the experiments a cell filled with different organic liquids which show strong Raman scattering (benzene, nitrobenzene, toluene, etc.) was put inside an optical resonator which also contained a ruby and a Q-switch. The frequency-displaced radiation had all the characteristics of stimulated emission: line narrowing, little beam spread, and the presence of a threshold.

The processes occurring in experiments of this kind can be described in terms of stimulated Raman emission, which is described in detail in §§ 38 and 40. We recall than an electromagnetic wave of frequency ω_1 acting on a molecular system having an eigenfrequency ω_0 produces negative susceptibility at a frequency $\omega_2 = \omega_1 - \omega_0$. We can now explain schematically the results of the experiments under discussion. The high-intensity light of frequency ω_1 (hundreds of kilowatts) obtained from a Q-switched ruby laser excites a medium (for example, an organic liquid) so that the latter is in a condition to amplify light of a frequency $\omega_2 = \omega_1 - \omega_0$ (ω_0 is one of the frequencies of the vibration spectrum of the molecules in the liquid). Under certain conditions this system is unstable (when all the losses, including the output radiation, are neutralized). This is achieved by making the mirrors of the ruby laser good reflectors at the frequency ω_2 as well as at ω_1. This kind of device, bearing in mind that it operates at a frequency ω_2, may be called a combination laser or Raman laser.‡

It has been shown in § 38, Vol. 1, that the negative susceptibility at the frequency $\omega_2 < \omega_1$ depends in the first place on the intensity of the pumping signal at the frequency ω_1 (the signal from the ruby laser), and in the second place on the matrix elements of the molecular system which determine the spontaneous Raman emission (see § 38). Therefore the self-excitation threshold at the frequency ω_2 will be determined by the intensity of the laser signal (the laser being the pumping source of the Raman laser), whilst the substances suitable for a Raman laser are those which have suitable symmetry (for example, central) which allows spontaneous Raman emission. Simple estimates which we shall make below show that fairly high powers in

† Section 56.7 has been written by E. G. Yashchin.

‡ Following the liquid Raman laser similar devices were made using solids: diamond and several others.

a very narrow range of frequencies are necessary for the self-excitation of a Raman laser. This is why the experimental discovery of stimulated Raman emission became possible only after the invention of the laser, although the spontaneous Raman effect was discovered as early as 1928.

Let us make some simple theoretical estimates of the operation of a Raman laser.

We shall assume that the signal from the ruby laser (the pumping signal for the Raman laser) is monochromatic. We expand the electromagnetic field inside the resonator in terms of the complex eigenfunctions $\boldsymbol{E}_\nu$ and $\boldsymbol{H}_\nu$ of the resonator (see § 39.4):

$$\boldsymbol{E} = \sum_\nu a_\nu \boldsymbol{E}_\nu \, e^{-i\omega_\nu t}, \tag{56.5}$$

$$\boldsymbol{H} = \sum_\nu a_\nu \boldsymbol{H}_\nu \, e^{-i\omega_\nu t}, \tag{56.6}$$

with the normalization condition

$$\int (\boldsymbol{E}_\nu \cdot \boldsymbol{E}_{\nu'}) \, dV = \int (\boldsymbol{H}_\nu \cdot \boldsymbol{H}_{\nu'}) \, dV = 4\pi\hbar\omega_\nu \delta_{\nu\nu'}. \tag{56.7}$$

The quantities $a_\nu a_\nu^*$ are the mean numbers of photons of energy $\hbar\omega_\nu$. Substituting the expansions (56.5) and (56.6) in Maxwell's equations (for the sake of simplicity we are discussing the isotropic case and are not taking dispersion of the medium into consideration) we obtain for the amplitudes $\boldsymbol{a}_\nu$

$$\frac{da_\nu}{dt} + \frac{1}{2}\frac{\omega_\nu}{Q_\nu} a_\nu = \frac{-i}{2\hbar\omega_\nu^2} \int \left(\ddot{\mathscr{P}}^{NL} \cdot \boldsymbol{E}_\nu\right) dV \, e^{i\omega_\nu t}, \tag{56.8}$$

into which we have introduced the attenuation phenomenologically (by means of the Q-factor Q_ν). We substitute in (56.7) the expression for the polarization produced by a two-quantum process

$$\mathscr{P}^{NL} = \sum_\nu \chi(\omega_\nu, |a_1|^2) \, a_\nu \boldsymbol{E}_\nu \, e^{-i\omega_\nu t}. \tag{56.9}$$

Here $\chi(\omega_\nu, |a_1|^2)$ is the Raman susceptibility, i.e. the susceptibility at the frequency ω_ν in the presence of a field of frequency ω_1 and amplitude a_1 ($\nu = 1$ is the pumping field). As we have already seen, $\chi'' < 0$ if $\omega_2 < \omega_1$ and $\varrho_2 < \varrho_1$.

From equations (56.7), (56.8) and (56.9) it is not hard to obtain the self-excitation condition of the laser in the form

$$Q_\nu^{-1} \leqq 4\pi \, |\chi''(\omega_\nu, |a_1|^2)|. \tag{56.10}$$

It is easiest, of course, to excite the oscillations at a frequency ν where the Q-factor has the highest value ($\omega_\nu \approx \omega_2 = \omega_t - \omega_0$).

Let us estimate the intensity of the pumping signal when self-excitation of the system occurs. For this we use (40.19) for the susceptibility produced by the Raman process. As a result we obtain the following estimate:

$$4\pi |\chi''| \sim \frac{4\pi d^2 T_2}{\hbar} N \frac{d^2 |E_1|^2}{\hbar^2 \omega^2}. \tag{56.11}$$

Here d, T_2, ω are respectively the characteristic values of the matrix elements of the dipole moment, the relaxation time τ_{mn} and the frequency differences, whilst N is the population difference of levels 1 and 2 ($\omega_{21} = \omega_0$). For liquids used in a Raman laser this quantity is of the order of 10^{22} cm^{-3}. If we take for the estimate $d \sim 10^{-18}$, $T_2 \sim 10^{-11}$ sec, $\omega \sim 10^{15}$ sec^{-1} and $Q \sim 10^7$, then the self-excitation condition (56.10) is satisfied with a pumping field of the order of 3 kV/cm. In practice the required field is far higher since our estimate is very approximate.

In conclusion we point out certain interesting effects which have been made possible by the use of stimulated two-quantum processes. If the substance taking part in a stimulated process is not excited, negative susceptibility can appear only at the frequency $\omega_2 < \omega_1$ (Stokes stimulated emission). However, if the substance is excited in some way ($\varrho_2 > \varrho_1$) self-excitation can be achieved at the anti-Stokes component $\omega_2 > \omega_1$ (see § 38). In addition a laser can be made in principle (see the same section), when $\omega_0 = \omega_1 + \omega_2$ and the substance is excited so that $\varrho_2 > \varrho_1$. For this resonators must be provided both the frequencies ω_1 and ω_2.

To explain this, we have from (38.35)

$$\chi''(\omega_1, |a_2|^2) < 0, \tag{56.12}$$

$$\chi''(\omega_2, |a_1|^2) < 0, \quad \text{if} \quad \varrho_2 > \varrho_1.$$

We can have, for example, the following state of affairs. We feed in a signal of frequency ω_1 of high enough intensity so that the self-excitation condition is satisfied at the frequency ω_2

$$4\pi |\chi''(\omega_2, |a_1|^2)| \geqq Q_2^{-1}. \tag{56.13}$$

In this case self-excitation need not occur at the frequency ω_1 and the signal at the frequency ω_1 will only be amplified.

The process of stimulated Raman emission in a Raman laser may be accompanied by a whole cascade of parametric excitations. In fact a signal of the frequency $\omega_2 = \omega_1 - \omega_0$, if it reaches a sufficient magnitude, interacts with the pumping signal ω_1 so that a polarization component appears at a frequency

$$\omega_3 = 2\omega_1 - \omega_2 = \omega_1 + \omega_0. \tag{56.14}$$

The possibility of this kind of process follows from an analysis of the expression for the density matrix in the same approximation (the third) in which the two-quantum emission is obtained [see (38.22)]. It is easy to see that the corresponding cross-susceptibility of this kind of process contains not the square of the modulus of the signal at the frequency ω_1, but the square of the amplitude, so if we are dealing with travelling waves then the z-dependence of the cross-susceptibility and also the z-dependence of the polarization at the frequency $2\omega_1 - \omega_2$ will be determined by the wave vector $2\boldsymbol{k}_1 - \boldsymbol{k}_2$.

Therefore for this process to proceed efficiently we must satisfy the following synchronism condition which expresses the equality of the phase velocities of the electromagnetic wave of frequency ω_3 and the polarization at the frequency $\omega_3 = 2\omega_1 - \omega_2$:

$$\boldsymbol{k}_3(\omega_3) = 2\boldsymbol{k}_1 - \boldsymbol{k}_2. \tag{56.15}$$

It is interesting to note that the parametric interaction of four waves under discussion differs from that discussed at the beginning of Chapter IX in that the first has a resonant nature, i.e. the condition

$$\omega_1 - \omega_2 = \omega_0 \tag{56.16}$$

is satisfied.

This process introduces additional losses at the frequency ω_2† which cannot be ignored in the more precise theory of a Raman laser. Lastly we note that a parametric process of the type (56.14) may occur both with an increase and with a decrease in frequency. It is apparently these processes which explain the additional colours that appear when an intense laser beam is passed through an appropriate medium.

57. The Kinetics of Oscillation Processes in Solid-state Lasers

57.1. The Phenomenon of Self-modulation

With rare exceptions existing solid-state lasers can work only when pulsed. This limitation is due to the high level of pumping power required to reach the oscillation threshold, and the severe thermal conditions of the laser. Xenon lamps are used for pumping pulsed lasers. The flash of light accompanying the discharge of the bank of capacitors through the lamp is of the form shown in Fig. XII.24a. Generally the duration of the flash is 0·1–1 msec. The start of oscillation in the laser is delayed after the lamp is fired until a sufficient number of ions is built up in the metastable level. The lag may be several hundred microseconds.

† In the more rigorous theory we must also allow for the losses at the frequency ω_1 caused by the positive susceptibility $\chi(\omega_1\ |a_2|^2)$.

Because the overall duration of the flash is comparable with the life of the metastable state, transients may occupy a considerable part of the oscillation time. Since there is no general non-stationary theory we must have recourse to a single-mode model of a maser oscillator which assumes that the working substance is concentrated into a small volume in order to estimate the possible nature and duration of the transients. We already know the equations of motion for this kind of system: they are (51.1). All that remains is to determine the value of the time parameter T_1. For this we turn to equations (54.1), allowing for the stimulated emission field by the addition of the term $W_{21}(N_1 - N_2)$ in the right-hand side of the equation for $\dot{N}_2$. Assuming that $\dot{N}_3/N_3 \ll T_{32}^{-1}$ we obtain the following equation from (54.1):

$$\dot{D}_{21} = -\left[\frac{1}{2}\frac{(T_{32}^{-1} - T_{31}^{-1})\,W_{31}}{T_{32}^{-1} + T_{31}^{-1} + W_{31}} + T_{21}^{-1} + \frac{1}{2}W_{31}\right] \times (D_{21} - D'_{21}) - 2W_{21}D_{21}. \tag{57.1}$$

Here $D_{21} = (N_2 - N_1)\,V_s$ is the population difference of the working levels, D'_{21} is the same quantity in the absence of oscillation. For a material of the ruby type $T_{32} \ll T_{31}$, $W_{31} \ll T_{32}^{-1}$ and the first coefficient in the right-hand side of (57.1) simplifies to

$$\dot{D}_{21} = -\,(T_{21}^{-1} + W_{31})\,(D_{21} - D'_{21}) - 2W_{21}D_{21}. \tag{57.2}$$

When there is no oscillation the population difference is given [see (54.2)] by

$$D'_{21} = \frac{W_{31}T_{21} - 1}{W_{31}T_{21} + 1}\,NV_s\,. \tag{57.3}$$

It follows from this that the condition for equal populations will be $W_{31} = T_{21}^{-1}$, and for the relaxation time in (51.1a) we must take

$$T_1 = (T_{21}^{-1} + W_{31})^{-1}. \tag{57.4}$$

Therefore the relaxation time depends upon the pumping power, and since we must have $D_{21} \ll NV_s$ for oscillation we can take it that $T_1 \simeq T_{21}/2$.

For a rough estimate of the frequency of the transient oscillations Ω, and their attenuation time τ, we can use the linear approximation developed in § 51.3. From the form of the roots of the characteristic equation (51.23) we can conclude that

$$\Omega = \sqrt{K_1(n' - 1)}/T_1;\quad \tau = 2T_1/n'. \tag{57.5}$$

Putting $K_1 = T_1/T_c = 10^6$ and $n' = 2$ (double the oscillation threshold), $T_{21} = 2\times 10^{-3}$ sec, we obtain $\Omega = 10^6$ sec^{-1}, $\tau = 10^{-3}$ sec, and when the

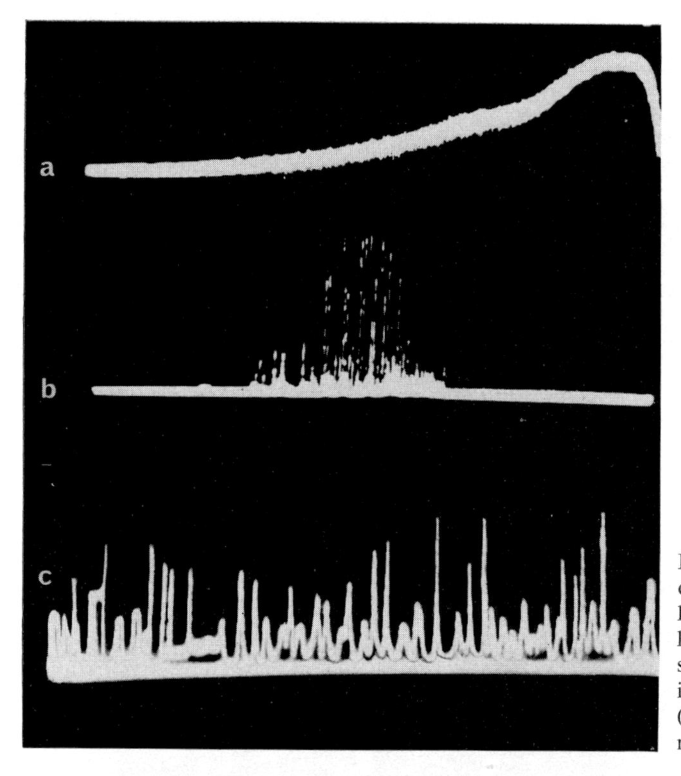

FIG. XII.24. Time curve of pumping (a) and oscillation (b), (c) of a ruby laser. The duration of the scan in cases (a) and (b) is 800 μsec and in case (c) is 230 μsec. The scan runs from right to left.

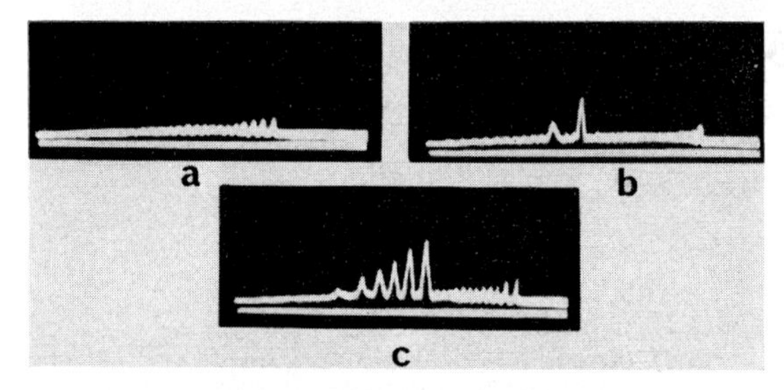

FIG. XII.25. Transients in a $CaF_2:U^{3+}$ laser (Sorokin and Stevenson, 1961). Case (a) corresponds to the minimum and case (c) to the maximum energy of the pumping pulse; scan from right to left.

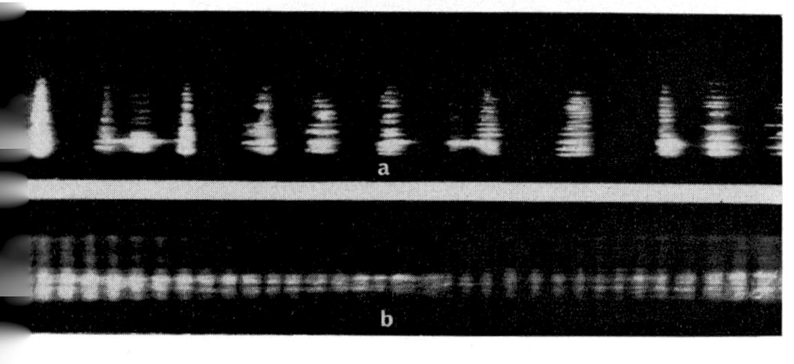

FIG. XII.27. Time scan of structure of the near field of a ruby laser; the crystal temperatures are (a) 220°K; (b) 120°K. The rates of both scans are the same.

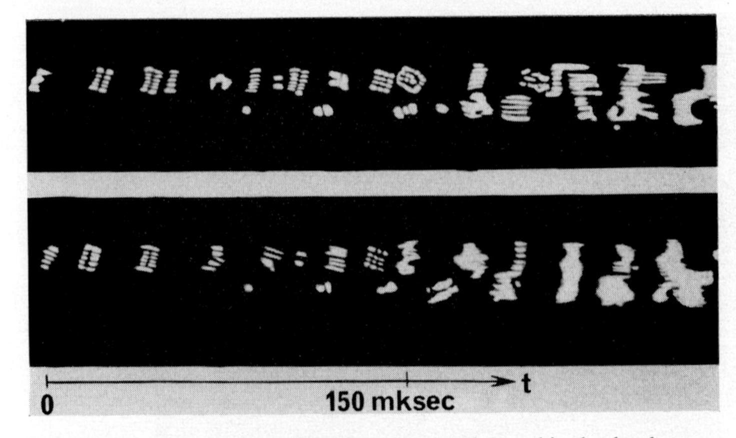

FIG. XII.28. The sequence of transverse modes observed in the development of the oscillation process (Hughes and Young, 1962). The initial portions of two successive flashes are shown (1 mksec ≡ 1 μsec).

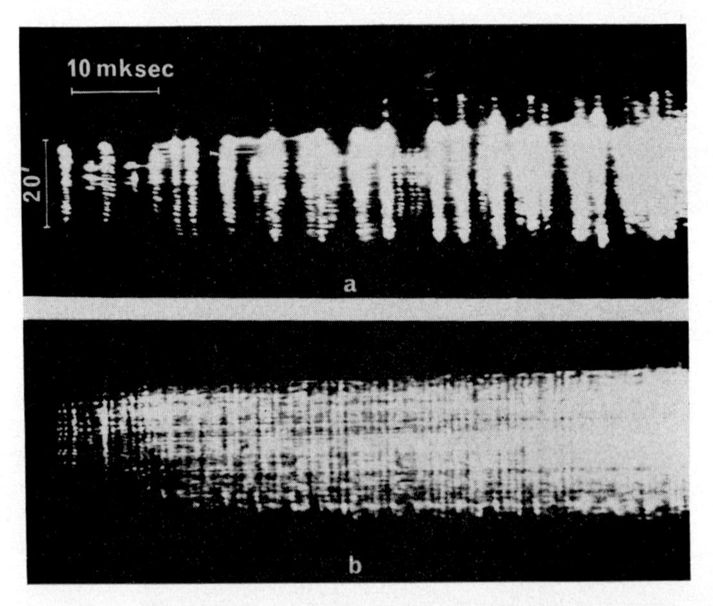

FIG. XII.29. The evolution of the angular distribution from a laser during oscillation: (a) 290°K; (b) 120°K (Korobkin and Leontovich, 1963); (1 mksec ≡ 1 sec).

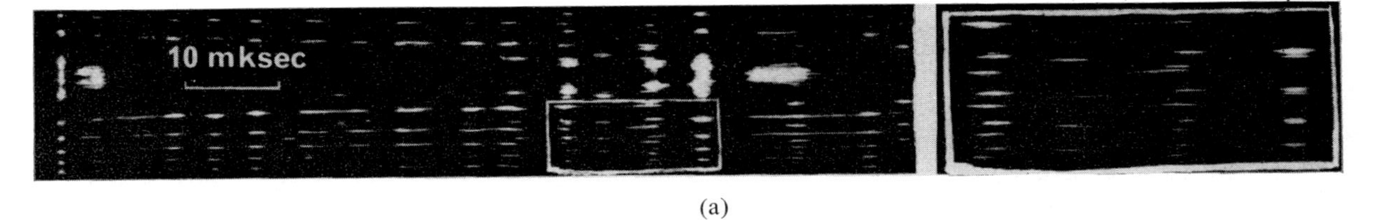

(a)

(b)

FIG. XII.30. The time scan of the spectrum of a ruby laser for different crystal temperatures: (a) 300°K (the frame covers a section where rapid modulation of the output can be seen); (b) 120°K. (1 mksec $\equiv$ 1 μsec).

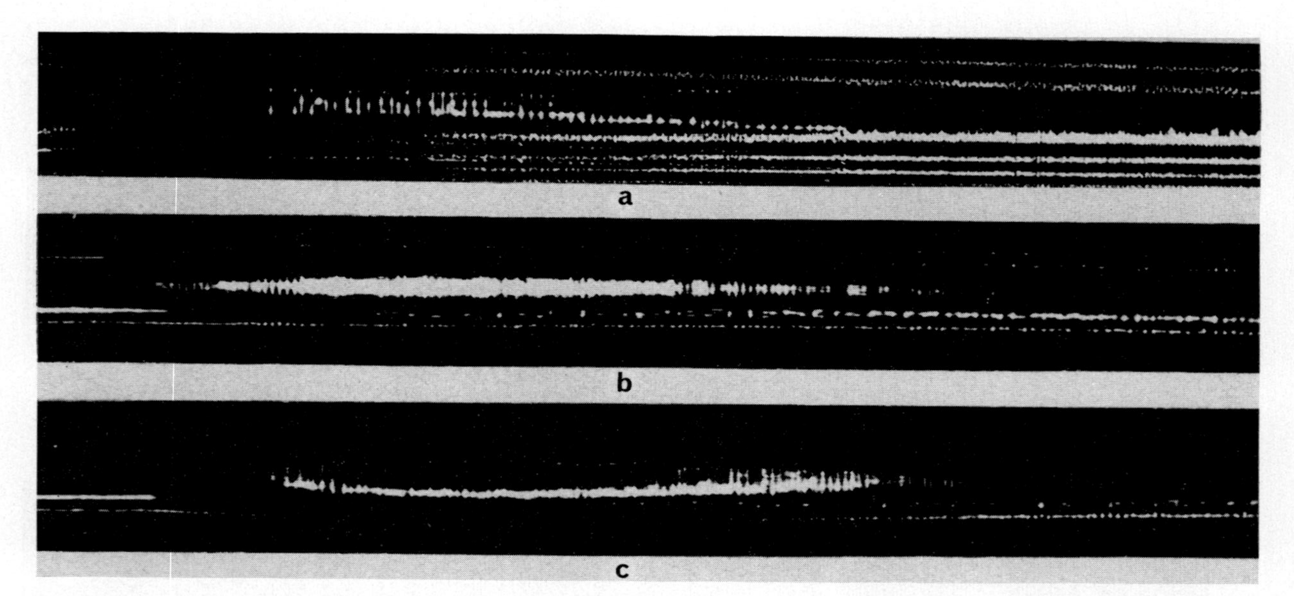

FIG. XII.31. Effect of uniform pulsed magnetic field on the frequency spectrum of a ruby laser for $T = 120°K$ (Kubarev and Piskarev, 1964). Speed 1·0 mm/μsec. The strength of the field in the pulse and the lag between applying the field and the beginning of laser action are (a) 3·5 kOe, 20 μsec; (b) 3·5 kOe, 500 μsec; (c) 1·7 kOe, 600 μsec.

threshold is exceeded by 1% ($n' = 1{\cdot}01$) $\Omega = 10^5$ sec^{-1}, $\tau = 2 \times 10^{-3}$ sec.

Let us move on to experimental data, starting with a ruby laser. An oscillogram of the oscillation process is shown in Fig. XII.24b and c. The pumping light and the laser output are recorded simultaneously on two independent photomultiplier channels whose outputs are connected up to a double-beam oscillograph. This kind of system is convenient for measuring the lag.

The picture of self-modulation ("spiking") which we see in Fig. XII.24 agrees in only one respect with the single-mode two-level model: the mean repetition rate of the small peaks agrees fairly well with Ω. The duration of a peak is 0·5 μsec. But on the other hand the variations in the amplitude of the peaks and the time between them do not display any regularity and the oscillations as a whole show no tendency to be damped. Qualitatively, the nature of the process is the same from one flash to the next but it is not reproducible in detail. The irregularity and irreproducibility of the oscillation process indicated that some sort of statistical element plays a definite part. We must, indeed, point out that the degree of irregularity depends on the temperature: it is far less at liquid-nitrogen temperatures than at room temperature.

By using the same experimental technique we can show the synchronous nature of the oscillation throughout the crystal. To do this we must cover the output mirror with an opaque screen, leaving two holes in it. The ray coming from each hole hits an individual photomultiplier and then the oscillation pictures are compared on the screen of a double-beam oscillograph. Experiments of this kind have shown that the variations in amplitude occur in synchronism in both channels. Only the ratio of the amplitudes may vary from peak to peak (Galanin *et al.*, 1962).

The phenomenon of self-modulation can be observed in the great majority of solid-state lasers. The exception is $CaF_2:Sm^{2+}$ which generates a smooth pulse. This result can be understood if we remember that the life of the metastable level is 2×10^{-6} sec (Kaiser *et al.*, 1961) and an estimate using the (57.5) for $n' = 2$ gives $\Omega = 2 \times 10^7$ sec^{-1}, $\tau = 2 \times 10^{-6}$ sec. The period of the oscillations $2\pi/\Omega$ and the attenuation time are of the same order, i.e. during a flash there may be several oscillations whose frequency is apparently beyond the range of the measuring equipment described by Kaiser *et al.* (1961).

A large number of working substances can be listed which generate the same random sequence of peaks as ruby, among them being calcium tungstate doped with Tm^{3+}, Ho^{3+}, Er^{3+}, neodymium and ytterbium glasses, etc. But there are also examples of a different kind. Amongst them is the transient in a $CaWO_4:Nd^{3+}$ laser. The output amplitude varies only for a brief initial period, being followed by the stationary self-oscillation mode which ends on the cessation of pumping. This process agrees with the results of the analysis of the single-mode model of a quantum oscillator.

Some examples of transients in a $CaF_2:U^{3+}$ laser at different pumping powers are shown in Fig. XII.25. The appearance of new series of damped oscillations indicates that the oscillation conditions change discontinuously. It is important to remember that this is connected with the self-excitation of completely different types of oscillation. The nature of the different forms of self-modulation can be understood only by a careful study of the kinetics of the process. A knowledge merely of the time variation of the oscillation amplitude is quite insufficient; information on the variation of the spatial distribution and spectrum of the output is necessary.

57.2. The Rapid Variations in the Structure of a Ruby Laser Beam and in the Spectrum of the Output

A photograph of a laser beam made with an exposure equal to the duration of the flash is the superposition of a succession of different pictures. Figures XII.17–21† are photographs of this kind. We must use high-speed photography to record the instantaneous intensity distributions and their changes. The high-speed photographic recorder (HPR) used for this purpose is a camera in which the film is stationary and the image of the object being photographed is moved rapidly along the film by reflection from a rotating

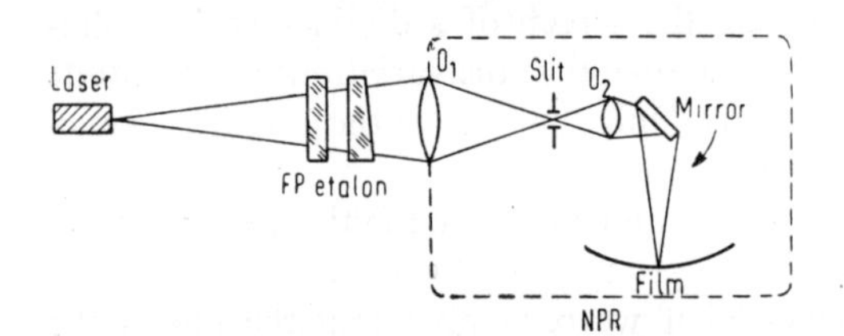

FIG. XII.26. Arrangement for HPR recording of the time resolved spectral distribution.

mirror. The HPR can operate both as a "slow-motion camera" (frame filming) and as a photographic recorder, in which case the shutter remains open for the whole of the filming time. The second method allows us to obtain continuous data, so it is preferable.

The form of the photographic recorder is similar to that of an ordinary camera and depends on the task it has to perform. If it is necessary to study the near-field distribution, the image of the output mirror of the laser is formed on the surface of the film. In investigations of the angular distribution and frequency spectrum the film should be in the focal plane of the objective. An arrangement for recording the spectrum from a laser is shown in Fig. XII.26.

† Excluding Fig. XII.19.

Examples of time scans of the image of the end of an oscillating ruby are shown in Fig. XII.27. The photographs cover only part of the duration of the laser pulse. The image on the film is limited by the narrow inlet slit of the HPR. The vertical bands on the photograph are caused by the "spiked" nature of the oscillation and the horizontal ones by the mosaic structure of the lasing resions in the crystal. It can be seen from these pictures that laser action starts and dies away synchronously in the whole crystal although the configuration of the lasing region does not remain the same. The latter fact can be seen clearly in photographs made with a wide input slit (Fig. XII.28). As a rule a structure corresponding to excitation of the simplest modes of the resonator is observed in the first peaks. As oscillation develops the structure becomes more complex, the oscillating region becomes larger, and the obvious regularity of the picture disappears. The degree of randomness in the near-field structure is determined not only by the duration of the oscillation but also by the pumping power and the quality of the crystal (Leontovich and Veduta, 1964; Stickley, 1963; Hughes and Young, 1962).

The most important conclusion that follows from the experimental data is that the transverse structure of the emission field changes from peak to peak. This is also confirmed by data on the angular distribution (Fig. XII.29), which also changes during laser action. A comparison of the photographs in Fig. XII.27a and b indicates that the dependence of the nature of the self-modulation on temperature is important; at 220 °K the intensity variations are far more random than at 120°K. In the latter case the output is quasi-continuous in nature.

Examination of the structure of the output field distribution of a laser shows the instability of its transverse modes; investigations into changes in the frequency spectrum lead to a similar conclusion about the longitudinal modes. Examples of time scans of a laser spectrum taken under different conditions are shown in Fig. XII.30. The horizontal bands are caused by different orders of interference in the Fabry–Perot etalon, and their fine structure by the discrete nature of the spectrum of the longitudinal modes of the resonator. In Fig. XII.30a, which covers the start of oscillation, we can see the tendency for the wavelength to increase with time. This is brought about by a discontinuous transition from one longitudinal mode to another. The fact that this structure is reproduced in each order indicates that the spectrum is independent of the direction of emission (within the beam, of course).

In Fig. XII.30a we can see splitting of the spectrum into two components separated by an interval of 0·38 cm^{-1} due to transitions to different sublevels of the ground state. Stimulated emission appears initially at the transition $\bar{E}(^2E) \rightarrow \pm\frac{1}{2}(^4A_2)$, and then (in the present case after 30 μsec) at the transition $\bar{E}(^2E) \rightarrow \pm\frac{3}{2}(^4A_2)$. There is another curious detail to be seen in Fig. XII.30a—emission with a modulation period of about 0·3 μsec occurs

50 μsec after the start of the process. This is approximately an order of magnitude less than the ordinary distance between peaks.

When the crystal is cooled to a temperature of 120°K (Fig. XII.30b) oscillation can also be observed initially on two lines displaced relative to each other. In this case, however, the mean oscillation frequency is constant throughout the pulse. The set of longitudinal mode frequencies emitted changes from peak to peak.

The HPR can be used to study the behaviour of a laser when the external conditions are varied. In Fig. XII.31 we can see how the output spectrum depends on the strength of the magnetic field in the crystal (Kubarev and Piskarev, 1964). The pulsed magnetic field is applied 20 μsec after the start of oscillation (Fig. XII.31 a). In this case oscillation at the transition $\bar{E}(^2E) \rightarrow \pm\frac{1}{2}(^4A_2)$ stops and is renewed again only after 20 μsec, but at different frequencies. The splitting of the spectrum and the nature of its dependence on the magnetic field indicate that the effect is connected with the removal of level degeneracy. The intense component covering the frequencies of 3–4 axial modes can be identified with the $-\frac{1}{2}\bar{E}(^2E) \rightarrow -\frac{3}{2}(^4A_2)$ transition. Two frequencies are generated around the field maximum (3·6 kOe). Other examples differing in the lags and magnetic field strengths are shown in Fig. XII.31 b and c. In all the cases there is a sharply defined retuning of the laser when the field is applied: oscillation at the $\bar{E}(^2E) \rightarrow \pm\frac{1}{2}(^4A_2)$ transition ceases and the $-\frac{1}{2}\bar{E}(^2E) \rightarrow -\frac{3}{2}(^4A_2)$ transition is stably excited. It should be noted that at room temperature a magnetic field of several thousand oersteds has no noticeable effect on the oscillation.

The above facts lead to a definite conclusion about the possible nature of the chaotic spiking of the output of solid-state lasers. Let us imagine a crystal placed in a resonator and evenly illuminated by a powerful enough lamp. Until oscillation starts the excited ions are evenly distributed throughout the crystal. Then let stimulated emission into one of the axial modes appear. This disturbs the uniform distribution of the active ions since the number of stimulated transitions is large near the antinodes of the field, whilst there are only spontaneous transitions at the nodes. As a result at the start of a peak a variation of the active ion density with a period of $\lambda/2$ is set up along the wave (Fig. XII.32). This kind of distribution is obviously unsuitable for exciting the same type of oscillation but it may be suitable for an adjacent mode which differs in the number of wavelengths along the resonator axis. Finally, the original mode cannot withstand the competition and gives place to a new one and the process has no time to reach a steady state. It will be seen that the hypothesis is based essentially on the system having many modes. If it is true, then in a travelling-wave laser the spiking should be attenuated even when using the same working substance as in a resonator laser. Numerous experiments (Röss, 1963; Tang *et al.*, 1963) prove that changing to the travelling-wave method really does change the nature of the self-modulation. The latter becomes damped, i.e. takes the form of a real transient from the state of rest to the state of stationary self-oscillations. Although the hypothesis does not contradict the known experimental facts it cannot be considered as the final answer until it has been supported by mathematical calculations.

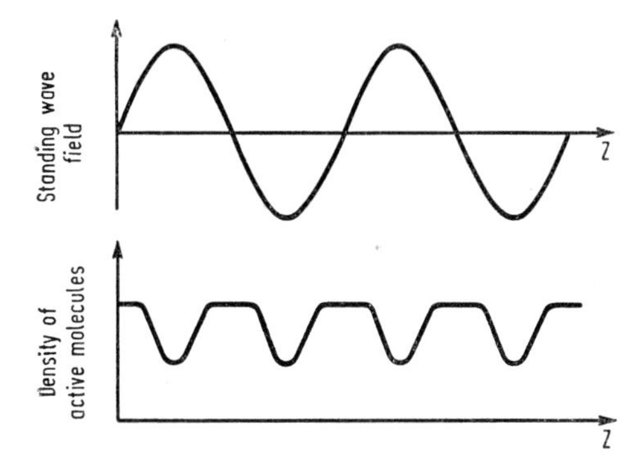

FIG. XII.32. The density distribution of active molecules in a standing wave.

57.3. Heterodyning of Light emitted by a Laser

The resolution of optical instruments used in detailed studies of laser spectra is often far from satisfactory for this purpose. We can hardly expect the capabilities of such equipment to be improved significantly in the near future; we must therefore apply new physical methods to the problem of spectrum analysis. One of these is the use of optical heterodyning.† This is based on the fact that a photocathode is a non-linear element with a square-law characteristic: the photocurrent is proportional to the intensity of the incident light. If such a cathode is subjected to the simultaneous action of two monochromatic light signals of different frequencies, beat frequencies appear in the photocurrent spectrum. The beat frequency of oscillations that cannot be resolved by optical methods lies in the radiofrequency band. The problem is therefore reduced to analysing radiofrequency spectra, which can be carried out with much higher accuracy. As the mixing element we can use the cathode of a vacuum tube if, of course, the beat frequency of the light signals falls within the passband of the tube. Photomultipliers, klystrons and travelling-wave tubes have been successfully used for this purpose. The latter has been used to study the beats between the different spectral components of a ruby laser (McMurtry and Siegman, 1962). The ray was directed onto the cathode of the travelling-wave tube. With a crystal length of 14 cm the adjacent longitudinal modes of the laser are 606 MHz apart and components which are multiples of this difference can be expected in the beat spectrum. Since the nominal bandwidth of the travelling-wave tube in the experiment described was limited to 2000–4000 MHz all the

† The idea of light heterodyning is due to Gorelik (1947, 1948).

harmonics from 1800 to 4200 MHz were observed. If we compare the oscillogram of the signal at the output of the travelling-wave tube with the oscillogram of the output from the laser, it is not hard to see their marked similarity (Fig. XII.33).

Every peak in the output from the travelling-wave tube has a corresponding peak in the laser output. There is, however, no reverse correspondence and this means that from peak to peak the set of longitudinal modes excited in the laser changes. Heterodyning of the output of a laser is not only a precise method for investigating its spectrum but may also be looked upon as a method for receiving and demodulating light. The method is very convenient as a basis of superheterodyne light-wave receivers.

57.4. Controlling a Laser by modulating its Parameters

Controlling the oscillation process in a laser is an important problem which inevitably arises when we wish to make practical use of optical maser oscillators. The most obvious example is the application of lasers to communications since the transmission of information inevitably presumes modulation of the transmitter. Another example is increasing the power of a pulsed laser by spoiling the Q-factor of the resonator.

Which elements of a laser can be controlled? In principle they all can, and the choice of the method of modulation is determined by the nature of the actual problem. The operation of a laser can be most easily controlled by means of the pumping system. A simple alteration of the power of the exciting light affects the number of active ions and thus the amplitude of the oscillations. The time taken to establish an oscillatory state corresponding to a new pumping level is of the same order of magnitude as T_1. For ruby this is 10^{-3}–10^{-4} sec, i.e. the pumping section of the laser has a very slow response and in practice its use for purposes of modulation is excluded.

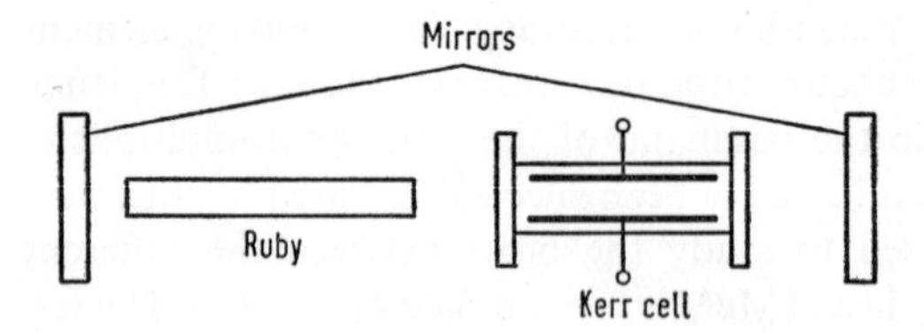

FIG. XII.34. Diagram of a laser with Q-spoiling.

Parameters which can easily be changed are the Q-factor of the resonator and its eigenfrequencies. In this case the maximum possible modulation frequency is determined by the time taken to establish oscillations in the resonator $T_c = Q/\omega$, i.e. of the order of 10^{-7}–10^{-9} sec. Lastly we can affect the working transition itself, using the Zeeman or Stark effect.

Control of the Q-factor is achieved by a shutter placed inside the optical resonator. In this case the resonator is formed by external mirrors. The shutter partly or completely screens one of the mirrors, the degree of screening being variable. The simplest mechanical modulator is an opaque disk with a hole (Basov, Zuev and Kryukov, 1962). The disk's axis of rotation is fixed outside the resonator parallel to its axis. At the times when the disk interrupts the coupling between the mirrors the system ceases to be a resonator and maximum Q is restored only when the hole crosses the axis. While the resonator is inoperative oscillation cannot start and pumping produces an accumulation of a large number of excited ions. When the resonator is restored to normal (the hole moves into the space between the mirrors) the system is in a strongly excited state well above the threshold. The oscillation process develops very vigorously and as a result the laser emits a short but powerful pulse. To obtain maximum power the time the shutter is open should be less than the oscillation rise time. From this point of view mechanical modulators are not very good since they provide an actuation time of no better than 10^{-6}–5×10^{-7} sec, which is insufficient.

Electric modulators have higher speeds. One of them is based on the Kerr effect (see, for example, Landsberg, 1957). A Kerr cell filled with, e.g. nitrobenzene, is put in the resonator between the ruby and the mirror (Fig. XII.34). The electric field in the cell is applied in a direction at 45° to the plane of polarization of the light generated by the crystal. The length of the cell and the strength of the field are selected so that the phase difference between the ordinary and extraordinary rays on passing through the cell is $\pi/2$. In this case the plane of polarization of a ray returned to the ruby after reflection from the mirror and passing through the cell twice is rotated 90° relative to the incident direction. If the ruby rod has a 90° orientation and therefore generates plane polarized light rotation of the plane of polarization by 90° is equivalent to total screening of the mirror. For the widely used crystals with 60° orientation which emit partly polarized light there will not be total screening and the oscillation threshold, as is shown by the results obtained by McClung and Hellwarth (1962), rises by only 20%. In a laser with a 0° ruby, or with other working substances emitting unpolarized light, the modulator can be used with a polarizer. This introduces considerable losses so is less useful.

When the electric field is removed from the Kerr cell its action ceases and the maximum Q-factor of the resonator is restored. The off time in the work of McClung and Hellwarth (1962) was 2×10^{-8} sec and the delay in starting oscillation was 2×10^{-7} sec.

Double refraction can be caused artificially by applying an electric field not only to liquids but also to certain crystals (the so-called Pockels effect). This effect takes place when light is propagated in the direction of the applied electric field. Amongst the crystals suitable for use in modulators which use this phenomenon are the dihydrogen phosphates of potassium and ammonium. (KDP and ADP). Both the Kerr cell and crystals of this type cause modultion frequency of the laser as well as amplitude modulation since the applied field causes a change in both the ε of the medium and the electrical length of the resonator (Barnes, 1962).

The losses in the resonator can be increased by introducing into it a second ruby rod which is not exposed to the exciting light. Q-spoiling is caused by resonance absorption in the additional rod. The absorption can be removed in a very short time by using the Stark or Zeeman effects which shift the transition frequency of the second crystal. In order to estimate whether this method can be used let us take some figures for the R_1 line of a ruby: the Stark coefficient is $1{\cdot}8 \times 10^5$ Hz cm/V and the Zeeman coefficient is 2·6 MHz/Oe. This means that the frequency displacement of the R_1-line by the width of the line itself (10^{11} Hz at 300°K) requires an electric field with a strength of 6×10^5 V/cm or a magnetic field of 4×10^4 Oe. Therefore the method can be recommended only for cooled specimens ($\Delta f \sim 10^{10}$ Hz). We note that this method can also be used to called a modulation of the number of active ions.

The Zeeman effect can be used for modulating a laser, without introducing an additional crystal into the resonator, by the action of a field on the oscillating crystal itself; in this case the field must necessarily be non-uniform. The resulting inhomogeneous broadening of the spectral line reduces the oscillation amplitude and may stop it altogether. An experiment of this type is described by Nedderman *et al.* (1962). A single-turn copper conductor surrounds the central part of the ruby rod. This turn produces a field in the centre of the rod 4 times stronger than at the ends. At room temperatures inhomogeneous broadening starts to produce an effect with a field strength at the centre of the rod of 25 kOe. At liquid-nitrogen temperature this figure drops to 6 kOe. If the non-uniform magnetic field is oscillatory the flashes of stimulated emission occur during the small time intervals when the magnetic field is close to its minimum.

The control element may be an ultrasonic cell (DeMaria and Gagosz, 1962) placed inside the resonator of the laser. If the length of the standing ultrasonic wave is greater than the cross-section of the beam the trajectory of the latter is distorted by refraction. The magnitude of the instantaneous deflection of the ray depends on the amplitude and phase of the ultrasonic oscillation. When there is no ultrasound the laser generates an ordinary random sequence of spikes. The ultrasonic oscillations synchronize the self-modulation process since the times that are favourable for laser action coincide with the zero phase of the ultrasonic oscillations.

Synchronization can also be achieved with a Kerr cell modulated by a sine-wave voltage. Oscillograms of the output of a laser in the synchronization mode with different cell modulation frequencies are given in Fig. XII.35.

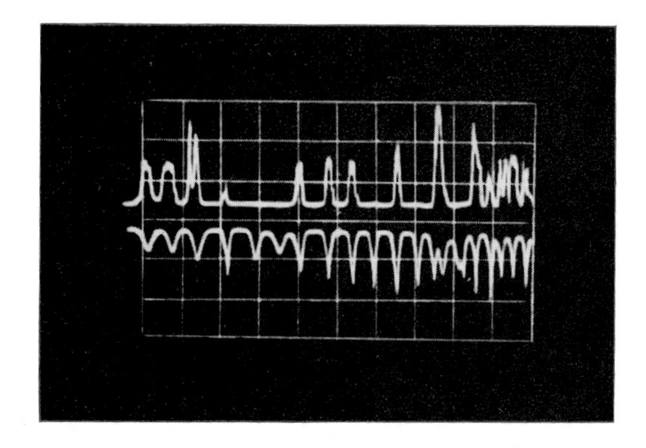

FIG. XII.33. Beats between spectral components of a ruby laser output obtained with a travelling-wave tube (McMurtry and Siegman, 1962).

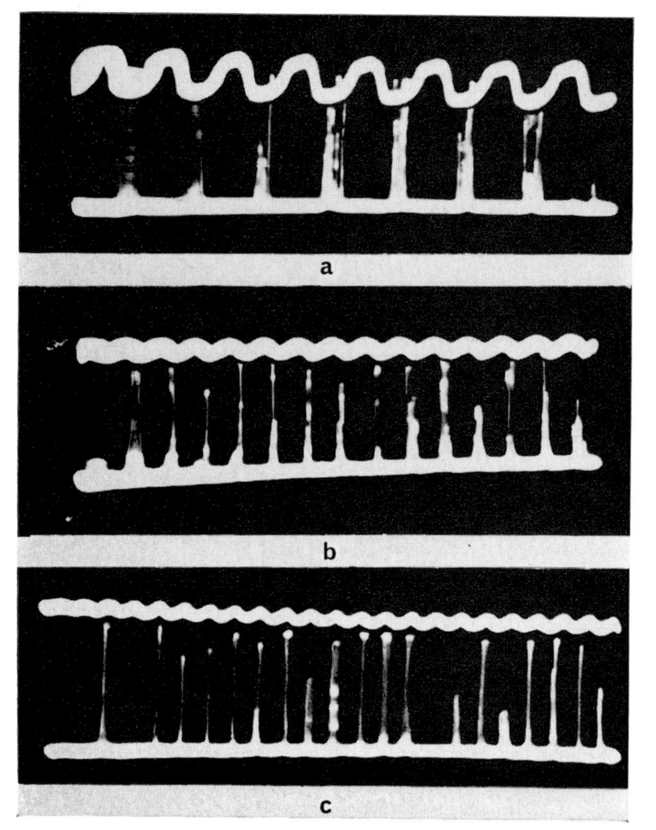

FIG. XII.35. Synchronization of the peaks of a ruby laser with sine-wave modulation of a Kerr cell located inside the resonator. The frequency of the alternating voltage across thecells is:
(a) 25 kHz;
(b) 100 kHz;
(c) 200 kHz.

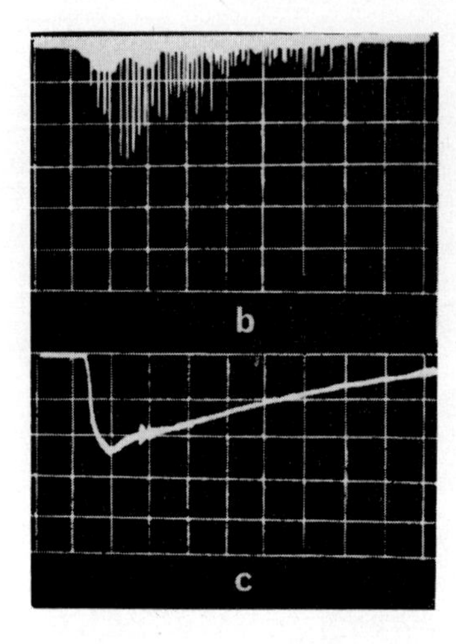

FIG. XII.36. (b) Spiked output when there is no feedback, and (c) smooth output pulse when there is feedback (Marshall and Roberts, 1962).

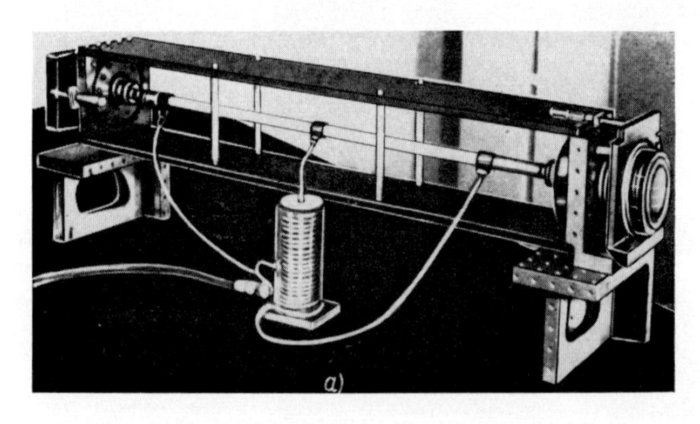

FIG. XII.39a. Helium-neon laser with internal mirrors (Bennett, 1962).

The effect is most clearly defined under conditions when the modulation period is close to the "natural" mean interval between peaks.

The electro-optical shutter can also be used for producing a smooth output pulse. To do this it is sufficient to introduce the appropriate feedback circuit from the receiving photoelectric cell to the Kerr cell (Fig. XII.36).

Of the modulating systems listed lasers with Q-switching are extensively used; they provide brief pulses of very high power (up to hundreds of

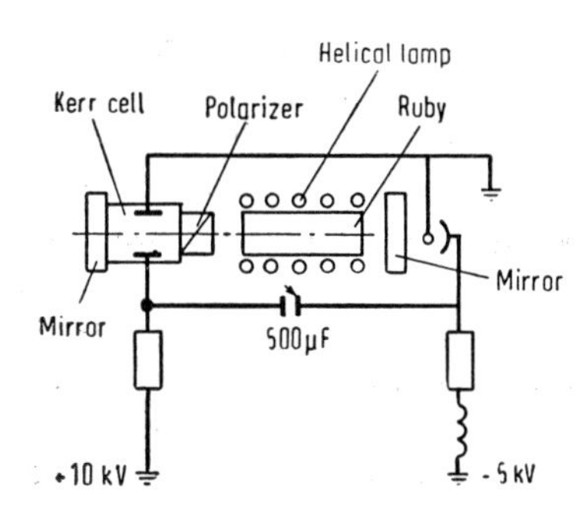

FIG. XII.36. Arrangement of a laser with an external feedback circuit.

megawatts). High powers are important in all cases when non-linear effects are to be exploited; in particular for doubling the output frequency, for producing the combination frequencies of the outputs from two different lasers and for pumping Raman lasers.

58. Gas Lasers

58.1. The Spectra of Inert Gases. Level Notations

The use of chemically active gases and vapours as the working substances of lasers causes additional technical difficulties which limit the types of materials suitable for making containers; inert gases are the most convenient from this point of view. Naturally, they were also used in the first experimental investigations (Javan *et al.*, 1961; Javan, 1961; Bennett, 1961; Herriott, 1961). The ground state of the atoms of inert gases, excluding helium, is characterized by the np^6 configuration of the outer electron shell. The lowest of the excited states has the configuration $np^5(n+1)s$. Higher excited states correspond to the configurations $np^5(n+1)p$, $np^5(n+2)s$, etc.

The arrangement and structure of the levels corresponding to these configurations are shown diagrammatically in Fig. XII.37.

Different systems of notations for the energy levels of these atoms are found in published papers. We very often find the system corresponding to

Russell-Saunders or *LS*-coupling (Landau and Lifshitz, 1963). We recall that *LS*-coupling is valid in cases when the spin–orbital coupling is small and the state of the atom is characterized by the total orbital moment L and the total spin moment S. Spectral terms are denoted by different letters which give the values of L:

$$\begin{aligned} L &= 0,\ 1,\ 2,\ 3\ldots, \\ &\quad S,\ P,\ D,\ F\ldots \end{aligned}$$

At the top left is shown the multiplicity $2S+1$ of the term, and at the bottom right the value of the total moment J. For example, 3^3P_2 denotes the state with $L=1$, $S=1$, $J=2$ and the principal quantum number of the excited electron $n=3$ (the case of *LS*-coupling is discussed in Appendix III). Russell-Saunders coupling is found in very many atoms, e.g. in the atoms of elements of the alkali group, oxygen, etc.

Sometimes *LS* notation is also applied to the terms of inert gases. With the exception of helium, however, there is little justification for this. Racah (1942) has shown that for inert gases the *jl* type of coupling is more characteristic, i.e. the coupling between the orbital moment of $\boldsymbol{l}$ of the excited electron and the total moment $\boldsymbol{j}$ of the rest of the atom. In this situation the state of the atom is characterized by a set of quantities such as $\boldsymbol{k} = \boldsymbol{j} + \boldsymbol{l}$,

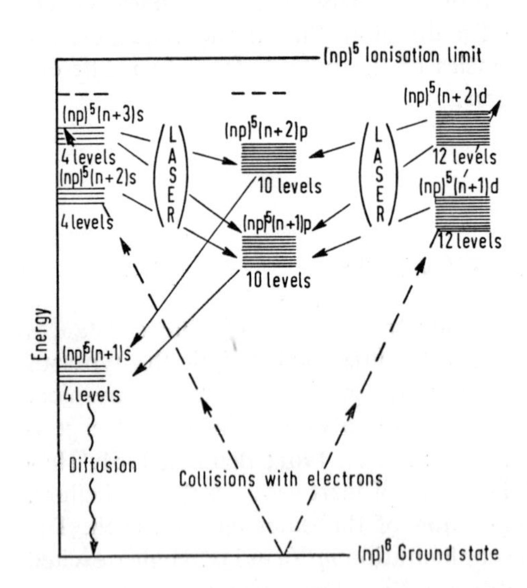

FIG. XII.37. Generalized diagram to show the arrangement and structure of the levels of heavy inert gases.

the orbital moment of the excited electron $\boldsymbol{l}$ and the total moment of the atom $\boldsymbol{J} = \boldsymbol{k} + \boldsymbol{s}$. The corresponding quantum numbers are used in the Racah system of indices. In the Racah system the notation of a term looks like the following: $3p'|\frac{3}{2}|_1$, which means that the excited electron is in a shell with $n = 3$ and has $l = 1$; the rest of the quantum numbers have the values $k = \frac{3}{2}$ and $J = 1$. The prime shows that the terms have $j = \frac{1}{2}$;† the absence of the prime shows that $j = \frac{3}{2}$.

The Paschen notation, which is often used for the terms of neon and heavier inert gases, is semi-empirical in nature and has no special physical

TABLE XII.2. RELATIVE PROBABILITIES OF $2s$–$2p$ TRANSITIONS IN NEON

Racah	LS	Paschen	$4s[^3/_2]^0_2$ 4^3P_2 $2s_5$	$4s[^3/_2]^0_1$ 4^3P_1 $2s_4$	$4s'[^1/_2]^0_0$ 4^3P_0 $2s_3$	$4s'[^1/_2]^0_1$ 4^1P_1 $2s_2$
$3p'[^1/_2]_0$	3^1S_0	$2p_1$	–	w	–	$^1/_9{}^l$
$3p'[^1/_2]_1$	3^3P_1	$2p_2$	w	w	$^1/_9{}^l$	$^2/_9{}^l$
$3p[^1/_2]_0$	3^3P_0	$2p_3$	–	$^1/_9$	–	w^l
$3p'[^3/_2]_2$	3^3P_2	$2p_4$	w	w	–	$^5/_9{}^l$
$3p'[^3/_2]_1$	3^1P_1	$2p_5$	w	w	$^2/_9{}^l$	$^1/_9{}^l$
$3p[^3/_2]_2$	3^1D_2	$2p_6$	$^1/_2{}^l$	$^1/_{18}$	–	w^l
$3p[^3/_2]_1$	3^3D_1	$2p_7$	$^1/_{18}$	$^5/_{18}$	w^l	w
$3p[^5/_2]_2$	3^3D_2	$2p_8$	$^1/_{18}{}^l$	$^1/_2{}^l$	–	w
$3p[^5/_2]_3$	3^3D_3	$2p_9$	$^7/_9{}^l$	–	–	–
$3p[^1/_2]_1$	3^3S_1	$2p_{10}$	$^5/_{18}$	$^1/_{18}$	w	w

Note. The letter w denotes transitions which give weak lines although they are forbidden in the jl-coupling scheme. The index l denotes that laser action is obtained at the given transition.

meaning; its conciseness justifies its use. The first excited state with the configuration $np^5(n+1)s$ is denoted by $1s$ and its four sublevels are numbered in order of decreasing energy from $1s_2$ to $1s_5$. The next excited state $np^5(n+1)p$, denoted by $2p$, consists of ten sublevels from $2p_1$ to $2p_{10}$, etc.

The selection rules for systems with jl-coupling allow transitions with

$$\begin{aligned} \Delta J &= 0, \pm 1, \\ \Delta k &= 0, \pm 1, \end{aligned} \tag{58.1}$$

excluding $J = 0 \to J = 0$ and $k = 0 \to k = 0$. These rules are confirmed in general by spectroscopic studies, although there are exceptions. Sometimes weak lines are observed which are forbidden by (58.1). This merely indicates that the idealization of jl-coupling is not equally good for all states, but for inert gases it gives the most satisfactory results. Table XII.2 gives the probabilities of transitions between the $2s$ and $2p$ levels calculated by Koster and Statz (1961) on the basis of the jl-coupling scheme. The notations of the levels are given in all systems for convenience.

† If the excited electron is removed from the atom, the ion is in the $2P_{1/2}$ state.

TABLE XII.3. SOME TRANSITIONS IN GASES AT WHICH LASER ACTION IS OBSERVED

λ (μ)	Transition		Optimum pressure of components (torr)	Amplification (%/m)	Laser power (mW)	References	Remarks
	Paschen or *LS*	Racah					
Neon (He–Ne mixture)							
0·5940	$3s_2$–$2p_8$	$5s'[^1/_2]^0_1$–$3p[^5/_2]_2$			0·4	White and Rigden, 1963 a	Laser action at each of the visible transitions (excluding 0·6328 μ) is observed when the competing transitions are swamped. The power corresponds to single-mode (TEM_{00}) oscillations
0·6046	$3s_2$–$2p_7$	$5s'[^1/_2]^0_1$–$3p[^3/_2]_1$			0·5	White and Rigden, 1963 a	
0·6118	$3s_2$–$2p_6$	$5s'[^1/_2]^0_1$–$3p[^3/_2]_2$			3·0	White and Rigden, 1963 a	
0·6293	$3s_2$–$2p_5$	$5s'[^1/_2]^0_1$–$3p'[^3/_2]_1$			0·3	White and Rigden, 1963 a	
0·6328	$3s_2$–$2p_4$	$5s'[^1/_2]^0_1$–$3p'[^3/_2]_2$	0·5/0·1		46	White and Rigden, 1963 a	
0·6351	$3s_2$–$2p_3$	$5s'[^1/_2]^0_1$–$3p[^1/_2]_0$			0·3	White and Rigden, 1963 a	
0·6401	$3s_2$–$2p_2$	$5s'[^1/_2]^0_1$–$3p'[^1/_2]_1$			1·0	White and Rigden, 1963 a	
0·7305	$3s_2$–$2p_1$	$5s'[^1/_2]^0_1$–$3p'[^1/_2]_0$			0·6	White and Rigden, 1963 a	
1·0798	$2s_3$–$2p_7$	$4s'[^1/_2]^0_0$–$3p[^3/_2]_1$	} 1/0·1				
1·0845	$2s_2$–$2p_6$	$4s'[^1/_2]^0_1$–$3p[^3/_2]_2$					
1·1143	$2s_4$–$2p_8$	$4s[^3/_2]^0_1$–$3p[^5/_2]_2$					
1·1177	$2s_5$–$2p_9$	$4s[^3/_2]^0_2$–$3p[^5/_2]_3$					
1·1390	$2s_5$–$2p_8$	$4s[^3/_2]^0_2$–$3p[^5/_2]_2$				Javan *et al.*, 1961	Only in afterglow
1·1409	$2s_2$–$2p_5$	$4s'[^1/_2]^0_1$–$3p'[^3/_2]_1$					
1·1523	$2s_2$–$2p_4$	$4s'[^1/_2]^0_1$–$3p'[^3/_2]_2$	1/0·1	12	20	Javan *et al.*, 1961	
1·1601	$2s_2$–$2p_3$	$4s'[^1/_2]^0_1$–$3p[^1/_2]_0$					
1·1614	$2s_3$–$2p_5$	$2s'[^1/_2]^0_0$–$3p'[^3/_2]_1$	} 1–2/0·1–0·2	2	1	Javan *et al.*, 1961	
1·1767	$2s_2$–$2p_2$	$4s'[^1/_2]^0_1$–$3p'[^1/_2]_1$					
1·1985	$2s_3$–$2p_2$	$4s'[^1/_2]^0_1$–$3p'[^1/_2]_1$				Javan *et al.*, 1961	
1·2066	$2s_5$–$2p_6$	$4s[^3/_2]^0_2$–$3p[^3/_2]_2$				Javan *et al.*, 1961	
1·5231	$2s_2$–$2p_1$	$4s'[^1/_2]^0_1$–$3p'[^1/_2]_0$		6	3		

3·3913	$3s_2$–$3p_4$	$5s'[^1/_2]^0_1$–$4p'[^3/_2]_2$	0·5/0·1			
			Helium			
2·0603	7^3D–4^3P		8	5	3	Patel, Faust *et al.*, 1962
			Neon			
1·1523	$2s_2$–$2p_4$	$4s'[^1/_2]^0_1$–$3p'[^3/_2]_2$	0·06	4	1	
2·1019	$4s''''_1$–$2p_6$	$4d'[^5/_2]^0_2$–$4p[^3/_2]_2$	0·2	3	1	Patel, Faust *et al.*, 1962
5·40	$5d_5$–$4p_7$	$5d[^1/_2]^0_1$–$5p[^3/_2]_1$				
	or	or				
	$5d_4$–$4p_6$	$5d[^7/_2]^0_3$–$5p[^3/_2]_2$				
			Argon			
1·6180	$2s_5$–$2p_3$	$5s[^3/_2]^0_2$–$4p'[^3/_2]_2$	0·05			Patel, Faust *et al.*, 1962
1·6941	$3d_3$–$2p_6$	$3d[^3/_2]^0_2$–$4p[^3/_2]_2$		3	0·5	Patel, Faust *et al.*, 1962
1·793	$3d_5$–$2p_6$	$3d[^1/_2]^0_1$–$4p[^3/_2]_2$				
	or	or	0·035			
	$3d_6$–$2p_7$	$3d[^1/_2]^0_0$–$4p'[^3/_2]_1$				
2·0616	$3d_3$–$2p_3$	$3d[^3/_2]^0_2$–$4p'[^3/_2]_2$		3	1	Patel, Faust *et al.*, 1962
			Krypton			
1·6900	$3d_5$–$2p_{10}$	$4d[^1/_2]^0_1$–$5p[^1/_2]_1$	0·07			Patel, Faust *et al.*, 1962
1·6936	$3d_1''$–$2p_7$	$4d[^5/_2]^0_2$–$5p[^3/_2]_1$	0·05			Patel, Faust *et al.*, 1962
1·7843	$3d_6$–$2p_{10}$	$4d[^1/_2]^0_0$–$5p[^1/_2]_1$	0·07			Patel, Faust *et al.*, 1962
1·8185	$3s_1''''$–$2p_2$	$4d'[^5/_2]^0_2$–$5p'[^3/_2]_2$	0·07			Patel, Faust *et al.*, 1962
1·9211	$4s_5$–$3p_8$	$8s[^3/_2]^0_2$–$6p[^5/_2]_2$	0·035			Patel, Faust *et al.*, 1962
2·1165	$3d_3$–$2p_7$	$4d[^3/_2]^0_2$–$5p[^3/_2]_1$	0·035	3	1	Patel, Fanst *et al.*, 1962
2·1902	$3d_3$–$2p_6$	$4d[^3/_2]^0_2$–$5p[^3/_2]_2$	0·035	3	1	Patel, Faust *et al.*, 1962
2·5234	$3d_5$–$2p_6$	$4d[^1/_2]^0_1$–$5p[^3/_2]_2$				
			Xenon			
2·0261	$3d_2$–$2p_7$	$5d[^3/_2]^0_1$–$6p[^3/_2]_1$	0·02	10	5	Patel, Faust *et al.*, 1962
5·5738	$3d'_4$–$2p_8$	$5d[^7/_2]^0_4$–$6p[^5/_2]_3$	0·01	very strong		

TABLE XII.3 (CONT.)

λ (μ)	Transition: Paschen or *LS*	Transition: Racah	Optimum pressure of components (torr)	Amplification (%/m)	Laser power (mW)	References	Remarks
			Xenon (He–Xe mixture)				
2·0261	$3d_2$–$2p_7$	$5d[^3/_2]^0_1$–$6p[^3/_2]_1$	5/0·02	120	10	Patel, Faust *et al.*, 1962	
2·3193	$3d'_1$–$2p_9$	$5d[^5/_2]^0_3$–$6p[^5/_2]_2$		Very weak			
2·6269	$3d''$–$2p_9$	$5d[^5/_2]^0_2$–$6p[^5/_2]_2$					
2·6511	$3d_2$–$2p_5$	$5d[^3/_2]^0_1$–$6p[^1/_2]_0$	1/0·02	Strong			
3·1069	$3d'_1$–$2p_6$	$5d[^5/_2]^0_3$–$6p[^3/_2]_2$		Very weak			
3·3667	$3d''$–$2p_7$	$5d[^5/_2]^0_2$–$6p[^3/_2]_1$	1/0·01	Strong			
3·5070	$3d_4$–$2p_9$	$5d[^7/_2]^0_3$–$6p[^5/_2]_2$	1/0·01	Very strong			
3·6788	$3d_5$–$2p_{10}$	$5d[^1/_2]^0_1$–$6p[^1/_2]_1$		Weak			
3·6849	$3d''$–$2p_6$	$5d[^5/_2]^0_2$–$6p[^3/_2]_2$	0·3/0·04	Strong			
3·8940	$3d_4$–$2p_8$	$5d[^7/_2]^0_3$–$6p[^5/_2]_3$		Very weak			
3·9955	$3d_6$–$2p_{10}$	$5d[^1/_2]^0_0$–$6p[^1/_2]_1$	1/0·01	Strong			
5·5738	$3d'_4$–$2p_8$	$5d[^7/_2]^0_4$–$6p[^5/_2]_3$		Very strong			
7·3147	$3d_3$–$2p_7$	$5d[^3/_2]^0_2$–$6p[^3/_2]_1$					
9·0040	$3d_3$–$2p_6$	$5d[^3/_2]^0_2$–$6p[^3/_2]_2$					
9·7002	$3d_5$–$2p_7$	$5d[^1/_2]^0_1$–$6p[^3/_2]_1$					
12·263	$3d_6$–$2p_7$	$5d[^1/_2]^0_0$–$6p[^3/_2]_1$					
12·913	$3d_5$–$2p_6$	$5d[^1/_2]^0_1$–$6p[^3/_2]_2$					
			Mercury (He–Hg mixture)				
1·5295			0·5–10/20			Paananen *et al.*, 1963	

			Caesium				
7·1821	$8^2P_{1/2}$–$8^2S_{1/2}$			170	0·025	Rabinowitz *et al.*, 1962	
			Atomic oxygen (He–O_2 mixture)				
0·84462	3^3P_2–3^3S_1		0·35/0·014	3	1	Bennett *et al.*, 1962	
			Atomic oxygen (Ar–O_2 mixture)				
0·84462	3^3P_2–3^3S_1		1·3/0·036	3	1	Bennett *et al.*, 1962	
			Molecular nitrogen (N_2)				
0·8663–0·8710	$B^3\Pi_g$–$A^3\Sigma^*_u$	Band 2–1		4	0·5 W	Mathias and Parker, 1963b	In each of the bands laser action occurs at several discrete frequencies. Laser action only in pulsed high-voltage discharge
0·8844–0·8908		Band 1–0					
1·0449–1·0505		Band 0–0		4			
1·2303–1·2347		Band 0–1		4			
0·7580–0·7620		Band 3–1		2			
0·7700–0·7750		Band 2–0		2	100 W		Power in pulse
			Carbon monoxide (CO)				
0·5591–0·5604	$B^1\Sigma$–$A^1\Pi$	Band 0–3		2	0·8 W	Mathias and Parker, 1963a	See remark above
0·6062–0·6075		Band 0–4		2	8 W		
0·6595–0·6614		Band 0–5		2	4 W		

Some of the transitions in atomic gases in which laser action has been achieved in one way or another are listed in Table XII.3. The overwhelming majority of the results are obtained in inert gases excited by a gas discharge.

Figure XII.38 shows the levels of neon and helium which are relevant to the operation of a He–Ne laser. A feature of the neon spectrum is that the levels corresponding to $(m + 1)$ *s*-configurations lie lower than the *md*-configurations. The opposite situation obtains in all the heavier inert gases.

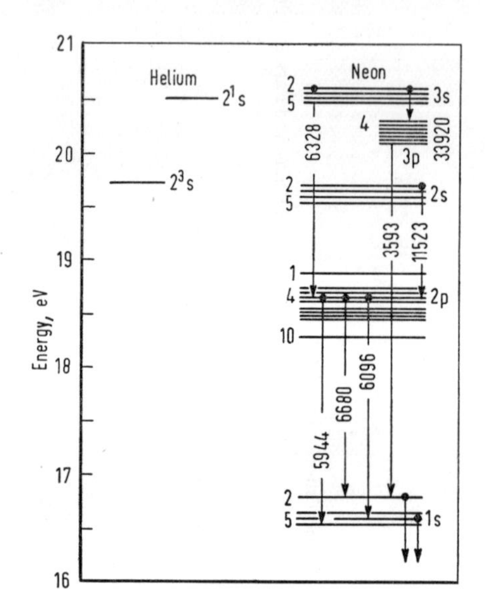

FIG. XII.38. The positions of the metastable levels of helium and the levels of neon which take part in the operation of a laser.

58.2. Design of a Gas-discharge Laser

An idea of the design and external appearance of the first model of a gas laser operat ingwith a neon–helium mixture is given in Figs. XII.39a, b. A radio-frequency oscillator was used to excite a discharge in the gas. The resonator was formed by a pair of high-quality plane mirrors placed inside the gas-filled chamber. A coefficient of reflection of 98·9% with a transparency of 0·3% was achieved by covering the surface of a quartz optical flat with 13–15 quarter-wave dielectric layers with alternating indices of refraction. The reflecting properties of the mirror coatings were high only at the required wavelength. The surface of the mirrors was kept flat with an accuracy of up to 0·01 λ. The relatively small amplification of the working medium imposed these very exacting requirements on the mirrors.†

† It should be noted that in modern designs of He–Ne lasers laser action can be achieved with lower-quality mirrors. Even a resonator made from two metalized mirrors is suitable for the $2s_2$–$2p_4$ transition.

A device designed by Javan, Bennett and Herriott was the first continuous optical maser oscillator. Of the five wavelengths at which simultaneous laser action was observed (1·118, 1·153, 1·16, 1·199 and 1·207 μ) the most intense emission, with an output power of the order of 10 mW, occurred at the wavelength of 1·153 μ, which corresponds to the $2s_2$–$2p_4$ transition. Many other forms of laser operate on this transition. The width of the polar diagram of a laser with a plane-parallel resonator does not exceed 1′. The whole dynamic range of the laser from the self-excitation threshold to saturation is covered with pumping powers in the 25–75 W range.

The gas laser design described has obvious disadvantages. The internal situation of the mirrors leads to complications in the design of the vacuum section of the device and considerably limits the experimental flexibility. Changing the mirrors or introducing additional elements such as diaphragms, lenses, etc., inside the resonator is a rather laborious affair. Another practical difficulty is connected with the form of the mirrors: a resonator with plane mirrors requires very careful adjustment. Lack of parallelism of only 4″ has an effect on the output.

Far more practical are resonators with spherical mirrors which do not require such careful setting up (see Appendix II). In a working laser it has

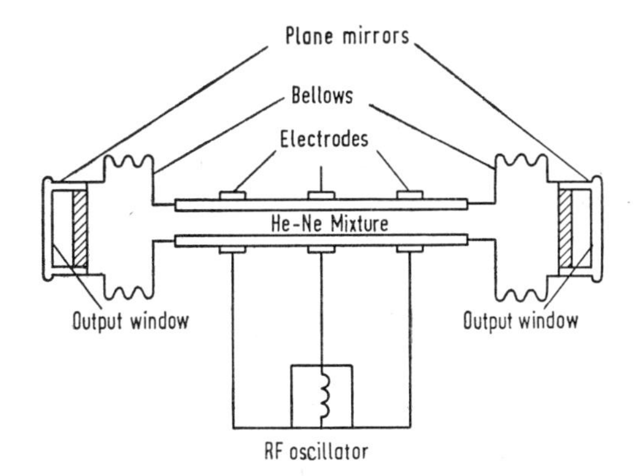

FIG. XII.39 b. Layout of a helium-neon laser with internal mirrors (Bennett, 1962).

been found that rotation of the spherical mirror by 25′ from its central position has no effect on the laser action. A widely used design of gas laser has external mirrors, which may both be spherical, or one spherical and one plane (Fig. XII.40a) (Brangaccio, 1962; Rigrod *et al.*, 1962). The optical windows at the ends of the discharge tube are orientated at the Brewster angle to the axis of the tube, since in this case the reflection of one component of polarization is negligibly small. There is therefore a preferred direction of polarization in the output (in the plane of incidence) for which the losses are reduced to a minimum. Besides a high-frequency discharge, a low-frequency discharge or a d.c. discharge can be used to excite the working gas. In the latter case a heated cathode may be used.

The dimensions of the gas-discharge tube are not critical: the length is determined by the amplification achievable per metre of the discharge. The less the amplification the longer must be the discharge for the overall amplification to compensate for the losses. The radius of the tube is one of the parameters affecting amplification; experimentally it has been found that $G \sim 1/r$. This dependence is apparently connected with the life of the metastable $1s$ state. Since de-excitation of the $1s$ state is caused by collisions

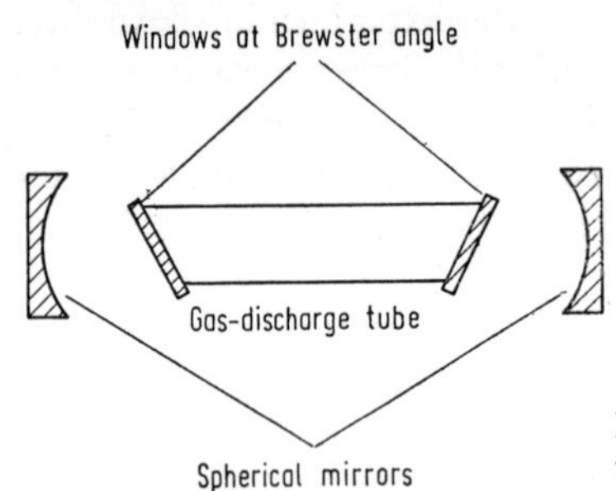

FIG. XII.40a. Layout of a gas laser with external spherical mirrors.

with the walls the effective lifetime of this state is the diffusion time, which is proportional to the radius of the tube. In present-day gas-discharge lasers the internal diameter of the tubes used is 3–10 mm. The length depends on the amplification and varies from tens of centimetres to several metres.

In order to avoid excessive heating of the tube it should be made of a transparent material: glass or quartz. The latter is preferable since its absorption is lower. In addition, unwanted impurities absorbed by the walls which might contaminate the working mixture can be removed more easily from quartz. No problems arise in selecting materials for end windows and mirrors for lasers operating in the visible and near infrared regions down to 2 μ. In the longer-wave region glass and quartz lose their transparency and other materials, such as BaF_2, have to be used.

In a He–Ne laser oscillation can be excited at a large number of wavelengths. Under favourable conditions laser action appears at several transitions simultaneously (White and Rigden, 1963a; 1963b). For some purposes it is necessary to suppress oscillation at all transitions except one; for others, on the contrary, we have to create conditions favourable for simultaneous oscillation. In the latter case we must ensure minimum resonator losses at the working frequencies. This is not hard to do if the frequencies are sufficiently close and are not outside the limits of the high-reflectivity region of the interference mirrors. If, however, they are in different spectral regions special hybrid mirrors are needed which have a high coefficient of reflection in both regions.

In order to ensure oscillation at a single wavelength we can place inside the resonator of the laser, between the discharge and the mirror, some dispersing or selectively absorbing component. Rigden and White (White and Rigden, 1963a; 1963b) have used for this purpose a prism with Brewster incidence angles. The arrangement with two prisms shown in Fig. XII.40b made it possible to achieve laser action at several transitions of the $3s_2$–$2p$ group which could not be excited by simpler methods because of interference with laser action by the dominant transitions ($3s_2$–$2p_4$; $3s_2$–$3p_4$).

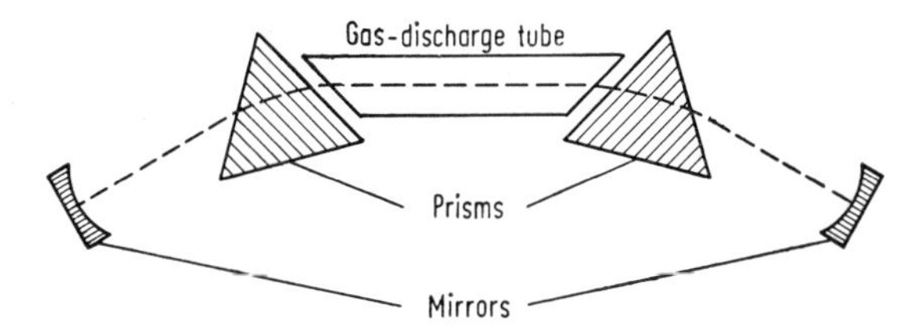

FIG. XII.40b. Prisms are placed inside the resonator to prevent laser action in transitions which compete with the selected ones (White and Rigden, 1963a).

58.3. *Remarks on the Excitation Mechanisms of Inert Gases*

If we produce a discharge in a pure gas, the populations of the different levels of the gas undergo changes because of collisions between the atoms and electrons in the discharge. In § 54.3 we have seen how this process takes place in an idealized gas consisting of three-level molecules. If we consider any real inert gas, the position will be somewhat different because of the presence of other levels. When an unexcited atom collides with an electron not all electron configurations are excited with equal probability. It has already been mentioned in § 54.3 that the excitation cross-section of a given level depends upon the matrix element of the corresponding transition. Because of this dependence certain states with np^5ms and np^5md configurations are most strongly excited in a gas discharge, whilst the np^5mp states cannot be excited directly by collision. In a number of cases conditions have been realized which made it possible to obtain stimulated emission at the ms–$m'p$ and md–$m'p$ transitions (see Table XII.3).

The population of the different levels which is achieved in a gas discharge depends on many factors such as intensity of the discharge, the density of the gas, the efficiency of the secondary processes, the dimensions of the discharge tube, etc. In order to discuss these basic features we take the example of neon, to which most attention has been devoted, remembering that exactly similar phenomena occur in other inert gases.

The energy levels of the neon atom are shown in Fig. XII.38. Laser action in pure neon is achieved on the single transition $2s_2$–$2p_4$ ($\lambda = 1{\cdot}1523\ \mu$). The $2s_2$ level becomes populated as a result of the process

$$\mathrm{Ne} + e \rightarrow \mathrm{Ne}(2s_2) + e, \tag{58.2}$$

whose cross-section is large since the Ne–Ne($2s_2$) transition is allowed; cascade transitions to $2s_2$ from higher levels may also play some part. Depletion of the $2s$ levels may occur either by a transition back to the ground state or by a transition into the $2p$ state. The first can be ignored since at ordinary pressures resonance absorption by one of the adjacent atoms causes the photon to disappear and the level population does not change.† Therefore the atom, after spending a certain time in the $2s$ state emits a photon and goes into the $2p$ state from which it reaches the $1s$ state by radiation. The possibility of creating a population inversion between $2s$ and $2p$ depends on the inequality of the mean livetimes of these states. For the $2s$ levels the lifetimes are between 0·1 and 0·2 μsec, whilst for $2p$ they are at least an order of magnitude less (Bennett, 1961).

The metastable $1s$ levels play a double role. On the one hand their presence ensures the necessary rate of depletion of the $2p$ levels by spontaneous emission. On the other hand these same levels under certain conditions are the reason for the increase in the density of Ne($2p$), i.e. they reduce the magnitude of the population inversion of the working levels. The process

$$\mathrm{Ne}(1s) + e \rightarrow \mathrm{Ne}(2p) + e \tag{58.3}$$

becomes more rapid as the gas density and the power of the discharge rise since there is an increase in the density of the $1s$ atoms and a reduction in their rate of diffusion towards the walls. This effect limits the power of the laser when operating continuously and determined the optimum value of certain parameters such as pressure and discharge current. It is also connected with the dependence of the amplification factor on the radius of the tube: $G \sim 1/r$, which was mentioned above.

The possibility of creating a population difference at any given transition depends upon the efficiency of the various excitation and de-excitation processes in the gas discharge. The comparatively small number of lasing transitions in pure gases leads us to the conclusion that under steady conditions it is possible to obtain the necessary population difference only in a very favourable combination of circumstances. Inversion is much more readily achieved, as is clear from Table XII.3, when an impurity gas is introduced into the discharge. Whilst stimulated emission has been observed at three transitions in pure neon, the number rises to twenty-two in a He–Ne mix-

† Transitions to the ground state from $2s_3$ and $2s_5$ are generally forbidden. We are dealing with the $2s_2$ and $2s_4$ levels.

ture. Likewise pure xenon lased only at two frequencies, but when mixed with He at seventeen.†

In a He–Ne mixture there is no doubt about the significance of the process of transfer of energy from He(2^3S) and He(2^1S) to an unexcited atom of neon in inelastic collisions. These levels of helium are metastable and in a discharge a large population is built up in them. It is true that the cross-section of the process

$$\text{He}(2^3S) + \text{Ne} \rightarrow \text{Ne}(2s) + \text{He} \tag{58.4}$$

is not large, which can be explained by the considerable energy difference between the interacting levels

$$E[\text{He}(2^3S)] - E[\text{Ne}(2s_3)] > kT.$$

The probability of the process

$$\text{He}(2^3S) + \text{Ne} \rightarrow \text{Ne}(2s_2) + \text{He} \tag{58.5}$$

is also low since the spin-conservation rule (Wigner's rule)‡ is broken. The cross-sections of some of the processes relevant to the operation of a He–Ne laser are given in Table XII.4.

Because of the small cross-section the widespread opinion about the efficiency of process (58.4) needs additional discussion. Data on the operation of a laser in the afterglow give information on this point. A pulsed discharge

TABLE XII.4. TOTAL CROSS-SECTIONS FOR ENERGY TRANSFER FROM METASTABLE HELIUM STATES (BENNETT, 1962a)

Type of collision	Cross-sections (10^{-16} cm^2)
He(2^1S)–Ne	4·1 ± 1
He(2^3S)–Ne	0·37 ± 0·05
He(2^3S)–A	6·6 ± 1·3
Ne($1s_5$)–A	2·6

is needed to observe this phenomenon. Immediately the discharge has ceased the generated power rises, and only after a certain time does laser action stop (Fig. XII.41) (Powers and Harned, 1963). The rise in the intensity can be explained by the rapid decay of the electron density and therefore by the

† The addition of helium into the gas discharge is essential for laser action in some other gases, besides the inert gases. As an example of a mixture of this type we can cite He–Hg, which also shows stimulated emission (Paananen *et al.*, 1963).

‡ In *LS* notations the singlet $^1P_1$ corresponds to the $2s_2$ level.

reduction in the rate of process (58.3).† The characteristic time of the subsequent attenuation of emission at $\lambda = 1{\cdot}153\ \mu$ is the same as the decay time of the metastable 2^3S state of helium. Under afterglow conditions neon atoms may be excited into $2s$ levels exclusively by process (58.4).

The dominant process in populating the $3s_2$ level of neon is

$$\mathrm{He}(2^1S) + \mathrm{Ne} \rightarrow \mathrm{Ne}(3s_2) + \mathrm{He}. \tag{58.6}$$

It has been shown experimentally that when the discharge conditions are changed the densities of $\mathrm{He}(2^1S)$ and $\mathrm{Ne}(3s_2)$ change identically (White and Gordon, 1963).

The process shown in (58.4) is not the only way in which the helium present in the discharge has an effect on the populations of the neon levels. The helium also causes an increase in the density of the electrons without reducing their mean velocity (Bennett, 1962). This increases the importance of (58.2) which is obviously comparable with the part played by (58.4).

In a He–Xe mixture processes of resonance transfer of excitation in an

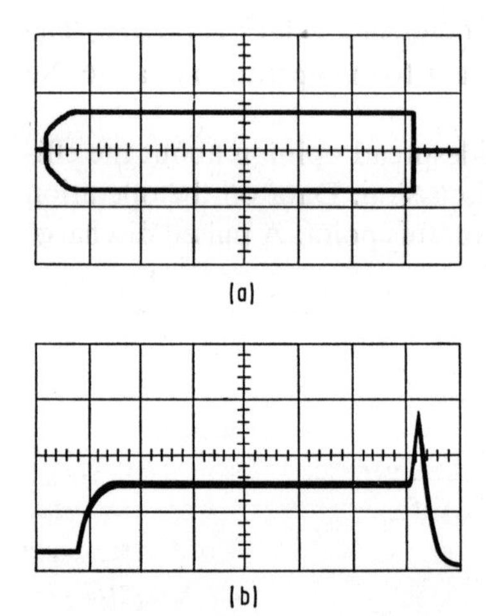

FIG. XII.41. (a) Variation of current in a pulsed high-frequency discharge; (b) stimulated emission accompanying the pulsed discharge.

inelastic collision are excluded since the energy of $\mathrm{He}(2^3S)$ is much greater than the ionization energy of xenon. For this reason the reaction

$$\mathrm{He}(2^3S, 2^1S) + \mathrm{Xe} \rightarrow \mathrm{Xe}^+ + \mathrm{He} + e, \tag{58.7}$$

which increases the electron density, is of importance. If we take the case of the $5d|\tfrac{3}{2}|_1^0\text{–}6p|\tfrac{3}{2}|_1$ line, with $\lambda = 2{\cdot}0261\ \mu$, the increase in the output power is in quantitative agreement with the increase of electron density (Aisenberg,

† The more favourable conditions for laser action in the afterglow make it possible to observe some transitions which do not lase in a continuous discharge.

1963): both are doubled when helium is added. This means that the dominant $5d|\frac{3}{2}|_1^0$ excitation mechanism of the xenon level is collisions with electrons. There is no similar information about other transitions. It is possible that cascade transitions or recombination processes are also of significance here.

It should be pointed out that laser action occurs, in inert gases heavier than neon, as a rule at the $d \rightarrow p$ transitions and not the $s \rightarrow p$ as in the case of neon. There are a number of ways of explaining this: in the first place in heavy inert gases the *d*-levels are lower than the *s*-levels and require less energy for excitation. In the second place the excitation cross-sections of the *d*-levels in the collision of an atom with an electron are an order of magnitude greater than the excitation cross-sections of the *s*-levels. Lastly the probability of the $d \rightarrow p$ transition is always greater than $s \rightarrow p$.

In Table XII.3 we can find more than one pair of transitions connected with a common initial or final level. In all such cases the lasing processes cannot be discussed at each of these transitions independently since they reduce the population differences not only between the particular pair of levels but also between the levels of the alternative transition. If the transition probabilities differ considerably, the laser action may in general be suppressed or significantly reduced in the weak line. For example, the $3s_2$–$2p_4$ transition in neon is far more probable than the other $3s_2$–$2p$ transitions. Oscillations at the frequency of the dominant transition (λ = 6328 Å) reduce the Ne($3s_2$) concentration so much that the remaining density is insufficient to satisfy the self-excitation conditions at the other frequencies in the visible region. The laser arrangement shown in Fig. XII.40b was used to obtain stimulated emission at the other frequencies. In its turn laser action at the $3s_2$–$2p_4$ transition is strongly quenched by the $3s_2$–$3p_4$ transition (λ = 3·39 μ). This intense, easily excited laser action in the infrared region leads to significant impoverishment of the $3s_2$ level, a decrease in N_{3s2}–N_{2p4} and therefore a decrease in the red emission. The latter can be considerably increased by placing in the resonator a piece of glass which absorbs infrared, and thus prevents laser action in this region. Transitions of the $2s$–$2p$ group are connected in the same way. White and Rigden (1963b) have observed in a He–Ne mixture simultaneous laser action at many transitions. They altered the lasing conditions by adjusting the discharge current. The changes in the relative intensities of five different components of the stimulated emission are shown in Fig. XII.42; these curves illustrate very well what has been said above. The drop in the intensity of the visible emission is accompanied by a rise in the infrared emission at λ = 1·15 μ ($2p_4$ is the common lower level) and at λ = 3·39 μ ($3s_2$ is the common upper level). Emission due to the $2s_2$–$2p_5$ transition at λ = 1·1409 μ, on the other hand, drops sharply. It should be pointed out that under ordinary conditions laser action in this line is completely quenched by $2s_2$–$2p_4$.

One-way transfer of energy in collisions between different kinds of molecules also occurs in Ne–O_2 and A–O_2 mixtures (Bennett, 1962a; Bennett *et al.*,

1962). However, the oxygen excitation process differs from the processes in mixtures of noble gases which have been discussed so far. When the metastable Ne(1*s*) atom collides with an unexcited O_2 molecule it transfers its

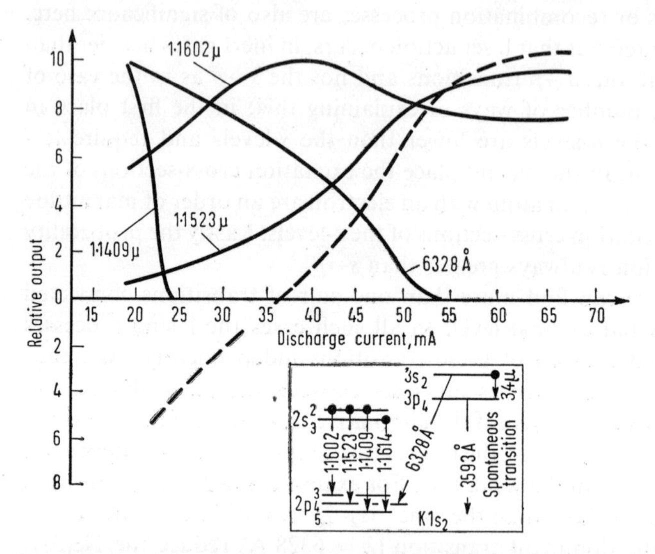

FIG. XII.42. The variation in the relative output of a He–Ne laser at various transitions which show simultaneous laser action (White and Rigden, 1963b). The level of oscillations at 3·39 μ is deduced indirectly from the amount of spontaneous emission at $\lambda = 3593$ Å.

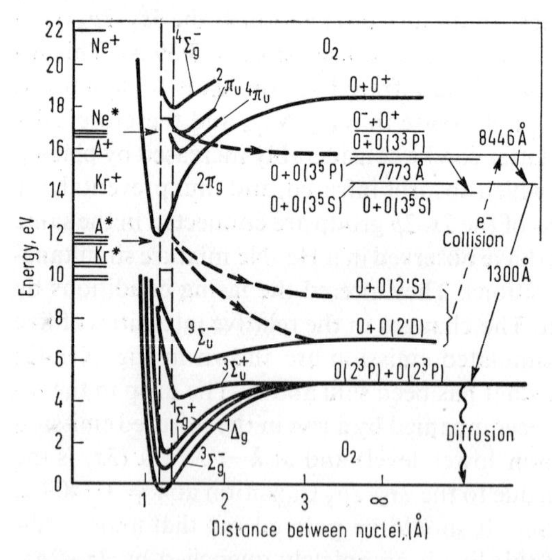

FIG. XII.43. Diagram of the excited states of an O_2 molecule and O atoms to explain the dissociation of an oxygen molecule in a collision with neon and A atoms (Bennett *et al.*, 1962).

energy to it, bringing the latter into an excited unstable state. Then the molecule dissociates and at least one of the atoms released is in an excited state (Fig. XII.43).

The results of the reaction depend on the initial energy of the Ne atom and the following alternatives are possible:

$$Ne(1s_3, 1s_4) + O_2 \to O(3^3P) + O(2^3P) + Ne, \quad (58.8)$$

$$Ne(1s_5) + O_2 \to O(3^3S) + O(2^3P) + Ne. \quad (58.9)$$

The excess population of the 3^3P level of atomic oxygen is quite sufficient to satisfy the self-excitation conditions at the frequency of the 3^3P–3^3S transition. The optimum pressures of the components of the mixture are given in Table XII.5.

The cross-section of a process of the type (58.8) depends on the energy difference of the levels of the colliding particles, but this dependence is less critical than for processes of the (58.4) type. A difference of the order of 1–2 eV, i.e. several tens of kT, is permissible, whilst in (58.4) reactions it

TABLE XII.5. VALUES OF VELOCITY-AVERAGED CROSS-SECTIONS OF COLLISONS OF Ne^*–O_2; A^*–O_2; Kr^*–O_2 (in units of 10^{-15} cm^3) (BENNETT, 1962)

Initial state of inert gas atom	Ne	A	Kr
$1s_2$	4	1·8	1·6
$1s_3$	2·8	1·8	1·8
$1s_4$	2·8	1·6	2·2
$1s_5$	0·9	1·2	1·9

should not exceed kT. This can be explained by the fact that as a result of the reaction three particles are formed and the conservation laws are more easily satisfied than in the case of two particles.

The energies of the metastable states of argon and krypton are insufficient for direct excitation of the $O(3^3P)$ level. Nevertheless, in an A–O_2 mixture the multi-stage process

$$A^* + O_2 \to \begin{Bmatrix} O(2^3P) + O(2^1S) \\ O(2^1D) + O(2^1D) \\ O(2^3P) + O(2^1D) \end{Bmatrix} + A, \quad (58.10)$$

$$O(2^1D, 2^1S) + e \to O(2^3P) + e \quad (58.11)$$

which provides the same amplification as (58.8) and (58.9) occurs quite efficiently. Ne–O_2 and A–O_2 lasers emit a wavelength of 8556 Å. It is curious to note that the frequency of the stimulated emission is slightly shifted towards the short-wave side of the maximum of the 3^3P–3^3S fluorescence line. This shift can be explained by the presence of an unidentified absorption line in the immediate vicinity of 8446 Å.

58.4. Caesium Laser with Optical Pumping

The caesium-vapour laser (Rabinowitz *et al.*, 1962) is unique among gas lasers. The difference is in the method of pumping. The caesium vapour is contained in a closed glass tube (Fig. XII.44), is heated to a temperature of

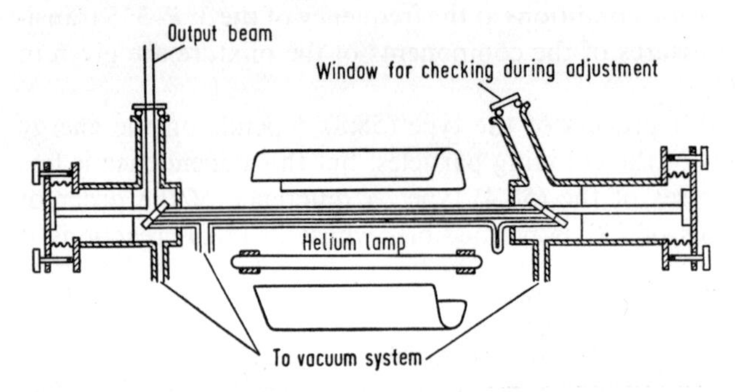

FIG. XII.44. Layout of a caesium laser with optical excitation (Rabinowitz *et al.*, 1962).

448 °K and excited by the light of a helium gas-discharge lamp. The Cs levels essential to the operation of the laser are shown in Fig. XII.45. The resonance transition† of the caesium atoms $6S_{1/2}$–$8P_{1/2}$ is very close in frequency to the

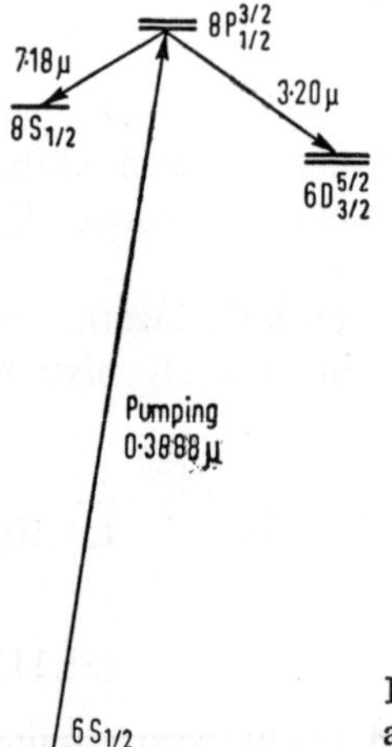

FIG. XII.45. Caesium levels relevant to the operation of a laser.

† In optics resonance transitions are allowed transitions between the ground state and any of the excited levels of the substance.

emission line of the helium lamp 3^3P–2^3S. A caesium laser pumped by a helium lamp can, in principle, lase at wavelengths of 3·20 μ ($8P_{1/2}$–$6D_{3/2}$) and 7·18 μ ($8P_{1/2}$–$8S_{1/2}$). In practice laser action has been obtained only on the latter which can be explained by the low self-excitation threshold. The total power emitted by the laser was 50 μW.

58.5. Pulsed Operation of Gas-discharge Lasers

As has already been pointed out above, processes take place in a gas-discharge plasma that lead to population of both the upper and lower working levels of the laser. Equation (58.3) is an example of the second type of process. The part it plays becomes particularly noticeable with large discharge powers when the density of the 1*s* atoms of helium is high. The 1*s*–2*p* transitions are the basic factor limiting the amplification of a He–Ne mixture, and the power of the laser, when operated continuously. Different ways of overcoming this difficulty can be devised. One of them depends on reducing the life of the 1*s* state, which can be achieved by using narrow-bore tubes or adding a suitable impurity.

The second way relies on attaining the optimum electron velocity distribution by selecting the appropriate discharge conditions (George, 1963). The 1*s* and 2*p* levels are separated by an interval of 2–3 eV. It is obvious that electrons with these energies are the most effective in producing reaction (58.2). The excitation potential of helium is about 22·5 eV. The electron distribution function for a helium pressure of the order of 1 torr has its maximum around 5–6 eV. This situation is far from being optimal. It can be improved by applying a pulsed electric field to the discharge region in addition to the constant field.. If the duration of the pulse is not greater than the time of free flight of the electrons, the latter will be accelerated and the time-averaged density of fast electrons rises at the expense of the slow ones. In practice it is simpler to produce a sinusoidally varying field than a pulsed one; the former is achieved by simultaneously applying constant and alternating voltages to the electrodes of the gas-discharge tube. In this way the power of a He–Ne laser (length of discharge 80 cm, diameter 7 mm, frequency of alternating voltage 30MHz) has been increased by 35% ($\lambda = 6328$ Å), 20% ($\lambda = 1{\cdot}1523\,\mu$) and 24% ($\lambda = 3{\cdot}39\,\mu$).

In many cases a definite gain in power can be obtained by changing to pulsed operation of the laser. The time variation of the different processes is not the same under non-equilibrium discharge conditions. At definite time intervals when the rates of processes that prevent an increase in the inverted population difference are low the amplification of the medium rises. The effect can be further reinforced by the increased intensity of the discharge conditions. This method has been used with success in many cases both for increasing the output power from substances already known to lase and for obtaining laser action in new substances.

We shall now give some results relating to a He–Ne laser (Byerly *et al.*, 1963). At a wavelength of 1·1523 μ under conditions of continuous high-frequency discharge the device has an output power of 8 mW. In pulsed operation with a field strength of several hundred W/cm† the pulse power is 50 times greater and the mean power reached 25% greater. The repetition rate of the pulses is limited by the time taken to depopulate the neon 1*s* levels, and the heating of the tube.

The use of pulse techniques has made it possible to obtain powerful laser action with the diatomic gases N_2 and CO as working substances (Mathias and Parker, 1963a and 1963b). In nitrogen stimulated emission in pulses about 1 μsec long has been observed on a large number of infrared lines in the first positive band. Laser action in carbon monoxide is in lines of the Ångstrom band in the visible part of the spectrum. The distinguishing feature of the second laser is that it lases only when the discharge current is increasing. Both lasers require gas pressures over 1 torr and pulse voltages of 20–30 kV with currents of several tens of amperes. The characteristic output powers are of the order of a watt during the pulse. At present there are still no data on the processes in molecular gases which will allow the necessary population differences to be created. It can only be assumed that laser action breaks down because the lower levels of the transitions fill up.

58.6. Oscillation Spectrum of a Gas Laser

Far higher optical homogeneity is inherent in gases than in solids. This results in the observed regularity in the structure of the beam emitted by a gas laser and its controllability. In particular it is easy to choose conditions which give laser action in a particular mode. The intensity distributions across the beam in single-mode operation are shown in Fig.XII. 46. Examples of multi-mode structures are also shown. The photographs were produced by a He–Ne laser operating at $\lambda = 6328$ Å. A particular beam structure is obtained by altering the angle between the mirrors of the resonator, which leads to a change in the loss associated with a given mode. The second controlling factor is the excitation power. Stable single-mode operation is characteristic of excitation conditions just above threshold.

Simultaneous laser action at several frequencies is a fairly obvious feature of a gas laser and is due to the inhomogeneous nature of the broadening of the spectral line. The Doppler effect makes the major contribution of the linewidth under the operating conditions of gas laser. When there is inhomogeneous broadening the effect of saturation shows up in a particular manner. Let the amplification of the gas take the form shown in Fig.XII.47. While $G < G_{\min}$ there is no laser action. As the number of active molecules rises the

† With a tube length of about 1 m, a voltage of the order of 20 kV is applied to the electrodes.

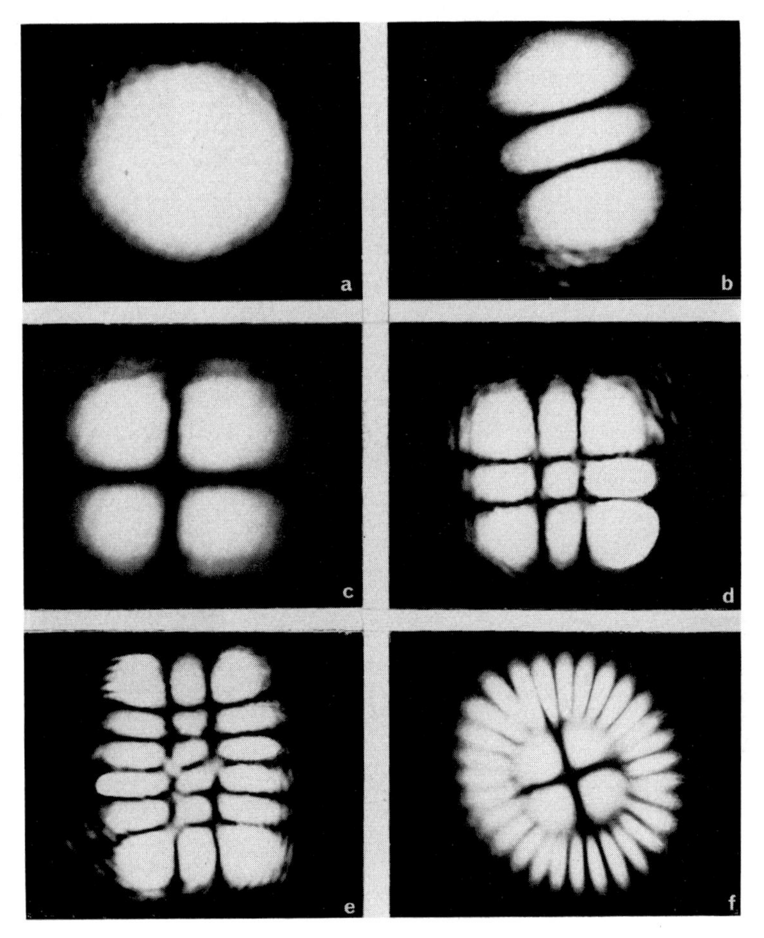

FIG. XII.46. Some intensity distributions across the beam emitted by a He–Ne laser ($\lambda = 6328$ Å).

amplification increases and when $G(\omega_1)$ exceeds G_{min} at ω_1 laser action starts. On a further increase in the excitation the amplification $G(\omega_1)$ will not rise, but it continues to increase in the rest of the line and under certain conditions modes are excited at frequencies ω_2, ω_3, etc. The presence of laser action at one frequency does not prevent the excitation of oscillations at other frequencies, although the oscillations are not entirely independent. The point is, roughly speaking, that the atoms concentrated near ω_2 make a contribution to the dispersion of the medium at ω_1. The formation of a hole in the

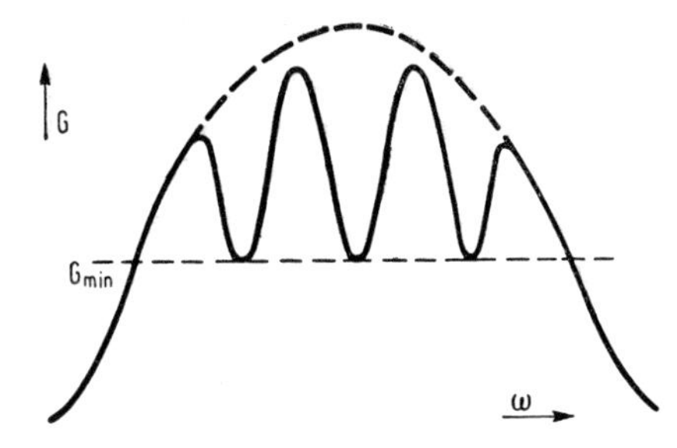

FIG. XII.47. "Hole-burning" in an inhomogeneously broadened line.

distribution function at ω_2 leads to a change in the phase shift at ω_1 and thus to a shift in the oscillation frequency. In its turn the hole at ω_1 also affects the oscillations at the frequency ω_2. The effect of "hole-burning" and the interaction of the holes has been described and evaluated by Bennett (1962b). It must be pointed out that the mobility of the gas molecules leads also to rapid spatial smoothing of the density of the active atoms, which excludes the possibility of a spontaneous transition from the generation of one set of modes to another as is the case in solid-state lasers.

The linewidth of a gas laser is extremely small; the optical methods of spectral investigation ordinarily used are useless for measuring it. The only way of obtaining information of the linewidth at present is frequency conversion by photoelectric devices. The frequencies of adjacent longitudinal modes are separated by an interval of the order of $c/2L$, i.e. for $L = 1$ m, $f_{p+1} - f_p = 150$ MHz. It is therefore convenient to use photomultipliers as the non-linear element.

The linewidth is affected by a number of factors. First of all there are fluctuations caused by spontaneous emission, which determine the natural width of the line; the actual width is far greater. It is due to mechanical vibrations of the mirrors caused by microphony. Thermal expansion of the structural components of the laser during operation also reduces the long-term stability. These mechanical and thermal effects have almost identical effects on the frequencies of the various oscillating modes, so they have little effect on the spectrum of the beats between different modes of a single laser. By beating together two such modes the natural linewidth of a gas laser can be measured. The results of Javan *et al.* (1962), which were made on the 1·153 μ line of a helium-neon laser, give a relative linewidth of 10^{-14}, i.e. an actual width of about 3 Hz.

The oscillations of two independent lasers must be combined to determine the total linewidth. The above authors indicate in their paper that microphony broadens the line to 100 kHz.

The relative thermal frequency drift of a laboratory device with an invar structure is 10^{-9} in 100 sec.

The frequency stability and simple tuning of gas lasers give them distinct advantages over existing solid-state devices. These qualities are desirable an many applications of coherent optical oscillators, particularly in laser communication systems. Sound and television images have already been transmitted by the use of gas lasers; their range is, however, limited to relatively short distances because of the low power output of the laser. The very high-frequency stability of gas lasers may lead to their use as frequency standards.

APPENDIX II

Laser Resonators

A.2. General Theory

A.2.1. Introductory Remarks

An electromagnetic field in the optical region interacts most efficiently with the active medium when it is put inside a resonator, just as in the microwave case. In the optical region, however, resonators similar to those used in the microwave band, whose dimensions are of the order of a wavelength, cannot be used. In a microwave resonator the natural frequencies within the working range are so widely spaced that in practice it is possible to discuss only a few modes, or even just one mode of oscillation. The important point is not only the difficulty of making resonators of microscopically small dimensions, and their small power handling capacity, but also the unfavourable scaling of the electrodynamic characteristics connected with the ohmic losses in the walls of the resonant cavity. In particular the Q-factor of the resonator decreases (in the case of the usual skin effect as $\sqrt{\lambda}$) when the wavelength and the dimensions of the resonant cavity are scaled down by the same factor. This in practice limits (on the short wave side) the application of resonator systems with dimensions of the order of a wavelength to the millimetre band, and necessitates the use, at shorter wavelengths, of multimode resonators whose dimensions are large compared with a wavelength; eigen-oscillations in the working range of the active medium are therefore of a high order.

Even in this case the use of the closed resonant cavities characteristic of the microwave region is unacceptable. This is because of the close spacing of the resonant frequencies of this kind of resonator when high-order modes are involved. The number of resonant modes of a closed cavity of volume V in a frequency interval $\Delta\omega$ is

$$\Delta N = \frac{\omega^2 V}{\pi^2 v^3} \Delta\omega, \qquad \text{(A.2.1)}$$

where v is the velocity of light in the substance filling the resonator.

It follows from (A.2.1) that the mean frequency interval between adjacent modes decreases inversely as the square of the frequency. In the case of the normal skin effect the Q-factor Q, whose order of magnitude is determined by the ratio of the resonator dimensions to the depth of penetration of the field into the metal (Vainshtein, 1957), is proportional to $\omega^{1/2}$. Therefore as the frequency rises the width of the resonance curve $\Delta\omega = \omega/Q$ increases as $\omega^{1/2}$. Therefore the resonance curves of high order modes of a closed cavity eventually overlap, i.e. at these frequencies the resonator loses its resonant properties. This is even truer of the optical region where various surface effects (roughness, oxide film, etc.) lead to absorption coefficients in the metal surface greater than those given by the skin effect formula.

Therefore for successful use of multi-mode resonators in the optical region we must find ways of decreasing the mode density, preferably reducing the energy losses from the resonant cavity at the same time. In principle there are several ways of doing this.

One of them consists of replacing the conducting metal walls of the cavity by dielectric boundaries and using the phenomenon of total internal reflection. Although the mode density of this kind of dielectric resonator is small compared with that of hollow cavities† their Q-factor may be much higher because of the extremely small losses connected with total internal reflection from a well-polished surface.

The mode density can be additionally reduced by changing to fine dielectric threads (fibre optics) whose transverse dimensions are only a few wavelengths. In this method, however, it is necessary to make the active substance in the form of fine fibres and there are considerable technical difficulties in doing this for substances which have a crystalline structure.

Another method, consisting of using open resonators, is now being widely used to reduce the mode density whilst retaining high Q-factors. The first suggestions concerning the use of open resonators of the Fabry–Perot interferometer type as oscillating systems at very short wavelengths came from Prokhorov (1958) and Dicke (1958). The theory of these resonators has been developed by Fox and Li (1960, 1961), Boyd and Gordon (1961), Boyd and Kogelnik (1962) and Vainshtein (1963a and 1963b). It is largely based on the laws of propagation of beams. These laws have also been studied by Goubau and Schwering (1961) as applied to so-called beam waveguides.

† For example, for a section of a circular dielectric cylinder placed between parallel metal planes at right angles to its axis the mode density is reduced $n/(n-1)$ times, where n is the refractive index of the dielectric.

A.2.2. Elementary Theory of a Fabry–Perot Resonator

Figure A.II.1 shows the open resonators most often used in lasers.

Although rigorous calculation of the characteristics of this kind of resonator necessitates the use of wave optics, an elementary theory which will give some idea of their features can be built up from geometrical optics. Let us take as an example the elementary theory of a Fabry–Perot resonator (Schawlow and Townes, 1958; Basov *et al.*, 1960; Genkin and Khanin, 1962; Kotik and Newstein, 1961).

In this theory it is assumed that the eigenwaves of the resonator are standing waves formed by superposition of uniform plane waves travelling in opposite directions. The energy losses have two main causes: absorption

(a) L (b) L

FIG. A.II.1. Open resonators: (a) with spherical mirrors; (b) with plane mirrors.

at the mirrors when the wave is reflected and radiation from the open side surface of the resonator. The radiation losses depend upon the angle θ between the direction of the wave vector and the normal to the reflector, dropping as θ decreases. For axial types of oscillation ($\theta = 0$) they are reduced to purely diffraction losses caused by spreading of the electromagnetic beam.

If we assume that during attenuation the structure of the field in the resonator does not change significantly,† then the decrease in the energy of the field in the resonator will be exponential with a certain characteristic time t_0 which determines the bandwidth of the resonator for the selected mode of oscillation, and therefore also the Q-factor of this mode:

$$Q = \omega t_0 . \tag{A.2.2}$$

Allowing for the losses on reflection leads to the Q-factor value (when the loss factor δ_r is a small quantity)

$$Q_r = \frac{kL}{\delta_r} = \frac{kL}{1-R}, \tag{A.2.3}$$

where k is the wave number for the medium filling the resonator; R is the energy reflection coefficient of the mirror.

† This means that this kind of field structure either relates to an individual non-degenerate mode of oscillation of the resonator or is a superposition of degenerate modes of oscillation and can therefore once again be looked upon as an eigen-oscillation of the system. It is the second factor that makes possible an approximate treatment of a section of a uniform plane wave, as is a mode of oscillation of the open resonator, when the dimensions of the mirrors are large and the ohmic losses in them are comparatively high.

In the optical range the Q-factor Q_r may reach comparatively high values, e.g. for $\lambda = 0{\cdot}5\ \mu$, $R = 0{\cdot}9$, $L = 10$ cm, $Q_r \sim 10^7$.

Using an approach based purely on geometrical optics the diffraction losses cannot, of course, be estimated. These losses will be calculated rigorously in § A.3. Here all we shall do is to use elementary diffraction theory to indicate the basic parameter that determines these losses. It is well known (Malyuzhinets, 1959) that diffraction near the edge of a screen can be looked upon as a process of spreading (diffusion) of the beam into the shadow region, this diffusion process covering a region of a size of the order of $2\sqrt{\lambda L}$ at a distance L from the screen (Fig. A.II.2), causing a relative increase of about $\sqrt{\lambda L}/a$ in the radius of the beam. The parameter $N = a^2/\lambda L$ will

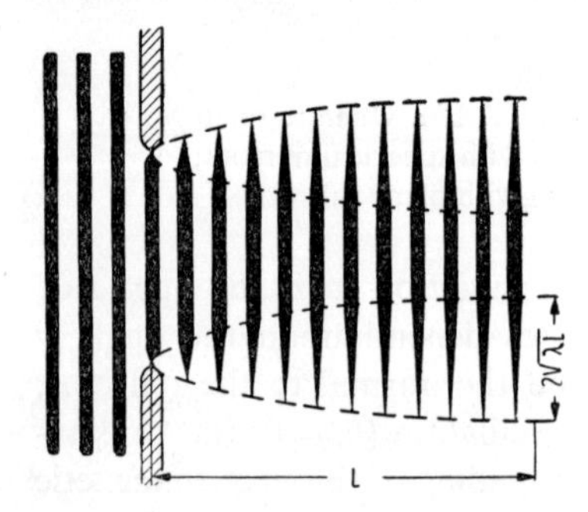

FIG. A.II.2. Diffraction of a wavefront at the edge of an aperture.

determine the magnitude of the diffraction loss factor $\delta_d(N)$ and therefore the corresponding Q-factor

$$Q_d = \frac{kL}{\delta_d(N)}. \tag{A.2.4}$$

From elementary considerations it is clear that the diffraction losses are a monotonically decreasing function of the parameter N.

As a rigorous calculation shows (§ A.3), in the optical resonators usually used this parameter is so great that the diffraction losses of axial modes need not be taken into consideration. When the values of N are not too high, however, the diffraction losses may be not only comparable with, but even greater than, the ohmic losses and it is necessary to allow for them. The dominance of the diffraction losses in fact means that the geometrical optics approach cannot be used and there is no sense in discussing this case within the framework of elementary theory.

A reduction of the Q-factor of actual Fabry–Perot resonators may be due to non-parallelism of the mirrors. It is not hard to estimate the corresponding losses, representing the field between the mirrors in the form of the superposition of two opposed beams of finite width $2a$ and calculating the amplitude transformation coefficient for one beam into another taking into consideration the dephasing introduced by the inclined mirror (Kotik and Newstein, 1961):

$$R = \frac{1}{2a}\int_{-a}^{a} e^{2i\beta kx}\,dx = \frac{\sin 2\Delta}{2\Delta},$$

where $\Delta = \beta ka$ is the advance in the phase difference in each of the wave-fronts at the edge of the mirror caused by tilting of the latter through the angle β.

With small symmetrical tilts ($2\Delta \ll 1$) the magnitude of the losses δ_β is

$$\delta_\beta = 1 - |R| \simeq \frac{4\Delta^2}{3} \tag{A.2.5}$$

and the Q-factor corresponding to them is determined by the relation

$$Q_\beta = \frac{3kL}{4\Delta^2}. \tag{A.2.6}$$

Fox and Li (1963) have shown that the expressions obtained agree quite closely with the results of a rigorous calculation made with an electronic computer.

When $Q_\beta > Q_r$ non-parallelism of the mirrors has no significant effect on the overall Q-factor. The condition $Q_\beta > Q_r$ imposes a limitation on the maximum permissible angles of tilt of the mirrors:

$$\beta < \beta_{\max} \sim \frac{(1 - R)^{1/2}}{ka}. \tag{A.2.7}$$

For the parameters $a \sim 0{\cdot}25$ cm, $R = 0{\cdot}99$, $\lambda \sim 10^{-4}$ cm, which are typical for a laser resonator, the permissible angle of tilt is of the order of 1″. We know that plane mirrors in gas lasers also have to be adjusted with an accuracy of seconds. The rough adjustment of the resonators for solid-state lasers can be explained by the optical inhomogeneity of the crystals used.

The modes that are a superposition of waves propagated at an angle θ to the axis of the resonator are damped more than the axial ones since for them there is a larger loss due to radiation through the side surface of the resonator. The order of magnitude of the Q-factor due these losses for circular mirrors with a radius a can be estimated from the expression (Basov *et al.*, 1960)

$$Q'' \sim \frac{ka}{2\theta}. \tag{A.2.8}$$

Just as in the case of tilt of the mirrors we can calculate the maximum angle θ_{max} which results in a decrease in the Q of the mode by a factor of more than two compared with Q_r:

$$\theta_{max} \sim \frac{ka}{2Q_r} = \frac{a}{2L}(1 - R). \tag{A.2.9}$$

From θ_{max} it is not hard to calculate the number of modes of the resonator per frequency interval $\Delta\omega$ with a Q-factor $Q > Q_r/2$. To do this it is sufficient to multiply expression (A.2.1), for the total number of oscillators in a volume $V = \pi a^2 L$, by $\theta^2_{max}/2$ (the relative share of high Q-factor non-axial types of oscillation). The reduction in the mode density achieved in an open resonator may be very considerable: for example, with $a = 0{\cdot}5$ cm, $L =$ 100 cm, $R = 0{\cdot}98$, $\theta^2_{max}/2 \sim 10^{-9}$ and with $a = 0{\cdot}5$ cm, $R = 0{\cdot}9$, $\theta^2_{max}/2 \simeq 10^{-6}$. Let us calculate the number of high-Q modes within the spectral line of the working substance, assuming that for the spectral lines of gases at room temperature $\Delta\omega/\omega = 10^{-6}$, and for paramagnetic crystals $\Delta\omega/\omega = 10^{-3}$. Assuming that the latter parameter relates to a solid-state laser and the former to a gas laser we have (for $\lambda = 1\mu$) $\Delta N \sim 10^5$ in the second and $\Delta N \sim 1$ in the first case. The lower values of θ_{max} and the correspondingly smaller number of modes within the spectral line of the working substance explain the more directional nature of theoutput of gas lasers as compared with actual solid-state lasers.

The elementary theory of a two-mirror resonator described above is quite a long way from being rigorous. Because of the boundary effects the eigenwaves differ from plane uniform waves. In addition, the elementary theory by its very nature cannot provide a correct estimate of the diffraction losses. A more closely argued analysis of the characteristics of open resonators will now be carried out on the basis of the wave theory of the propagation of electromagnetic waves.

A.2.3. Elements of Theory of Propagation of Beams

Let us examine some general laws concerning wavefronts in free space (Goubau and Schwering, 1961; Bondarenko and Talanov, 1964). The electromagnetic field in a broad beam propagated in a direction $\boldsymbol{k}_0 = k\boldsymbol{n}$ $(k = 2\pi/\lambda)$ can be written in the form†

$$\begin{aligned} \boldsymbol{E}(\boldsymbol{r}) &= \boldsymbol{E}_0(\boldsymbol{r})\, e^{i(\boldsymbol{k}_0 \cdot \boldsymbol{r})}, \\ \boldsymbol{H}(\boldsymbol{r}) &= \boldsymbol{H}_0(\boldsymbol{r})\, e^{i(\boldsymbol{k}_0 \cdot \boldsymbol{r})}, \end{aligned} \tag{A.2.10}$$

where $\boldsymbol{E}_0(\boldsymbol{r})$, $\boldsymbol{H}_0(\boldsymbol{r})$ are functions of the coordinates which change by a negligible amount in a distance λ; we shall consider only the electric field $\boldsymbol{E}$.

† The time dependence $e^{-i\omega t}$ is understood.

Substituting for $\boldsymbol{E}$ in the equations

$$\nabla^2 \boldsymbol{E} + k^2 \boldsymbol{E} = 0, \tag{A.2.11}$$

$$\operatorname{div} \boldsymbol{E} = 0 \tag{A.2.12}$$

changes this pair into the following:

$$-\frac{1}{2ik} \nabla^2 \boldsymbol{E}_0 = (\boldsymbol{n} \cdot \nabla) \boldsymbol{E}_0, \tag{A.2.13}$$

$$-\frac{1}{ik} \operatorname{div} \boldsymbol{E}_0 = (\boldsymbol{n} \cdot \boldsymbol{E}_0) \tag{A.2.14}$$

Let us take an arbitrary cylindrical system of coordinates with the z-axis along the direction of propagation of the beam $\boldsymbol{n}$. Remembering that $\boldsymbol{E}_0$ is a slowly changing function we can neglect the term $(1/2ik)(\partial^2 \boldsymbol{E}_0/\partial z^2)$ compared with $\partial \boldsymbol{E}_0/\partial z$ in equation (A.2.13) and $(1/ik)(\partial E_{0z}/\partial z)$ compared with E_{0z} in (A.2.14) and rewrite the equations (A.2.13), (A.2.14) in the form

$$-\frac{1}{2ik} \nabla_\perp^2 \boldsymbol{E}_0 = \frac{\partial \boldsymbol{E}_0}{\partial z}, \tag{A.2.15}$$

$$-\frac{1}{ik} \operatorname{div}_\perp \boldsymbol{E}_0 = E_{0z}. \tag{A.2.16}$$

Unlike the original Helmholtz equation (A.2.13) the first of these equations belongs to the parabolic class and has the same form as the standard thermal conductivity and diffusion equations. Equations of this kind are extensively used in diffraction theory, where they have been given the name of transverse diffusion equations. According to (A.2.15) the propagation of an electromagnetic wave is represented as the process of diffusion of a wave amplitude across the rays. Apart from the simplicity of this description, which has been developed in diffraction theory in papers by Malyuzhinets (1959), Malyuzhinets and Vainshtein (1961a and 1961b) and Fok and Vainshtein (1963), it also makes it possible to obtain comparatively simply asymptotic expressions in problems of diffraction by bodies whose dimensions are large compared with the wavelength. It is also useful in the study of optical, or, to be more precise quasi-optical,† resonators, which are

† The term quasi-optical, generally speaking, is more appropriate here since by optical systems we generally mean systems in which the propagation of electromagnetic waves is subject to the laws of geometrical optics, and wave optics is introduced only in exceptional cases: when studying diffraction effects, when investigating phenomena near caustics, in particular near the focus of an optical system, etc. The dimensions of the systems we are discussing (or of the regions in them occupied by the electromagnetic field) are as a rule such that the propagation of electromagnetic waves takes place under conditions where wave optics and Fresnel diffraction laws are appropriate. Moreover, only by taking diffraction phenomena into consideration can we provide a correct explanation of the features of the majority of the systems being studied.

discussed in the present chapter [and also of optical waveguides (Goubau and Schwering, 1961)].

The solution of equation (A.2.15), which describes a beam being propagated in the direction of the z-axis, can be written in two equivalent forms:

$$E_0(r) = \frac{k}{2\pi iz} \int\int_S E_S(x', y')\, e^{\frac{ik}{2z}[(x-x')^2+(y-y')^2]}\, dx'\, dy' \qquad \text{(A.2.17)}$$

or

$$E_0(r) = \int_{-\infty}^{\infty}\int_{-\infty}^{\infty} E_S(k_\perp)\, e^{i(k_\perp \cdot r) - i\frac{k_\perp^2 z}{2k}}\, d^2k_\perp, \qquad \text{(A.2.18)}$$

where $k_\perp$ is a vector with the components k_x, k_y.

The first of these forms gives the solution of the Cauchy problem for the equation (A.2.15) when we know the distribution of the field E_S in the cross-section of the beam at $z = 0$; the second defines the field $E_0(r)$ in terms of its angular distribution $E_S(k_\perp)$ in this cross-section. The relation (A.2.17) is the mathematical expression of the Huygens–Kirchhoff principle, according to which the field in a certain region can be determined from its values at the boundaries of the region. We notice that according to the uniqueness theorem of electrodynamics we can limit ourselves in (A.2.17) and (A.2.18) to discussing only the components orthogonal to the z-axis of the wave; the third component of the field can be determined from (A.2.16). According to (A.2.16) this component has a relative order of magnitude of λ/a, where a is the characteristic dimension of the region of significant change of the field in the cross-section of the beam.

In order to determine in greater detail the field of applicability of (A.2.17) and (A.2.18) it is simplest to turn to the rigorous solution of (A.2.11), which can be written in the form of a Fourier integral in plane waves:

$$E = E_0(r)\, e^{i(k_0 \cdot r)} = e^{ikor} \int\int E_0(k_\perp)\, e^{i((k-k_0)\cdot r)}\, d^2k_\perp, \qquad \text{(A.2.19)}$$

where the vector k has the components k_x, k_y, k_z and k_0 has the components 0, 0, k.

If the is broad enough ($a \gg \lambda$) and its angular distribution $E_0(k_\perp)$ is finite in only a small region $|k_\perp| \sim 2\pi/a \ll k$ near the origin, then expanding the difference $k - k_0$ in powers of $k_\perp/k$:

$$k - k_0 = k_\perp + \frac{k_0}{k}\left\{-\frac{k_\perp^2}{2k} - \frac{1}{8}\frac{k_\perp^4}{k^3} + \cdots\right\}, \qquad \text{(A.2.20)}$$

and taking only the first two terms in the expansion we arrive at (A.2.18). Therefore to estimate the range of applicability of this relation and of

(A.2.17), which is equivalent to it, the first of the terms which were neglected in the exponent in (A.2.19) must be small compared with π:

$$\frac{1}{8}\frac{k_\perp^4 z}{k^3} \ll \pi. \tag{A.2.21}$$

This condition can be rewritten in the form

$$p = \frac{\sqrt{\lambda z}}{a} \ll 2\frac{a}{\lambda}, \tag{A.2.21a}$$

from which it follows that (A.2.17) and (A.2.18) can be used even when the parameters p are appreciably greater than unity (when $a/\lambda \gg 1$).

A.2.4. The Integral Equation for the Field in a Quasi-optical Resonator

We shall now apply expression (A.2.17), obtained above, to the case of an open resonator formed by two identical mirrors placed at $z = +z_0 = \pm L/2$. Since (A.2.17) can be referred to any of the Cartesian components of the field E_x and E_y we shall in future discuss only one of them, denoting it by $\psi(x, y, z)$. The problem is thus completely reduced to a scalar form.

Let the field distribution $\psi(x, y, -z_0) \equiv \psi_S(x, y)$ be given at the plane $z = -z_0$. The field at z can be obtained from (A.2.17) by the substitution $z \to z + z_0$:

$$\psi(x, y, z) = \frac{k}{2\pi i(z+z_0)} \times \iint_S \psi_S(x', y')\, e^{\frac{ik}{2(z+z_0)}[(x-x')^2+(y-y')^2]}\, dx'\, dy'. \tag{A.2.22}$$

Here S is the region occupied by the beam at the section $z = -z_0$ (the surface of the mirror).

The reflecting surface at $z = z_0$ can be looked upon as a converter that transforms the phase of the wavefront by an amount $\varphi(x, y)$ determined by the shape of the mirror. For the sake of simplicity we shall assume that the region outside the mirrors is an ideal absorber. From symmetry considerations we require that the distribution of the field in the wavefront reflected from the second mirror should repeat the distribution of the field at $z = -z_0$ except for the introduction of an arbitrary constant factor p:

$$\psi(x, y, z_0)\, e^{-i\varphi(x,y)} = p\psi_S(x, y). \tag{A.2.22A}$$

Substituting the expression for $\psi(x, y, z_0)$ from (A.2.22) in this produces an integral equation for determining the eigenfunctions $\psi_{S\nu}$ and the eigenvalues

p_ν of an open resonator

$$\frac{k}{2\pi iL}\iint_S \psi_S(x', y')\, e^{\frac{ik}{2L}[(x-x')^2+(y-y')^2]-i\varphi(x,y)}\, dx'\, dy' = p\psi_S(x, y). \qquad (A.2.23)$$

From the eigenvalues p_ν the frequencies of the free oscillations of the resonator can be determined from the relation

$$p_\nu^2\, e^{2ikL} = e^{2iq\pi}, \qquad (A.2.24)$$

where q is an integer.

To conclude the section we point out that the possibilities of the general approach discussed here, based on the use of a parabolic equation, are not limited to systems with quasi-plane wavefronts. This description is also useful when studying diverging fronts, when diffusion of the wave amplitude takes place on spherical or cylindrical surfaces (Vainshtein, 1963b).

A.3. Resonators with Spherical and Plane Mirrors (Fox and Li, 1960 and 1961; Boyd and Gordon, 1961; Boyd and Kogelnik, 1962; Vainshtein, 1963a and 1963b)

A.3.1. The Resonator formed by Rectangular Confocal Spherical Mirrors

Let the surface S of the mirrors of a resonator be a rectangle with sides $2a$ and $2b$. We have a phase transformation of the following form:

$$\varphi(x, y) = \frac{k\varrho^2}{L}, \qquad (A.3.1)$$

where $\varrho^2 = x^2 + y^2$. This kind of phase transformation is produced by a spherical mirror of radius L.

Substituting (A.3.1) in (A.2.23) and separating the variables we obtain two integral equations of the same type:

$$\sqrt{\frac{k}{2\pi iL}}\int_{-a}^{a} \tilde{\xi}_S(x')\, e^{-\frac{ikxx'}{L}}\, dx' = p_x\tilde{\xi}_S(x), \qquad (A.3.2)$$

$$\sqrt{\frac{k}{2\pi iL}}\int_{-b}^{b} \tilde{\eta}_S(y')\, e^{-\frac{ikyy'}{L}}\, dy' = p_y\tilde{\eta}_S(y), \qquad (A.3.3)$$

where

$$\psi_S(x, y) = \xi_S(x)\, \eta_S(y), \quad p = p_x p_y,$$

$$\tilde{\xi}_S(x) = \xi_S(x)\, e^{\frac{ikx^2}{2L}}, \quad \tilde{\eta}_S(y) = \eta_S(y)\, e^{\frac{iky^2}{2L}}.$$

Because of the complete similarity of equation (A.3.2) and (A.3.3) we shall limit ourselves to investigating the first of them. Formally it describes the eigenfunctions of a two-dimensional resonator formed by a system of confocal cylindrical mirrors with a radius L.

By introducing the dimensionless variables and parameters

$$u = \sqrt{\frac{k}{L}}\, x, \quad u' = \sqrt{\frac{k}{L}}\, x', \quad l_a = \sqrt{\frac{k}{L}}\, a, \quad \tilde{p}_x = \sqrt{i} p_x,$$

equation (A.3.2) is brought to the form

$$\sqrt{\frac{1}{2\pi}} \int_{-l_a}^{l_a} \tilde{\xi}_S(u')\, e^{-iuu'}\, du' = \tilde{p}_x \tilde{\xi}_S(u). \tag{A.3.4}$$

We shall first apply (A.3.4) to finding the field in a resonator with infinitely wide mirrors $l_a \to \infty$. Physically this corresponds to the case when the dimensions of the mirror are considerably greater than the cross-sectional dimensions of the region in which the field is localized in the resonator. When $l_a \to \infty$ the left-hand side of the equation (A.3.4) is a Fourier transform of the function $\xi_S(u)$. Therefore the solution of (A.3.4) in this case should be functions which are Fourier pairs. It is known (Titchmarsh, 1937) that one such set of functions, which are orthogonal and their square is integrable in the interval $(-\infty, \infty)$,† are the Hermitian functions

$$\varphi_m(u) = H_m(u)\, e^{-\frac{1}{2}u^2}, \tag{A.3.5}$$

where $H_m(u)$ are Hermite polynomials:

$$H_0 = 1, \quad H_1 = 2u, \quad H_2 = 4u^2 - 2, \quad \ldots, \quad H_n(u) = (-1)^n\, e^{u^2} \frac{d^n\, e^{-u^2}}{du^n}.$$

The function $\varphi_m(u)$ satisfies equation (A.3.4) (when $l_a \to \infty$) if

$$\tilde{p}_{xm} = (-i)^m. \tag{A.3.6}$$

The condition for orthogonality of the Hermitian functions

$$\int_{-\infty}^{\infty} e^{-u^2} H_m(u) H_n(u)\, du = \begin{cases} 0, & m \neq n, \\ 2^n n! \sqrt{\pi}, & m = n \end{cases} \tag{A.3.7}$$

is equivalent to orthogonality of the eigenmodes of the resonator.

Equation (A.3.3) is treated in the same way. Since $|p_{xm}| = 1$ and $|p_{yn}| = 1$ the eigenfrequencies defined by (A. 2.24) are purely real:

$$kL = \pi q + \frac{\pi}{2}(m + n + 1). \tag{A.3.8}$$

† The latter condition ensures that the energy flux in the beam is finite.

The oscillations obtained can be designated as of the TEM_{mnq} type, since they have quasi-transverse polarization of the field.

Let us find the field distribution inside the resonator. It is the superposition of two waves travelling towards each other; we shall discuss the wave travelling in the positive direction of the z-axis. Substitution of ψ_S in the form of the product of Hermitian functions in (A.2.22) gives

$$\psi_{mn}(x, y, z) = \xi_m(x, z)\, \eta_n(y, z), \tag{A.3.9}$$

where

$$\xi_m(x, z) = \sqrt[4]{\frac{2}{1+\zeta^2}}\, H_m\left(u\sqrt{\frac{2}{1+\zeta^2}}\right) \exp\left[-\frac{u^2}{1+\zeta^2}\right]$$

$$\times \exp\left\{i\left[\frac{\zeta}{1+\zeta^2}u^2 - \left(\frac{\pi}{2}-\varphi\right)\left(\frac{1}{2}+m\right)\right]\right\}; \tag{A.3.10}$$

$$\zeta = 2z/L, \quad \tan\varphi = \frac{1-\zeta}{1+\zeta}.$$

The function $\eta_n(y, z)$ is also defined by (A.3.10) but with the substitutions $u \to v \equiv \sqrt{(k/L)}\, y$, $m \to n$. It follows from (A.3.9) that the amplitude distribution in the beam depends only on the arguments $u\sqrt{[2/(1+\zeta^2)]}$, $v\sqrt{[2/(1+\zeta^2)]}$, i.e., different planes in the beam they are similar in the to ζ=const. As is to be expected the amplitude distribution is symmetrical about the plane $\zeta = 0$. That part of the phase of the beam that depends on the transverse coordinates $x = \sqrt{(L/k)}\, u$, $y = \sqrt{(L/k)}\, v$ is an odd function of ζ. In the planes $z = \pm z_0 = \pm L/2$ the phase distribution is described by the functions $\pm(u^2 + v^2)/2$, which corresponds to spherical waves with a radius of curvature of $R_{\pm z_0} = \pm L$.

The equation of the phase surface passing through an arbitrary point $\zeta = \bar{\zeta}\,(z = \bar{z})$ on the axis of the resonator has, in accordance with (A.3.9), the form

$$\frac{kL}{2}(1+\zeta) + \frac{\zeta}{1+\zeta^2}(u^2+v^2) - \left(\frac{\pi}{2}-\varphi\right)(1+m+n)$$

$$= \frac{kL}{2}(1+\bar{\zeta}) - \left(\frac{\pi}{2}-\bar{\varphi}\right)(1+m+n), \tag{A.3.11}$$

where $\bar{\varphi} = \arctan[(1-\bar{\zeta})/(1+\bar{\zeta})]$.

From this we can obtain for points near the axis of the system, when the values of m and n are not too high, the relation

$$z - \bar{z} = \simeq \frac{\bar{\zeta}}{1+\bar{\zeta}^2}\frac{x^2+y^2}{L}, \tag{A.3.12}$$

corresponding to a spherical front with a radius of curvature

$$R_{\bar{z}} = \frac{L^2 + 4\bar{z}^2}{4\bar{z}}. \qquad \text{(A.3.13)}$$

It is easy to see that the radius of curvature has its lowest value $R_{\min} = L$ at the plane $\bar{z} = \pm L/2$ where the mirrors are located. In future we shall use (A.3.13) to determine the radii of curvature of the mirrors in non-confocal resonators.

Let us estimate the transverse dimensions of the beam for the fundamental mode TEM_{00q}. We shall take as the radius of the beam ϱ_s the value of the transverse coordinate when the amplitude of the wave has decreased by a factor of e. Then from (A.3.9) we have

$$\varrho s = \sqrt{\frac{L}{k}\left[1 + \left(\frac{2z}{L}\right)^2\right]}. \qquad \text{(A.3.14)}$$

The radius of the beam at the mirrors is $\varrho_s(\pm L/2) = \sqrt{2L/k}$, and at the centre of the resonator is $\varrho_s(0) = \sqrt{L/k}$, i.e. $\sqrt{2}$ times less than at the mirrors. If one of the mirrors is slightly transmitting, then from (A.3.14) we can also estimate the angular divergence of the outgoing beam by putting $z \gg L/2$:

$$\theta_{1/e} = \frac{2\varrho_s}{z} = \frac{4}{\sqrt{kL}}. \qquad \text{(A.3.15)}$$

The divergence of the beam is generally characterized not by a drop in amplitude by a factor of e but by the half-power level:

$$\theta_{0.5} = 2\sqrt{\frac{\lambda}{\pi L}\ln 2}. \qquad \text{(A.3.15a)}$$

Let us make an estimate of the beam divergence obtained in practice. With the dimensions typical for a gas laser of $L = 100$ cm, $\lambda = 10^{-4}$ cm we have $\theta_{0.5} \sim 10^{-3}$ radian $\approx 3{\cdot}5'$. Additional focusing of the beam which will ensure a plane wave at the output of the resonator can reduce the divergence by a factor of $\sqrt{2}$.†

Let us now return to the analysis of the integral equations (A.3.2), (A.3.3) that describe the field distribution in a resonator with confocal mirrors of finite dimensions. Integral equations of this type have been investigated in detail in Slepian and Pollak (1961) and in Landau and Pollak (1961 and

† This is not difficult to see if we look upon the outgoing wave as being created by the field distribution at the section $z = 0$, where the wavefront is half as wide as at the mirrors and is plane.

1962). Their solutions are real functions with a limited spectrum that are their own finite Fourier transform:

$$\tilde{\xi}_S(u) = S_{0m}(l_a^2, u/l_a), \quad \tilde{\eta}_S(v) = S_{0n}(l_b^2, v/l_b), \tag{A.3.16}$$

for the eigenvalues

$$\tilde{p}_{xm} = \sqrt{\frac{2}{\pi}}\, l_a(-i)^m R_{0m}^{(1)}(l_a^2, 1), \quad \tilde{p}_{yn} = \sqrt{\frac{2}{\pi}}\, l_b(-i)^n R_{0n}^{(1)}(l_b^2, 1), \tag{A.3.17}$$

where the parameter l_b is defined similarly to l_a: $l_b = \sqrt{(k/L)}\, b$. The functions $S_{0m}(l^2, u/l)$ and $R_{0m}^{(1)}(l^2, 1)$ are called respectively the angular and radial wave functions in a prolate spheroidal system of coordinates. These functions have been tabulated by Stratton *et al.* (1956). The real nature of the eigenfunctions $\tilde{\xi}_S(x)$ and $\tilde{\eta}_S(y)$ means that the phase distribution in the planes $z = \pm L/2$ has, just as with infinitely large mirrors, the nature of a spherical wave with a radius of curvature L, i.e. the surfaces of the equal phases and the surfaces of the mirrors coincide with each other. We can determine the eigenfrequencies of the resonator from (A.2.24) by substituting $p_\nu = i\bar{p}_{xm}\bar{p}_{yn}$:

$$k = k' - ik'' = \frac{1}{v}(\omega' - i\omega''), \tag{A.3.18}$$

$$k'L = \pi q + \frac{\pi}{2}(m + n + 1), \tag{A.3.19}$$

$$2k''L = -\ln T, \tag{A.3.20}$$

where

$$T = T_{mn} = \frac{4}{\pi^2}\, l_a^2 l_b^2 [R_{0m}^{(1)}(l_a^2, 1)]^2\, [R_{0n}^{(1)}(l_b^2, 1)]^2 \tag{A.3.21}$$

is the power transfer coefficient per reflection. Since with finite l_a and l_b $T < 1$, $k'' \neq 0$. Therefore the real parts of the eigenfrequencies of a confocal resonator with mirrors of finite size are the same as with infinitely large mirrors, but the finite size of the mirrors, of course, causes the appearance of an attenuation coefficient which depends on the mode and the dimensions of the resonator.

In accordance with (A.3.18) it is convenient to represent the eigenfrequency spectrum of a confocal resonator on a complex plane ω (Fig. A.II.3). The eigenfrequencies corresponding to one and the same longitudinal number q but different transverse numbers m and n lie on a curve which moves away from the real axis as the frequency ω' rises. Figure A.II.3 has two important features. In the first place it illustrates clearly the decrease in the mode density mentioned in § A.2: the least distance along the real axis between the frequencies of the oscillations is now $v\pi/2L$, which is many orders ($2\pi V/\lambda^2 L$ times) greater than the mean frequency interval between modes found from (A.2.1). In the second place we have mode degeneracy if we consider the agreement only of the real parts of the eigenfrequencies. The latter effect leads to simultaneous excitation in the resonator of modes with completely different structures if the means of excitation has no spatial selectivity with respect to the field distribution of the individual modes.

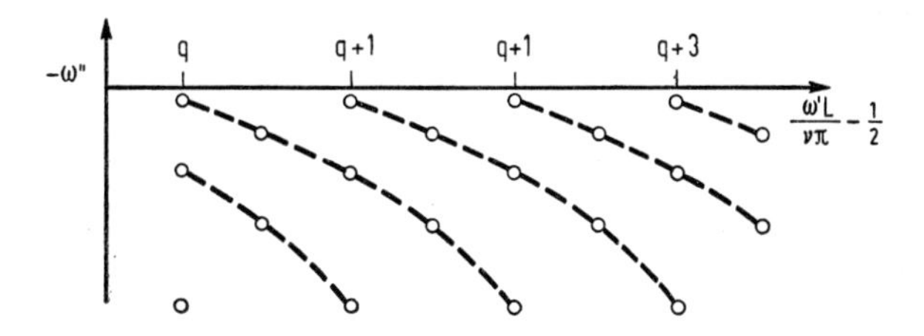

FIG. A.II.3. Frequency spectrum of a resonator with confocal mirrors.

Instead of by the attenuation coefficient ω'', the type of oscillation can be characterized by its Q-factor Q, which is connected with ω'' in the usual way:

$$Q = \frac{\omega'}{2\omega''} = -\frac{k'L}{\ln T}. \tag{A.3.22}$$

With transfer coefficients T close to unity

$$Q = \frac{k'L}{1 - T}. \tag{A.3.23}$$

The quantity $\delta_d = 1 - T$ is the diffraction loss factor. In the general case, if there are losses in the mirrors as well as diffraction losses, the former being caused by absorption or partial transparency of the reflecting surfaces and described by the loss factor $\delta_r = 1 - R$, then the Q-factor of a given type of oscillation will be

$$Q = \frac{k'L}{\delta_d + \delta_r}. \tag{A.3.24}$$

Figure A.II.4 shows the quantity $1 - |\tilde{p}_m|^2$ as a function of l^2 for $m = 0$, 1, 2 and Fig. A.II.5 shows δ_d as a function of $l^2 = l_a^2 = l_b^2$ for a few of the simplest modes of a confocal resonator with square mirrors. It can be seen from Fig. A.II.5 that as the mode number increases the diffraction losses rise, but for the fundamental mode they do not exceed 1% when $l^2 > 4$. As a comparison a coefficient of reflection $R \approx 0{\cdot}99$ can be obtained only by the use of high-quality multilayer dielectric coatings. Therefore in a confocal resonator with any available reflecting coating the losses in the mirrors will be greater than the diffraction losses provided the parameters l_a^2 and l_b^2 are more than a few units. In optical resonators the latter condition

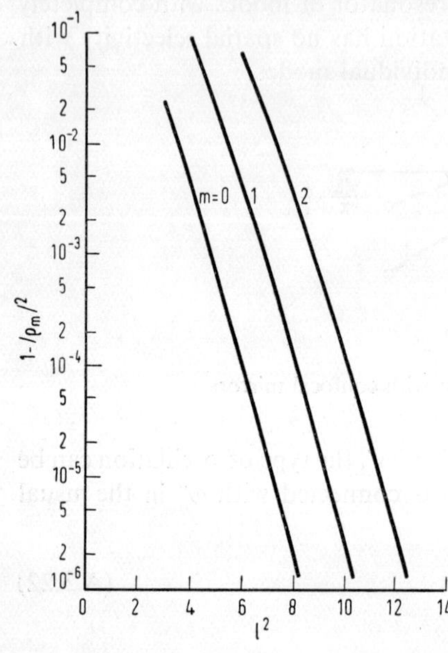

FIG. A.II.4. Diffraction losses of a confocal resonator with infinitely long cylindrical reflectors.

is easily satisfied, e.g. for $L = 100$ cm, $\lambda = 10^{-4}$ cm and $a = b = 0{\cdot}5$ cm, $l^2 = 50\pi$.

The ratio of the width of the resonance curve $\delta\omega = \omega/Q$ to the interval $\Delta\omega = v\pi/L$ between the frequencies of the axial modes

$$\frac{\delta\omega}{\Delta\omega} = \frac{1-R}{\pi}$$

is small when $R \sim 1$: in the example under discussion it is 0·006 when $R = 0{\cdot}98$, i.e. the axial modes are well resolved. The field distribution in the resonator at the surface of the mirrors is determined by the eigenfunctions $S_{0m}(l_a^2, u/l_a)$, $S_{0n}(l_b^2, v/l_b)$. The functions $S_{0m}(l^2, w)$ are orthogonal in the range $|w| < 1$. When $w^2 \ll 1$ (near the axis of the resonator) they are approximately the same as the Hermitian functions†

$$S_{0m}(l^2, u/l) \approx \frac{\Gamma\left(\frac{m}{2}+1\right)}{\Gamma(m+1)} H_m(u)\, e^{-\frac{1}{2}u^2}. \tag{A.3.25}$$

† The normalizing factor here is selected so that for even m

$$S_{0m}(l^2, 0) = \pm 1.$$

Therefore near the axis of the resonator the field can be found by using (A.3.25), and only when calculating the field at the edges of the mirrors is it necessary to have recourse to tables of spheroidal functions. Since when $l^2 \gg 1$ the field is localized a fairly long way from the edges we can in this case use (A.3.25) over the whole range of variation of the variable $u/l = x/a$.

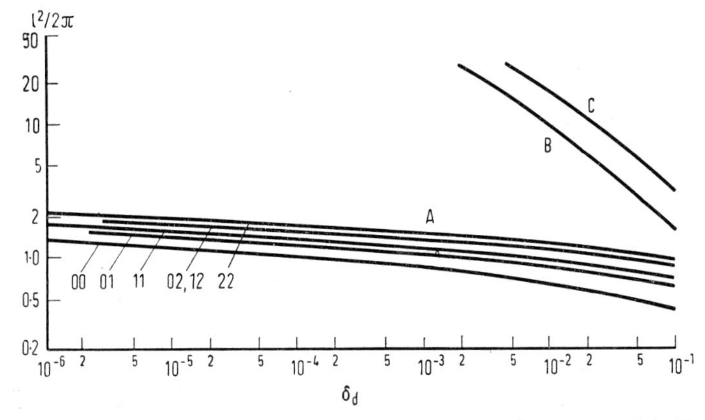

FIG. A.II.5. Diffraction losses of confocal (*A*) and plane-parallel (*B*, *C*) resonators: *A*—confocal reflectors of square cross-section; *B* and *C*—TEM_{00} and TEM_{10} modes respectively of a resonator with circular plane reflectors.

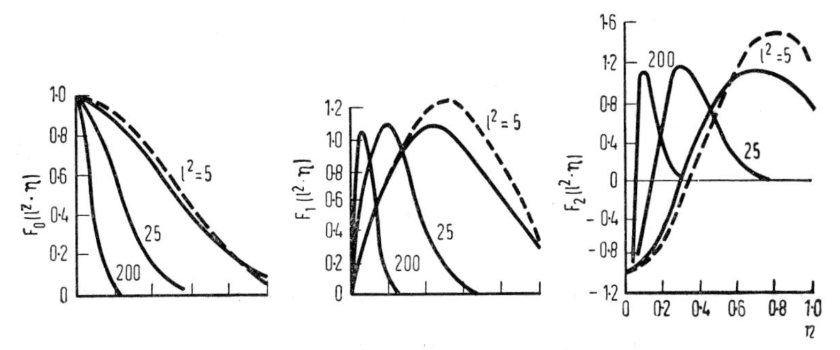

FIG. A.II.6. Amplitude distributions of the field $F_m(l^2, \eta = u/l)$ in the simplest modes of a confocal resonator.

Examples of the amplitude distributions of some simple modes for different values of l^2 are shown in Fig. A.II.6. The solid curves are calculated by use of the approximate formula (A.3.25). The dotted curve for $l^2 = 5$ is calculated from tables of spheroidal functions. A comparison of the solid and dotted curves for $l^2 = 5$ justifies the approximate calculation of the field distribution from (A.3.25), the accuracy of the calculation rising as l^2 rises. The current distribution at the mirrors for the simplest modes of a confocal resonator is shown diagrammatically in Fig. A.II.7.

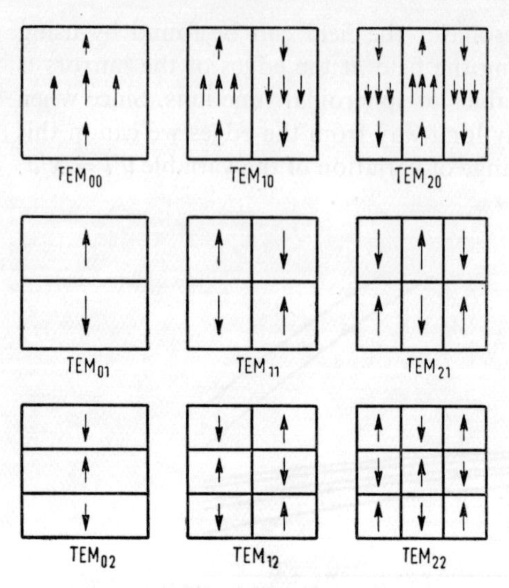

FIG. A.II.7. The current distribution at square mirrors for the simplest modes.

All the results obtained above can also easily be generalized to the case of resonators with mirrors of different cross-sectional dimensions (Boyd and Kogelnik, 1962). When this is done the eigenfrequencies [of the (A.3.19) type] remain unchanged; the eigenvalues (A.3.17) which, as we have seen, determine the diffraction losses in the resonator, are replaced by

$$\tilde{p}_{xm} = \sqrt{\frac{2}{\pi}}\, l_{a_1} l_{a_2} (-i)^m R_{0m}^{(1)}(l_{a_1} l_{a_2}, 1),$$
$$\tilde{p}_{yn} = \sqrt{\frac{2}{\pi}}\, l_{b_1} l_{b_2} (-i)^n R_{0n}^{(1)}(l_{b_1} l_{b_2}, 1), \tag{A.3.26}$$

where $l_{a_1,a_2} = \sqrt{(k/L)}\, a_{1,2}$, $l_{b_1,b_2} = \sqrt{(k/L)}\, b_{1,2}$ ($2a_1$, $2b_1$, and $2a_2$, $2b_2$ are respectively the dimensions of the first and second mirrors).

The eigenfunctions in this case are $S_{0m}(l_1 l_2, u/l_1)$ or $S_{0m}(l_1 l_2, u/l_2)$ depending on which mirror the field is determined.

A.3.2. Resonator with Circular Confocal Spherical Mirrors

For a resonator with confocal mirrors of circular cross-section (Boyd and Gordon, 1961; Boyd and Kogelnik, 1962) the general integral equation (A.2.23), by substitution of the variables

$$x = r\cos\varphi, \quad x' = r'\cos\varphi',$$
$$y = r\sin\varphi, \quad y' = r'\sin\varphi', \tag{A.3.27}$$

can be rewritten in the form

$$\frac{k}{2\pi iL}\int_0^a\int_0^{2\pi}\tilde{\psi}_S(r',\varphi')\,e^{-\frac{ik}{L}rr'\cos(\varphi-\varphi')}\,r'\,dr'\,d\varphi' = p\tilde{\psi}_S(r,\varphi), \tag{A.3.28}$$

where

$$\tilde{\psi}_S(r,\varphi) = \psi_S(r,\varphi)\,e^{\frac{ikr^2}{2L}}; \tag{A.3.29}$$

and a is the radius of the mirror.

Its solutions are

$$\tilde{\psi}_S(r,\varphi) = \tilde{v}_n(r)\,e^{in\varphi}. \tag{A.3.30}$$

The radial functions $\tilde{v}_n(r)$ can be determined from the integral equation

$$p_n\tilde{v}_n(r) = \frac{k}{Li^{n+1}}\int_0^a \tilde{v}_n(r')\,J_n\left(\frac{krr'}{L}\right)r'\,dr'$$

or in terms of the variables

$$u = \sqrt{\frac{k}{L}}\,r,\quad u' = \sqrt{\frac{k}{L}}\,r',\quad l = \sqrt{\frac{k}{L}}\,a,\quad p_n = \frac{\tilde{p}_n}{i^{n+1}}, \tag{A.3.31}$$

$$\tilde{p}_n\tilde{v}_n(r) = \int_0^l \tilde{v}_n(u')\,J_n(uu')\,u'\,du'. \tag{A.3.31a}$$

An integral equation of this type has been investigated by Fox and Li (1961) and Goubau and Schwering (1961). From its real solutions $v_{nm}(u)$ and the

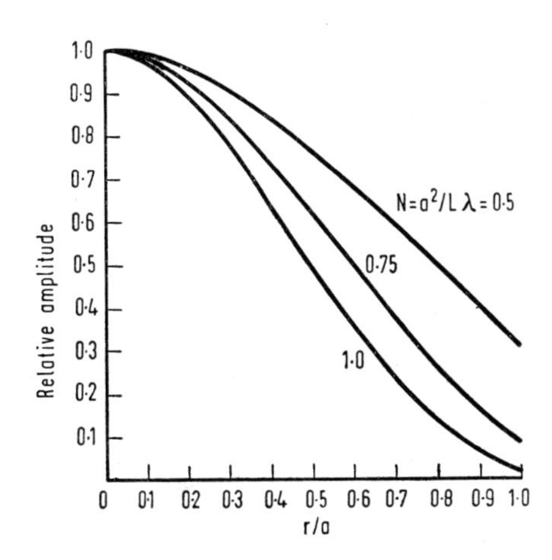

FIG. A.II.8. The distribution of the field amplitude in the TEM_{00} mode in a resonator with circular confocal mirrors.

eigenvalues $\tilde{p}_{nm} = |\tilde{p}_{nm}| (-1)^m$ from (A.2.24) there follows immediately the resonance frequency condition

$$k'L = \pi q + \frac{\pi}{2}(n + 1 + 2m), \tag{A.3.32}$$

which indicates, just as in the case of rectangular mirrors, that there is frequency degeneracy.

Figures A.II.8 and A.II.9 show the amplitude distribution as a function of the coordinates for the TEM_{00} and TEM_{10} modes respectively for different

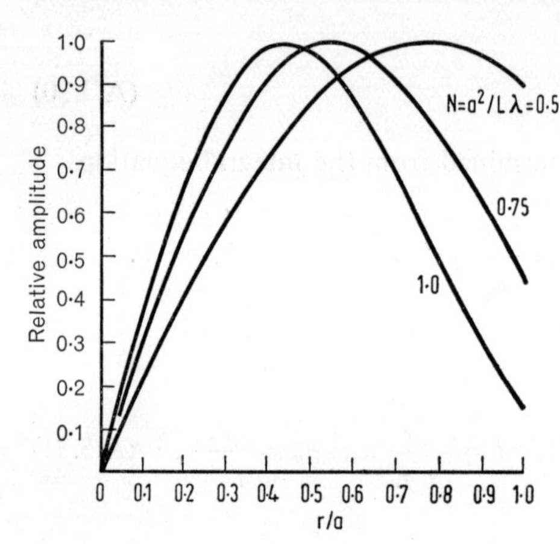

FIG. A.II.9. The distribution of the field amplitude in the TEM_{10} mode in a resonator with circular confocal mirrors.

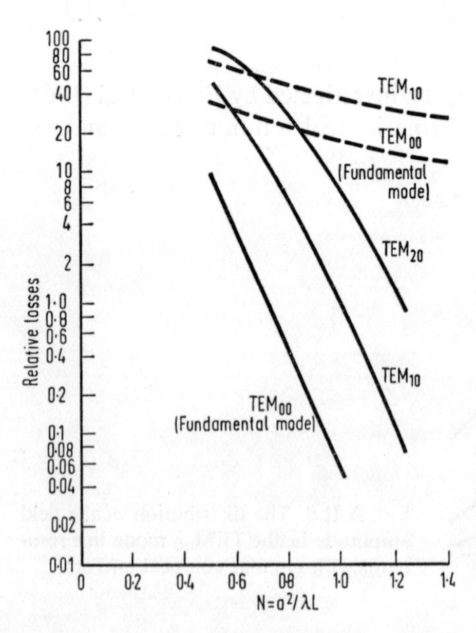

FIG. A.II.10. The diffraction losses of a resonator with confocal spherical (solid curves) and plane (dotted curves) mirrors of circular cross-section.

values of the parameter $N = l^2/2\pi = a^2/\lambda L$. The loss coefficient δ_d is shown in percent as a function of the same parameter in Fig. A.II.10. The current distributions at the mirrors are shown diagrammatically in Fig. A.II.11.

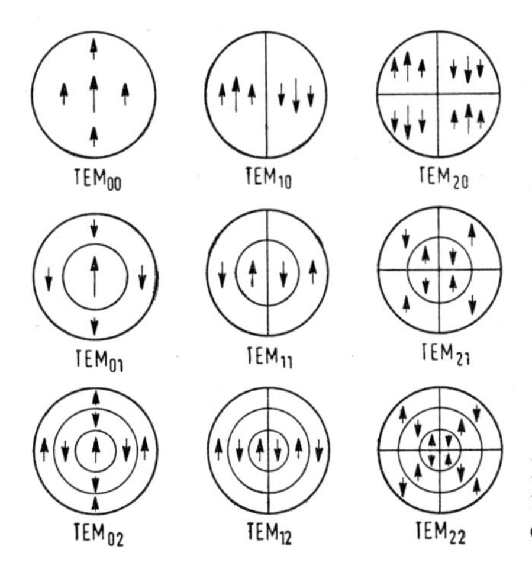

FIG. A.II.11. The current distributions in the simplest modes with mirrors of circular cross-section.

A.3.3. Non-confocal Resonators

Non-confocal systems may be treated by starting from the known structure of the field of a confocal resonator. We shall limit ourselves here to discussing resonators with rectangular mirrors, for which we have found above expressions for the field distribution across a wavefront. In a confocal

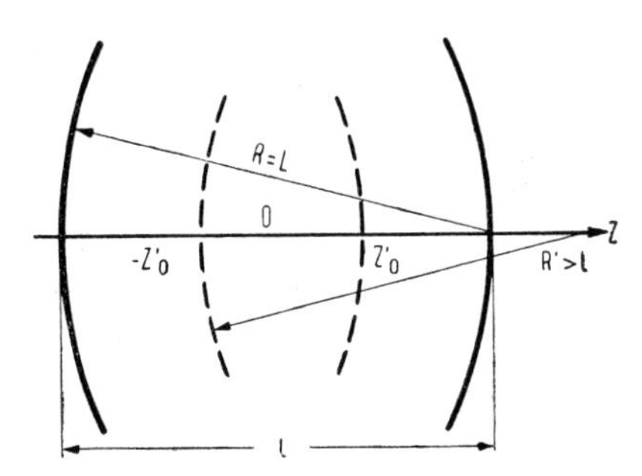

FIG. A.II.12. Wavefronts in a confocal resonator.

system with a distance L between the mirrors we can select two wavefronts crossing the axis of the resonator at the points $z = \pm z_0' = \pm L'/2$ (Fig. A.II. 12). From (A.3.13) the curvature of the wavefronts at these points is

$$R' = L\,\frac{1+\zeta_0^2}{2\zeta_0}, \tag{A.3.33}$$

where $\zeta_0 = \pm\, 2z_0'/L$.

If these wavefronts are replaced by reflecting surfaces, then from (A.2.24) and (A.3.9) the eigenfrequencies of the resonator obtained are given by

$$kL' = \frac{\pi}{2}\left[2q + (m+n+1)\left(1 - \frac{4}{\pi}\arctan\frac{L-L'}{L+L'}\right)\right]. \quad (A.3.34)$$

Since the second term in the brackets is now not necessarily an integer the degeneracy of which we spoke above is eliminated. There is still degeneracy, however, of the types of wave for which $m + n = \text{const}$. This degeneracy can be eliminated if as the reflecting surfaces we use mirrors with a double curvature R_1 and R_2, so that the distributions of the field with respect to each of the coordinates x and y correspond to confocal resonators with different distances L_1 and L_2 between the mirrors. In this case

$$kL' = \frac{\pi}{2}\left[2q + \left(m + \frac{1}{2}\right)\left(1 - \frac{4}{\pi}\arctan\frac{L_1 - L'}{L_1 + L'}\right) + \left(n + \frac{1}{2}\right)\left(1 - \frac{4}{\pi}\arctan\frac{L_2 - L'}{L_2 + L'}\right)\right], \quad (A.3.35)$$

so, generally speaking, the degeneracy is eliminated in all the indices.

Let there now be two mirrors with different radii of curvature $R^{(1)}$ and $R^{(2)}$ located at a distance d from each other. In order to find the confocal system in which these mirrors would coincide with the wavefronts we must solve the equations

$$R^{(1)} = L\frac{1+\zeta_1^2}{2\zeta_1}, \quad R^{(2)} = L\frac{1+\zeta_2^2}{2\zeta_2}, \quad \zeta_1 + \zeta_2 = \frac{2d}{L} \quad (A.3.36)$$

for the parameters ζ_1, ζ_2 and L. Eliminating ζ_1 and ζ_2 from these equations we have

$$2d = R^{(1)} + R^{(2)} \pm \sqrt{R^{(1)2} - L^2} \pm \sqrt{R^{(2)2} - L^2}. \quad (A.3.37)$$

Figure A.II.13 shows qualitatively the function $d(L)$ for $R^{(1)} < R^{(2)}$. It can be seen from the figure that only in the regions of the parameters

$$R^{(2)} < d < R^{(1)} + R^{(2)}, \quad 0 < d < R^{(1)} \quad (A.3.38)$$

does a confocal system exist in which the given mirrors coincide with the wavefronts. In the regions

$$R^{(1)} < d < R^{(2)}, \quad d > R^{(1)} + R^{(2)} \quad (A.3.39)$$

there is no such confocal system.

In these regions the resonator modes are rapidly attenuated because of radiation losses. This conclusion is confirmed by a purely ray treatment as well as the wave treatment (Vainshtein, 1963b). An arbitrary ray in a system with the parameters (A.3.39) is always unstable; when the number of reflections is high enough it may leave at as large a distance as one likes from the

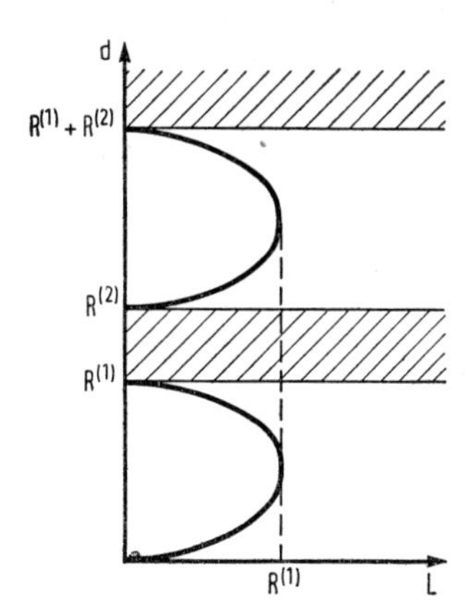

FIG. A.II.13. Dependence of d on L for $R^{(1)} < R^{(2)}$.

axis of the system (Boyd and Kogelnik, 1962). In a system with the parameters (A.3.38) it periodically oscillates in a space bounded by the surface of the caustic enclosed by the mirrors.

The radii ϱ_1 and ϱ_2 of the illuminated spot on the mirrors of a non-confocal system can be found starting with (A.3.14) by substituting in it the parameters of the equivalent confocal system.

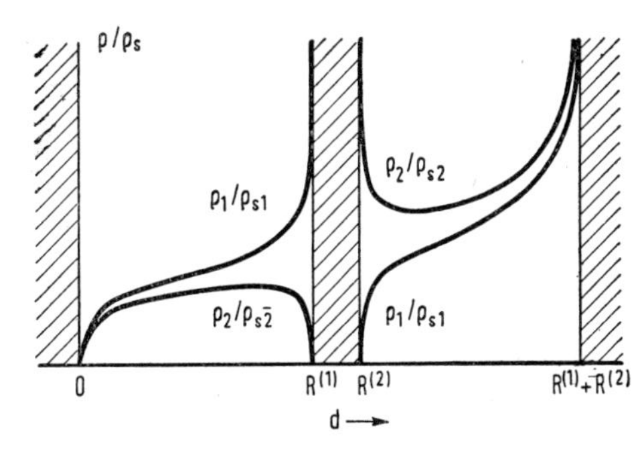

FIG. A.II.14. The size of the illuminated region on mirrors of a non-confocal resonator.

Two relations are obtained for determining ϱ_1 and ϱ_2:

$$\left(\frac{\varrho_1}{\varrho_2}\right)^2 = \frac{R^{(1)}}{R^{(2)}} \frac{R^{(2)} - d}{R^{(1)} - d}, \quad (\varrho_1\varrho_2)^2 = \left(\frac{\lambda}{\pi}\right)^2 \frac{R^{(1)}R^{(2)}d}{R^{(1)} + R^{(2)} + d},$$

whose roots are shown in Fig. A.II.14. In this figure $\varrho_{S1} = \sqrt{(R^{(1)}\lambda/\pi}$ and $\varrho_{S2} = \sqrt{R^{(2)}\lambda/\pi}$.

By a comparison with a confocal system the magnitude of the losses in a non-confocal system may readily be estimated. For simplicity let $R^{(1)} = R^{(2)} = R'$. Then the radius of the illuminated spot on the mirrors is

$$\varrho_s' = \left(\frac{\lambda d}{\pi}\right)^{1/2} \left[2\frac{d}{R'} - \left(\frac{d}{R'}\right)^2\right]^{-1/4}.$$

We notice that is smallest when $d/R' = 1$, in the non-confocal system. The parameter l^2 which determines the losses in a non-confocal system can be written in the form

$$l^2 = \frac{ka^2}{L} = 2\frac{a^2}{\varrho_s^2}, \tag{A.3.40}$$

where $\varrho_s = \sqrt{L\lambda/\pi}$ is the radius of the illuminated spot on the mirror.

By analogy with (A.3.40) we can define for a non-confocal system the parameter

$$l'^2 = 2\frac{a'^2}{\varrho_s'^2} = 2\pi\frac{a'^2}{\lambda d}\left[2\frac{d}{R'} - \left(\frac{d}{R'}\right)^2\right]^{1/2}, \tag{A.3.41}$$

where a' is the size of the mirror in the non-confocal resonator.

Then in view of the similarity noted above of the structures of the field at different cross-sections of the resonator we can assume that the losses in a non-confocal resonator will be approximately defined by the same relations as in a confocal one but with the substitution $l^2 \to l'^2$. It is easy to check that the parameter l'^2 reaches its maximum in the case of a confocal system: $d/R = 1$. From this we can conclude that a confocal resonator has smaller losses than all other resonators which can be formed by spherical mirrors of a given size.†

A.3.4. The Fabry–Perot Type of Resonator

Resonators with plane mirrors of finite dimensions can be treated by various methods (Fox and Li, 1961; Tang, 1962; Vainshtein, 1963a). One of them, which was developed by Fox and Li (1961), consists of direct solution by numerical methods of the integral equation (A.2.23) for $\varphi(x, y) = 0$.

By separating the variables in Cartesian coordinates (rectangular mirrors), or in cylindrical coordinates (circular mirrors) this integral equation is re-

† We can also prove the more general statement: with axially symmetrical apertures a system of confocal spherical mirrors has, in the simplest mode, the least radiation losses when compared with all other possible mirror configurations (Bondarenko and Talanov, 1964).

duced to the form

$$\frac{1}{\sqrt{2\pi}}\int_{-l}^{l} v(u')\, e^{\frac{i}{2}(u-u')^2}\, du' = \tilde{p}v(u) \tag{A.3.42}$$

or

$$\int_{0}^{l} J_n(uu')\, e^{\frac{i}{2}(u^2+u'^2)}\, v_n(u')\, u'\, du' = \tilde{p}_n v_n(u), \tag{A.3.43}$$

where $v(u)$ in (A.3.42) is one of the functions $\xi(x)$ or $\eta(y)$; $l = l_a$ or l_b and $\tilde{p} = \tilde{p}_x$ or $\tilde{p}_y$; $v_n(u)$ in (A.3.43) is the radial part of the field variation at the mirror: $\psi_S(u, \varphi) = v_n(u)\, e^{in\varphi}$; the rest of the notation is as before. In Fox and Li's paper (1961) the equations (A.3.42), (A.3.43) are solved by an iteration method on an electronic computer. Here the actual computation process actually simulates the transient regime of oscillations in a resonator as the result of multiple reflection from the mirrors of the original wave. A stationary distribution was achieved as the result of a large enough number of steps (of the order of 300). Examples of the stationary distributions of the amplitude and phase of the field $v_n(u) = |v_n(u)|\, e^{-i\psi(u)}$ at circular mirrors for the fundamental TEM_{00} and first asymmetrical TEM_{10} types of oscillation are shown in Figs. A.II.15 and 16 for different values of the parameter $N = a^2/\lambda L$.

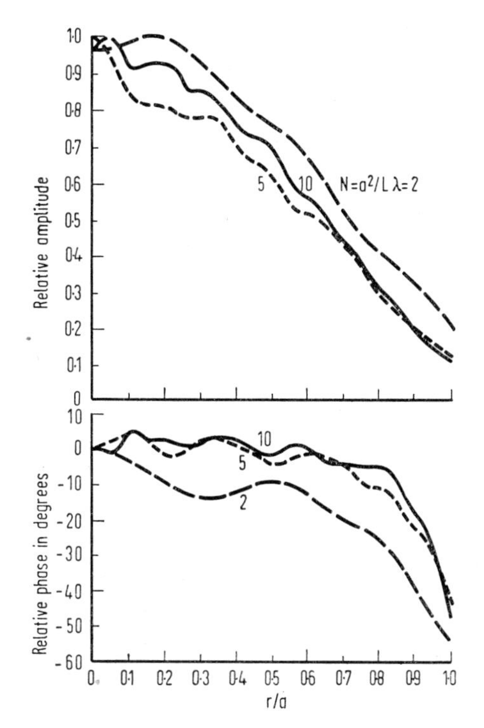

FIG. A.II.15. The distribution of the field amplitude and phase for the TEM_{00} mode at plane circular mirrors.

The considerable drop in amplitude of the field towards the edge of the mirror is characteristic, but the amplitude of the field at the edge for the same values of N is significantly higher than in a confocal system.

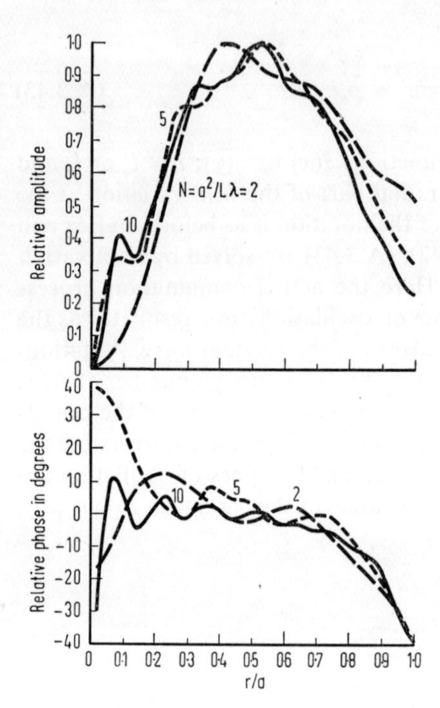

FIG. A.II.16. Field amplitude and phase distribution for the TEM_{10} mode at plane circular mirrors.

A feature of the phase distribution is the relatively small phase shift towards the edge of the mirror of not more than $\pi/4$, which for a beam diameter of tens of λ produces only slight distortion of the wavefront. The more complicated field distribution of the higher modes leads to an increase in the losses just as in the case of confocal resonators, and the dominant wave is TEM_{00}. This is illustrated by Fig. A.II.17, which shows the relative losses (per reflection) and the additional phase shift Δ defined by the condition

$$kL = \pi q + \Delta \tag{A.3.44}$$

for TEM_{00} and TEM_{10} modes in the case of circular mirrors. For identical modes the losses in a resonator with plane mirrors are several orders higher than in a confocal system; this can be seen from Fig. A.II.10 which shows the losses for the fundamental mode of a plane mirror resonator (dotted curve) as well as the loss factor of different types of oscillation of a confocal resonator for comparison.

Another method of calculation consists of plotting, starting from (A.3.42) or (A.3.43) the quadratic functional $\tilde{p}[v]$ or $\tilde{p}_n[v]$ which must have a stationary value in all physical cases (Tang, 1962). Substitution in this functional of approximate values of the field distribution (e.g. of the type $\cos (m\pi/2a)x$ (m odd) or $\sin (m\pi/2a)x$ (m even) in the case of rectangular mirrors) provides sufficiently accurate eigenvalues from equations (A.3.42) or (A.3.43).

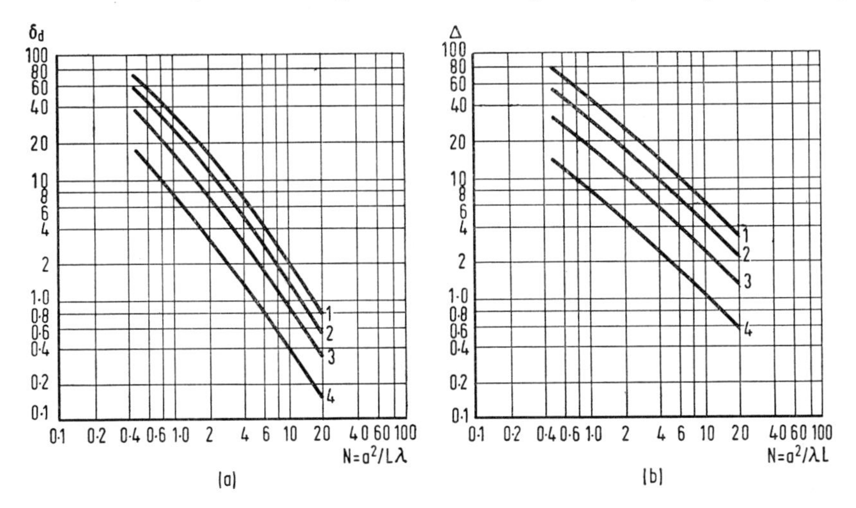

FIG. A.II.17. Relative losses in percent (a) and additional phase shift in degrees (b) for TEM_{00} and TEM_{10} modes: 1—circular mirrors, TEM_{10} mode; 2—strip mirrors, TEM_{10} mode; 3—circular mirrors, TEM_{00} mode; 4—strip mirrors, TEM_{00} mode.

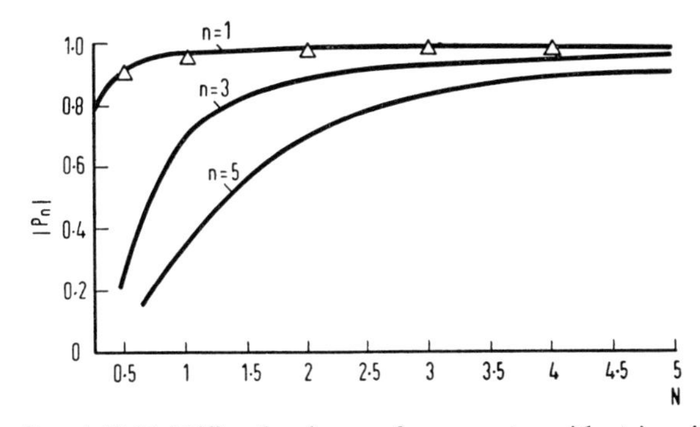

FIG. A.II.18. Diffraction losses of a resonator with strip mirrors calculated by the variational method. △—Fox and Li's (1961) results.

Figure A.II.18 shows for comparison the curves of the diffraction losses obtained by the method indicated and the results taken from Fox and Li's paper. It is easy to see that there is close agreement, the variational method requiring far less computation.

There is still one further method for investigating resonators with plane mirrors which has been developed in detail by Vainshtein (1963a). This method, which is based on the rigorous theory of the diffraction waves from the open end of a waveguide, allows us to obtain simple and self-evident relations for the field distributions and eigenfrequencies of resonators of this kind which agree closely with the numerical calculations of Fox and Li. In Vainshtein's theory the smallness of the diffraction losses from open-sided

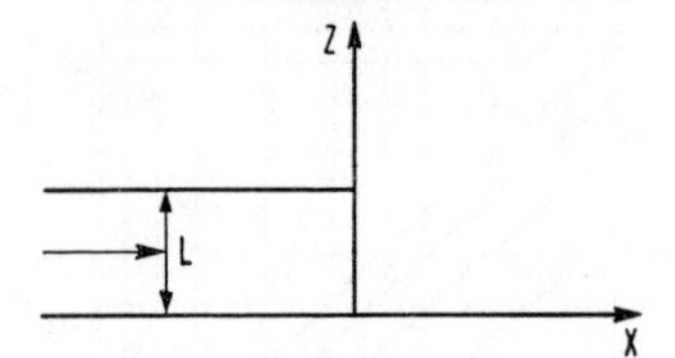

FIG. A.II.19. Plane waveguide with open end.

resonators can be explained by the fact that the side of the resonator is approached by a waveguide wave whose frequency only slightly exceeds the critical value, so that there is a slow change in the field across the reflecting plates of the resonator. It follows from the general theory (Vainshtein, 1953) that this kind of wave is hardly radiated at all and is not transformed into other types of wave, but is reflected with a reflection coefficient close to unity in absolute magnitude. As a result it is possible to take as the equivalent of an open resonator a section of a transmission line terminated by an impedance with a known reflection coefficient.

Let us first discuss some elementary aspects of the theory of wave diffraction at the open end of a broad plane waveguide (Fig. A.II.19). Let the wave number $k = 2\pi/\lambda$ be related to the distance L between the sides of the waveguide by the relation

$$kL = \pi(q + 2p), \tag{A.3.45}$$

where q is a large integer and $|p| < \frac{1}{2}$.

Let the open end of a waveguide be approached by a wave of a type E_{0q} or H_{0q} whose field depends upon the z coordinate and whose frequency, by virtue of (A.3.45), is close to its critical frequency (when $p = 0$ these frequencies are the same).

It follows from the asymptotic formulae of the rigorous theory obtained on the assumption of a slow change of the field along x† that in this case the distribution of the current density on the wall of the waveguide can be

† Changing to the asymptotic formulae of the rigorous theory of wave diffraction at the open end of a waveguide, as can easily be shown, is fully equivalent to changing from the wave equation to the transverse diffusion equation (A.2.15). This also determines the limits of applicability of the relations obtained below; they agree with the condition (A.2.21a) for the validity of the diffusion approximation.

written in the form

$$f(x) = A(e^{iW_0 x} + \sum_j R_{0,j} e^{-iW_j x}), \tag{A.3.46}$$

where A is the amplitude of the current in the incident wave; $R_{0,j}$ is the transformation coefficient (for $j \neq 0$) of the incident wave into a wave with the suffix $j(E_{0q-2j}$ or $H_{0q-2j})$, and $R_{0,0}$ is the reflection coefficient of the incident wave with respect to the current; W_j are the wave numbers of the waves H_{0q-2j} or E_{0q-2j} in a plane waveguide and are approximately $(W_j \ll k)$

$$W_j = \sqrt{\frac{k}{L}}\, s_j, \quad s_j = \sqrt{4\pi(j+p)}, \quad j = 0, \pm 1, \pm 2 \ldots \tag{A.3.47}$$

The coefficients R_{0j} can be expressed in terms of special functions $U(s, p)$ which have studied in detail and tabulated by Vainshtein (1953).

Expressions similar to (A.3.46) can be obtained in the case of an incident wave with a suffix $j \neq 0$.

Figure A.II.20 shows the absolute values of the coefficients R_{00}, R_{01}, R_{11} and R_{10} as a function of p for $|p| < \frac{1}{2}$. For small p a wave with a suffix $j = 0$ is strongly reflected, being all but transformed into waves with a suffix $j \neq 0$. The latter, on the other hand, are weakly reflected from the end $(|R_{1,1}| < 0{\cdot}1)$. Therefore modes with small radiation losses can exist only

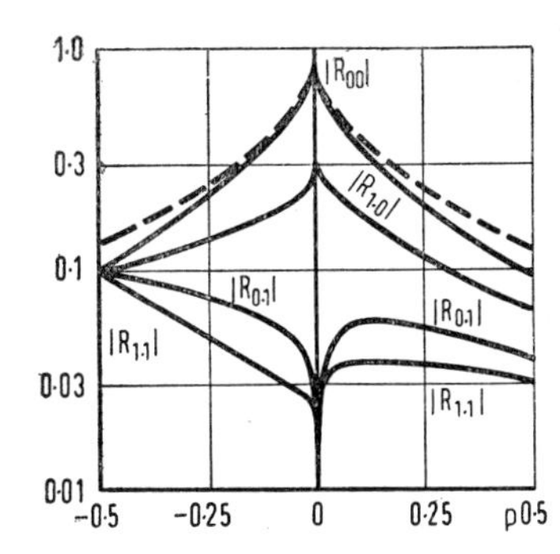

FIG. A.II.20. Absolute magnitudes of the current reflection and transformation coefficients.

when $j = 0$ and their characteristics will be determined by the reflection coefficient

$$R_{00} = -|R_{00}|\, e^{i\theta_{00}}, \tag{A.3.48}$$

whose phase θ_{00} is shown in Fig. A.II.21. It is an odd function of p, whilst $|R_{00}|$ is an even function of this argument.

For small $|p|$ the coefficient (A.3.48) can be written in the form

$$R_{00} = -e^{i\beta(1+i)s_0}, \tag{A.3.49}$$

where

$$\beta = 0{\cdot}824; \quad s_0 = \sqrt{4\pi p}.$$

It can be seen from Figs. A.II.20 and A.II.21, where the functions $|R_{00}|$ and θ_{00} calculated from the approximate formula (A.3.49) are shown by the dotted line, that it provides graphical accuracy for $|p| < 0{\cdot}05$ and is qualitatively applicable right up to $|p| \approx 0{\cdot}5$. Therefore the following calculations will be based on (A.3.49).

The asymptotic laws, despite the different behaviour of the current near a sharp boundary, are the same for waves of both polarizations E_{0q} and H_{0q}.

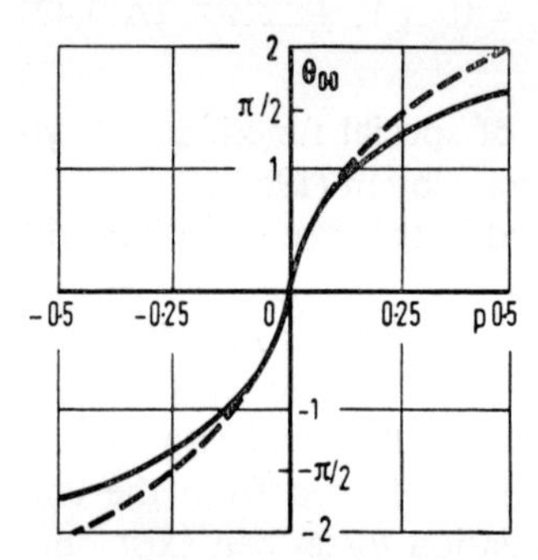

FIG. A.II.21. The phase of the reflection coefficient.

In addition, there is no current flowing to the outside of the walls in the asymptotic solution (A.3.46), so it is valid for waveguides with flanges, with walls of finite thickness, etc.

Let us apply the relations obtained to the case of actual resonators. Let us first discuss a resonator formed by two strip mirrors $2a$ wide located at a distance L from each other. We can obtain the eigenfrequencies of this kind of resonator by using the equivalent arrangement described above as a basis:

$$R_{00}^2\, e^{4iW_{0}a} = e^{2im\pi} \quad (m \text{ is an integer}). \tag{A.3.50}$$

Substituting in this the value of R_{00} from (A.3.49) and solving the relation obtained for s_0 we find

$$s_0 = \sqrt{4\pi p} = \frac{m\pi}{\beta(1+i)+M}, \quad m = 1, 2, 3, \ldots, \tag{A.3.51}$$

where the quantity $M = 2\sqrt{ka^2/L}$ is related to the parameter $N = a^2/\lambda L$ introduced by Fox and Li (1961) by $M = 2\sqrt{2\pi N}$. The unknown correction p that in accordance with (A.3.45) determines the eigenfrequencies of the resonator is

$$p = \frac{\pi m^2/4}{[\beta(1+i)+M]^2} \tag{A.3.52}$$

or, after separating the real and imaginary parts,

$$p = p' - ip'', \quad p' = \frac{\pi m^2}{4} \frac{M(M+2\beta)}{[(M+\beta)^2+\beta^2]^2},$$

$$p'' = \frac{\pi m^2}{2} \frac{\beta(M+\beta)}{[(M+\beta)^2+\beta^2]^2}. \tag{A.3.53}$$

The expressions (A.3.52) and (A.3.45) can be combined to give the eigenvalues of the resonator k_{mq}:

$$k_{mq}^2 = \left[\frac{m\pi}{2a\left(1+\beta\frac{1+i}{M}\right)}\right]^2 + \left(\frac{\pi q}{L}\right)^2. \tag{A.3.54}$$

The current distribution is the superposition of incident and reflected waves with a complex wave number (along x)

$$W_0 = \sqrt{\frac{k}{L}}\, s_0.$$

It is clear from symmetry that the resultant current distribution (longitudinal, i.e. along x, for E_{m0q} modes, or transverse for H_{m0q} modes) will be described by the functions

$$f(x) = \cos\frac{m\pi x}{2a\left(1+\beta\frac{1+i}{M}\right)}, \quad m = 1, 3, 5 \ldots,$$

$$f(x) = \sin\frac{m\pi x}{2a\left(1+\beta\frac{1+i}{M}\right)}, \quad m = 2, 4, 6 \ldots \tag{A.3.55}$$

For the modes where $m = 1$ and $m = 2$ these distributions are shown in Figs. A.II.22 and 23. It can be seen from a comparison of the curves in these

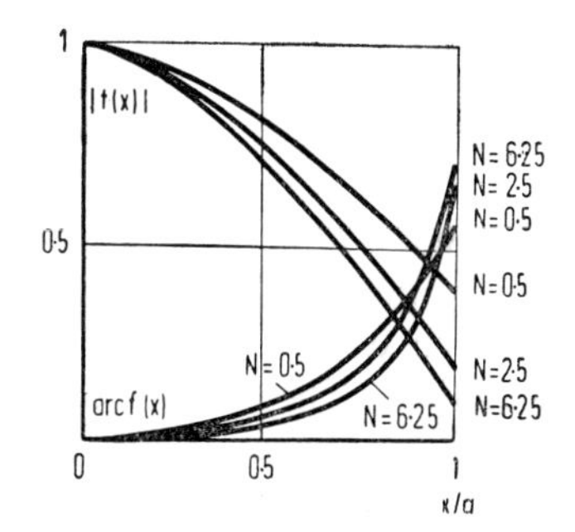

FIG. A.II.22. Function (A.3.55) for $m = 1$.

figures with the corresponding curves of Fox and Li (1961) that they differ only in that the latter show a slight wave variation. This difference can be explained by the fact that in finding the current distribution (A.3.55) waves with numbers $j \neq 0$ which appear, although with small amplitudes, when there is diffraction of a wave $j = 0$ at the edge of the resonator were not taken into consideration.

The corrections p' and p'' obtained above have a simple physical meaning: the quantity $\Delta = 2\pi p'$ is the additional (to $kL = \pi q$) phase variation in the

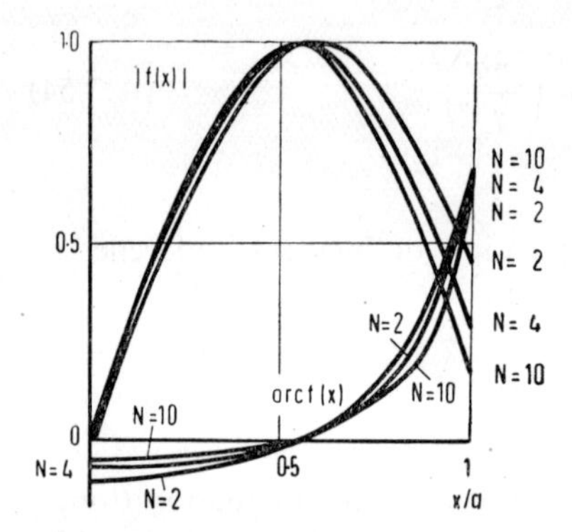

FIG. A.II.23. Function (A.3.55) for $m = 2$.

time $\tau = 2a/v$ taken for a wave to pass through the resonator, and the parameter

$$\delta_d = 1 - e^{-4\pi p''} \approx 4\pi p'', \tag{A.3.56}$$

which has been used earlier, gives the relative energy loss from the mode in the same time interval.

In the case under discussion

$$\Delta = \frac{\pi^2 m^2}{2} \frac{M(M + 2\beta)}{[(M + \beta)^2 + \beta^2]^2}, \tag{A.3.57}$$

$$\delta_d = 2\pi^2 m^2 \frac{\beta(M + \beta)}{[(M+\beta)^2 + \beta^2]^2}. \tag{A.3.58}$$

A comparison of (A.3.57) and (A.3.58) for $m = 1$ and $m = 2$ with the curves of Fox and Li (1961) shows that they lead to identical results within the accuracy of the graph. But (A.3.57) and (A.3.58) also make it possible to do a very simple calculation even for large values of m, which using Fox and Li's method would involve extremely laborious calculations. The absence of a fine structure in the current distribution has no significant effect on the calculation of the eigenfrequencies of the resonator.

Let us now discuss an open resonator formed by parallel rectangular mirrors:

$$|x| < a, \quad |y| < b, \quad z = \pm\frac{L}{2}.$$

It is clear from the general integral equation (A.2.23) that here, just as in the case of rectangular confocal mirrors, the variables x and y can be separated:

$$\psi_S(x, y) = f_a(x) f_b(y), \tag{A.3.59}$$

the problem being equivalent to discussing the corresponding strip resonator for each of the functions $f_a(x)$ and $f_b(y)$. Finally, we have the relations:

$$kL = \pi[q + 2(p_a + p_b)], \tag{A.3.60}$$

$$p_a = \frac{\pi m^2}{4(M_a + \beta + i\beta)^2}, \quad p_b = \frac{\pi n^2}{4(M_b + \beta + i\beta)^2}, \tag{A.3.61}$$

$$M_a = 2\sqrt{\frac{k}{L}}\, a, \quad M_b = 2\sqrt{\frac{k}{L}}\, b, \tag{A.3.62}$$

$$k_{mnq}^2 = \left[\frac{\pi m}{2a\left(1 + \beta\dfrac{1+i}{M_a}\right)}\right]^2 + \left[\frac{\pi n}{2b\left(1 + \beta\dfrac{1+i}{M_b}\right)}\right]^2 + \left(\frac{\pi q}{L}\right)^2. \tag{A.3.63}$$

The functions f_a and f_b that define the current distribution are:

$$f_a(x) = \cos\frac{m\pi x}{2a\left(1 + \beta\dfrac{1+i}{M_a}\right)}, \quad f_b(y) = \cos\frac{n\pi y}{2b\left(1 + \beta\dfrac{1+i}{M_b}\right)},$$

$$m, n = 1, 3, \ldots, \tag{A.3.64}$$

$$f_a(x) = \sin\frac{m\pi x}{2a\left(1 + \beta\dfrac{1+i}{M_a}\right)}, \quad f_b(y) = \sin\frac{n\pi y}{2b\left(1 + \beta\dfrac{1+i}{M_b}\right)},$$

$$m, n = 2, 4, \ldots$$

The general picture of the current distribution at the mirrors for different modes is just the same here as in a resonator with spherical mirrors (Fig. A.II.7).

An open resonator formed by circular mirrors can be investigated in a similar way if we use the reflection coefficient R_{00} (A.3.49), derived for a plane wave, for approximate calculation of the reflection of a cylindrical wave from the edge of a mirror with $r = a$. As a result the following relations are obtained:

$$p = \frac{\nu_{mn}^2}{\pi(M + \beta + i\beta)^2} \quad (n = 1, 2, \ldots, \quad m = 0, 1, 2, \ldots), \tag{A.3.65}$$

$$\Delta = 2\pi p' = 2\nu_{mn}^2 \frac{M(M + 2\beta)}{[(M + \beta)^2 + \beta^2]^2},$$

$$\delta_d = 4\pi p'' = 8\nu_{mn}^2 \frac{\beta(M + \beta)}{[(M + \beta)^2 + \beta^2]^2}, \tag{A.3.66}$$

$$k_{mnq}^2 = \left[\frac{\nu_{mn}}{a\left(1 + \beta\dfrac{1 + i}{M}\right)}\right]^2 + \left(\frac{\pi q}{L}\right)^2, \quad M = 2\sqrt{\frac{k}{L}}\, a. \tag{A.3.67}$$

The current distribution in either of the two independent Cartesian components can be described by the function

$$\psi_{Smn}(r, \varphi) = J_m\left(\frac{\nu_{mn} r/a}{1 + \beta\dfrac{1 + i}{M}}\right) \begin{matrix}\cos \\ \sin\end{matrix} (m\varphi), \quad \begin{matrix} m = 0, 1, 2, \ldots, \\ m = 1, 2, \ldots \end{matrix} \tag{A.3.68}$$

Here ν_{mn} are the roots of the Bessel function $J_m(x)$. The modes with $m \neq 0$ have fourfold degeneracy (twofold rotational degeneracy corresponding to selection in (A.3.68) of one of the functions $\sin m\varphi$ or $\cos m\varphi$, and two possible polarizations).

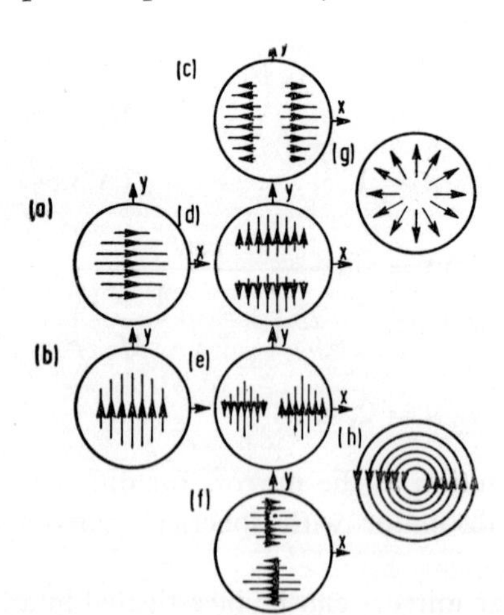

FIG. A.II.24. The current distributions at circular mirrors for TEM_{01q} and TEM_{11q} modes.

The first column of Fig. A.II.24 shows diagrammatically, allowing for rotational degeneracy, the current distribution on a mirror for TEM_{01q} modes polarized along the x (Fig. A.II.24a) or y (Fig. A.II.24b) axis; the second shows that of TEM_{11q} modes (allowing for rotational and polarization degeneracy). By superposition of modes polarized in different planes

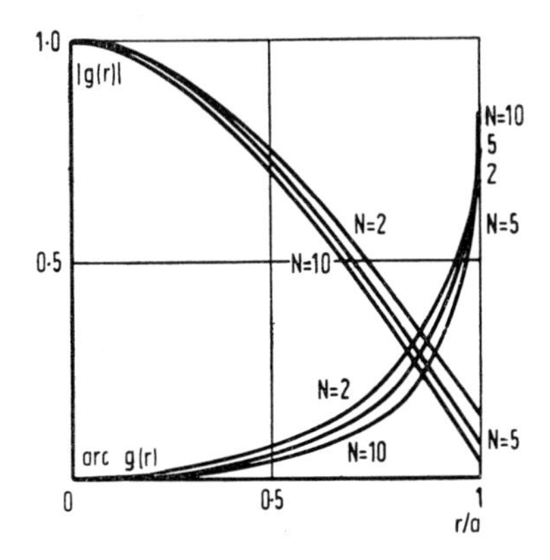

FIG. A.II.25. Function (A.3.68) for $m = 0$, $n = 1$.

we can obtain modes with radial or azimuthal currents having symmetry of rotation (Fig. A.II.24, column three). Figures A.II.25 and A.II.26 show the dependence of the modulus and phase of the function (A.3.68) for $m = 0$, $n = 1$ ($\nu_{01} = 2{\cdot}405$) and $m = n = 1$ ($\nu_{11} = 3{\cdot}832$). The graphs shown in Fig. A.II.17 for δ_a and Δ, with $m = 0$ and 1, $n = 1$, are practically equivalent to (A.3.66), whilst the smooth curves of the current distribution over the mirror (Figs. A.II.25 and A.II.26) repeat on the whole the general shape of the wavy curves shown in Figs. A.II.15 and A.II.16.

Just as in the case of the confocal resonators described above, the spectrum of the eigenfrequencies of resonators with plane mirrors is less dense than that of closed resonators. The reduction of the mode density can be seen clearly if we represent the complex frequencies defined by (A.3.52), (A.3.61) and (A.3.65) in the form of points on the plane of the complex variable ω (Fig. A.II.27). The points are located on straight half-lines originating at the points $\omega_q = \pi v q/L$ and making with the abscissae the angle

$$\psi = 2 \arctan \frac{\beta}{M + \beta}, \qquad \text{(A.3.69)}$$

where the value of M must be taken when $\omega = \omega_q$.

As the suffixes m and n increase the radiation loss of the modes increases.† The coordinates of the points in Fig. A.II.27 then rise far more rapidly than

† In fact the attenuation increases more rapidly than follows from Fig. A.II.27, since (A.3.49) for finite $|p|$ reduces the radiation losses (see Fig. A.II.20).

the difference of the abscissae of the adjacent points $\omega_{\mu q}$, $\omega_{\mu+1q}$ ($\mu = m, n$), which leads, starting at a certain $\omega_{\mu q}$ determined by the condition

$$\omega'_{\mu+1q} - \omega'_{\mu q} < \frac{\omega_{\mu q} + \omega_{\mu+1q}}{2},$$

to intersection of the resonance curves. As a result there remain only a few resonant modes around each frequency ω_{0q}. If the number of these modes is reduced to one ($m = n = 1$ for rectangular mirrors and $m = 0$, $n = 1$ for

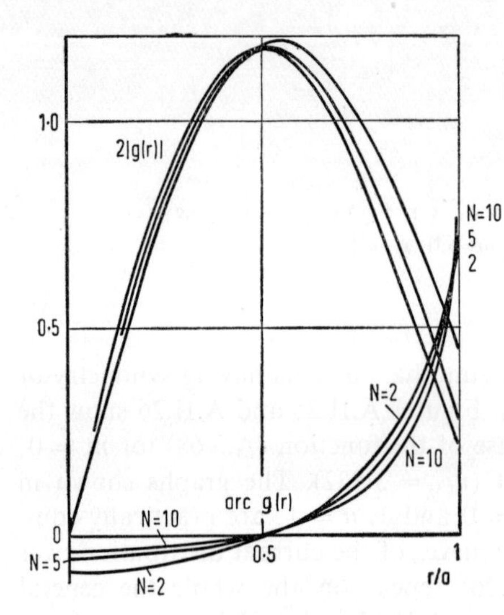

FIG. A.II.26. Function (A.3.68) for $m - 1$, $n = 1$.

circular mirrors), then a practically equidistant spectrum of the eigen-frequencies is obtained which is the same as the ideal case of a one-dimensional resonant system.

Although the diffraction losses, as has already been indicated, are much higher in the case of plane mirrors than in a confocal resonator, with the resonator dimensions used in practice they are nevertheless significantly less than the losses in the actual reflecting surfaces forming the resonator. For example, with $L = 100$ cm, $a = 1$ cm, $\lambda = 10^{-4}$ cm the quantity δ_d for TEM_{01} oscillations in a resonator with circular mirrors is 3×10^{-4}, in accordance with (A.3.66), whilst for the better dielectric multi-layer coatings $\delta_r \simeq 0{\cdot}01$.

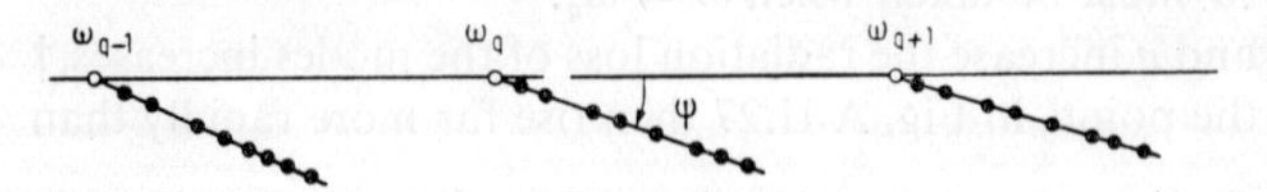

FIG. A.II.27. The spectrum of open resonators with plane mirrors.

Therefore the relative width of the resonance curve will be determined by the quantity δ_r:

$$\left(\frac{\delta\omega}{\omega}\right)_r = \frac{\delta_r}{kL}.$$

We compare this quantity with the frequency interval between adjacent lower order modes TEM_{01q} ($\nu_{01} = 2{\cdot}405$) and TEM_{11q} ($\nu_{11} = 3{\cdot}832$):

$$\left(\frac{\Delta\omega}{\omega}\right) = \frac{2\Delta p'}{kL}.$$

For these modes we have from (A.3.66) $2\Delta p' \simeq 2\times 10^{-3}$, which is several times less than δ_r. Therefore the transverse modes TEM_{01q} and TEM_{11q} are not resolved. However, if the resonator is filled with an active medium as in a laser, then as the oscillation threshold is approached adjacent modes may be resolved. In this case the difference of their diffraction Q-factors has a significant effect on the self-excitation conditions of the laser in different modes and on the ratio of the amplitudes of these modes when oscillating.

To conclude let us briefly examine the requirements which should be imposed on the accuracy of manufacture and adjustment of optical resonators. The permissible deviation of the surface of resonator mirrors from a flat surface or a sphere can be estimated by proceeding from the following considerations. A wave successively reflected from the resonator walls is attenuated e times in $n = -(\ln T)^{-1}$ reflections, where T is the transmission coefficient per reflection. If at some part the optical path differs from the mean by a quantity $\Delta\varphi_0$, then this leads after n reflections to an additional phase difference of

$$\Delta\varphi = n\,\Delta\varphi_0.$$

This phase difference is insignificant when $\Delta\varphi < \pi/2$, which at once leads to the requirement

$$\Delta\varphi_0 < \frac{\pi}{2n} = -\frac{\pi}{2}\ln T$$

or for T close to unity

$$\Delta\varphi_0 < \frac{\pi}{2}(1 - T). \qquad \text{(A.3.70)}$$

From (A.3.70) in particular follows the requirement for the permissible deviation ΔL of the surface of the mirrors from a plane or spherical form

$$\Delta L < \frac{\lambda}{4}(1 - T). \qquad \text{(A.3.70a)}$$

The smaller the loss coefficient $\delta = 1 - T$ the more rigid is this requirement.

When conditions (A.3.70) and (A.3.70a) are not satisfied there is strong coupling of the modes appropriate to an ideal resonator, transformation of lower order modes into higher ones and as a consequence to an increase in the radiation losses. In real resonators because of breaking of the conditions (A.3.70) and (A.3.70a) the structure of the various modes and their frequencies may prove to be essentially different from those in resonators of correct geometrical shape.

We have already spoken about the rigorous tolerances on non-parallelism of the mirrors when discussing the theory of a Fabry–Perot resonator. A system with spherical mirrors has undoubted advantages over a plane resonator in respect to adjustment. As slight tilt of spherical mirrors which does not take the beam beyond their edges has hardly any effect on the magnitude of their diffraction losses. The far lower sensitivity to accuracy of adjustment of a resonator with spherical mirrors (as opposed to a plane parallel resonator) can be explained by the strong concentration of the field near the axis of the resonator.

APPENDIX III

The Spectra of Paramagnetic Crystals

IN THIS appendix we shall discuss the theory of the spectra of paramagnetic crystals, which are extensively used as active substances in masers and lasers.

Paramagnetic crystals are dilute solid solutions of the ions of chemical elements with incomplete inner electron shells (generally elements of the iron or rare earth group) in the crystal lattice of various oxides, salts and other compounds. Strictly speaking, the paramagnetic ions are the active substance of a quantum device, and its operation is governed by their energy levels. The important properties of the active substance, however, are largely determined by the crystal lattice. The positions of the energy levels of free paramagnetic ions are considerably changed when they are put into a crystal. The crystal field partially or wholly splits the degenerate energy levels of the free ions and also alters their position. In an external magnetic field it is possible to obtain complete splitting of the levels, and by altering the strength of the field and the orientation of the crystal in the field the positions of the levels can be changed. An appropriate set of levels is selected for each quantum device, from among the many levels of a paramagnetic crystal.

At present theoretical calculations of the energy levels can be carried out for comparatively simple substances, e.g. for crystals containing paramagnetic ions; the calculations use the perturbation theory method (Landau and Lifshitz, 1963). The complete Hamiltonian of the paramagnetic ion in the crystal takes account of the interaction of all the particles in the ion with each other, and also, with the crystal field and the external magnetic field. Then all the terms of this Hamiltonian are taken into consideration in decreasing order of magnitude thus building up a detailed picture of the energy levels and the electron states corresponding to them.

A.4. The Hamiltonian of a Paramagnetic Ion in a Crystal

The complete Hamiltonian of a paramagnetic ion in a crystal in an external magnetic field can be approximately written in the form

$$\hat{H} = \hat{H}_0 + \hat{H}_{LS} + \hat{H}_{cr} + \hat{H}_{SS} + \hat{H}_H + \hat{H}_{nucl}. \tag{A.4.1}$$

Here

$$\hat{H}_0 \equiv \sum_{k=1}^{N} \left(\frac{\hat{p}_k^2}{2m} - \frac{Ze^2}{r_k} \right) + \frac{1}{2} \sum_{k=1}^{N} \sum_{j \neq k}^{N} \frac{e^2}{r_{kj}} \tag{A.4.2}$$

describes (in the non-relativistic approximation) the kinetic energy of N electrons in the ion, the Coulomb interaction of the nucleus with the electrons and the interaction of the electrons with each other;

$$\hat{H}_{LS} \equiv \lambda(\hat{\boldsymbol{L}} \cdot \hat{\boldsymbol{S}}) \tag{A.4.3}$$

describes the spin–orbit interaction electrons in the paramagnetic ion, i.e. the magnetic interaction of the total orbital ($\boldsymbol{L}$) and spin ($\boldsymbol{S}$) moments of the ion (Landau and Lifshitz, 1963).

$\hat{H}_{cr}$ is the Hamiltonian describing the interaction of the paramagnetic ion with the crystal surroundings; the actual form of $\hat{H}_{cr}$ will be discussed below;

$$\hat{H}_{SS} \equiv \varrho(\hat{\boldsymbol{L}} \cdot \hat{\boldsymbol{S}})^2 \tag{A.4.4}$$

describes the spin–spin interaction of the electrons in the ion (Landau and Lifshitz, 1963);

$$\hat{H}_H \equiv \beta((\hat{\boldsymbol{L}} + 2\hat{\boldsymbol{S}}) \cdot \boldsymbol{H}_0) \tag{A.4.5}$$

is the Zeeman term that allows for the interaction of the electron magnetic moments with the external magnetic field $\boldsymbol{H}_0$;

$\hat{H}_{nucl}$ describes the contribution of the nucleus† to the energy of the ion (the dipole–dipole interaction of the nuclear magnetic moment with the electron moments, the electric interaction of the nuclear quadrupole moment with the electrons and the energy of the nuclear magnetic moment in the external magnetic field).

The orders of magnitude of the energy corresponding to the terms of the Hamiltonian (A.4.1) can be estimated from experimental data and are:‡

$$H_0 \sim 10^5\ \text{cm}^{-1}\ \text{(including the term } \tfrac{1}{2}\sum_{k}^{N}\sum_{j\neq k}^{N} e^2/r_{kj} \sim 10^4\ \text{cm}^{-1}\text{)},$$

$$H_{LS} \sim \begin{cases} 10^2\ \text{cm}^{-1}\ \text{for the iron group,} \\ 10^3\ \text{cm}^{-1}\ \text{for rare earths and actinides,} \end{cases}$$

† Part of this contribution is already allowed for by the term $-Ze^2/r_k$ in (A.4.2).

‡ Here and later we are using the conventional units of energy (in cm^{-1}) used in spectroscopy; the ordinary energy values can be obtained from them by multiplying by $\hbar c$, and the frequency by multiplying by $c = 3 \times 10^{10}\ \text{cm sec}^{-1}$.

$H_{SS} \sim 1 \text{ cm}^{-1}$,
H_H depends on the magnitude of the applied field (it generally reaches $0{\cdot}1$–1 cm^{-1}),
$H_{\text{nucl}} \sim 10^{-1}$–$10^{-3} \text{ cm}^{-1}$ (in future we shall neglect this term).

The interaction energy of different paramagnetic ions with the crystal field H_{cr} may be more or less than the energy of the spin–orbit interaction. Correspondingly the perturbing action of the operator $\hat{H}_{\text{cr}}$ must be taken into consideration before or after allowing for the action of the operator $\hat{H}_{LS}$. In this connexion we can distinguish three cases:

1. *A strong crystal field* with an energy H_{cr} of the order of 10^4 cm^{-1}, i.e. of the order of the energy of the Coulomb interaction of the electrons. This case holds for paramagnetic centres strongly bound to the surrounding ions, when the chemical bond is ionic and not covalent in nature. It occurs typically for paramagnetic ions of the $4d$ and $5d$ groups and occasionally for ions of the iron group ($3d$)† (in certain crystals such as the cyanides). The perturbing action of a strong crystal field must be taken into consideration simultaneously with the Coulomb interaction $\left(\frac{1}{2}\sum_{k}^{N}\sum_{k\neq j}^{N} e^2/r_{kj}\right)$ and sometimes even before it.

2. *An intermediate crystal field* with an interaction energy greater than H_{LS} but less than the Coulomb interaction of the electrons, so the perturbing action of $\hat{H}_{\text{cr}}$ must be taken into consideration immediately after $\hat{H}_0$ and before $\hat{H}_{LS}$. This case corresponds to an ionic type of bonding and is particularly characteristic of elements of the iron group (incomplete $3d$ shell) in many crystals.

3. *A weak crystal field with an energy $\hat{H}_{\text{cr}} \ll H_{LS}$*, so that the action of the crystal field on the energy levels must be examined after allowing for the spin–orbital interaction, i.e. as a weak perturbation on the free ions. This case occurs for ions of rare-earth elements with an incomplete $4f$ shell which is well protected from the influence of the crystal field by the outer complete shells $5s^2$ and $5p^6$.

The case of a strong field is rarely met in a pure form so we shall take no further interest in it. We shall deal with the case of an intermediate field, and also discuss the case of a weak field.

Before we discuss the effect of the crystal field on the paramagnetic ion we shall briefly list the basic conclusions of the theory of the spectra of free atoms; in the next section we give some information from atomic spectroscopy which we shall need.

† See § A.5.

A.5. The States of a Free Many-electron Atom

Without taking electron spin and the magnetic interaction of the orbital moments into consideration, a free many-electron atom can be approximately described by the Hamiltonian:

$$\hat{H}_0 = \sum_{k=1}^{N} \left(\frac{\hat{p}_k^2}{2m} - \frac{Ze^2}{r_k} \right) + \frac{1}{2} \sum_{k=1}^{N} \sum_{j \neq k}^{N} \frac{e^2}{r_{kj}} \cong \sum_{k=1}^{N} \hat{H}_k, \tag{A.5.1}$$

where

$$\hat{H}_k = \frac{\hat{p}_k^2}{2m} + V_{CF}(r_k)$$

and the potential $V_{CF}(r_k)$ takes account of the interaction of the kth electron with the nucleus and the rest of the electrons in the central field (CF) approximation (Landau and Lifshitz, 1963).

The eigenfunctions of the single-electron problem with the Hamiltonian $\hat{H}_k$ are, in polar coordinates, of the form (Landau and Lifshitz, 1963; Condon and Shortley, 1935):

$$\psi_{nlm_l}(r_k, \theta_k, \varphi_k) = R_{nl}(r_k) Y_l^{m_l}(\theta_k, \varphi_k), \tag{A.5.2}$$

where $Y_l^{m_l}(\theta, \varphi)$ are spherical harmonics:

$$Y_l^{m_l}(\theta, \varphi) = (-1)^l \left[\frac{1}{4\pi} \frac{(2l+1)(l-|m_l|)!}{(l+|m_l|)!} \right]^{1/2} P_l^{m_l}(\cos\theta)\, e^{im_l\varphi}, \tag{A.5.3}$$

and

$$P_l^{m_l}(u) = \frac{(1-u^2)^{m_l/2}}{2^l \cdot l!} \frac{d^{l+m_l}}{du^{l+m_l}} (u^2 - l)^l.$$

To describe the symmetry properties which the electron wave function acquires when the *spin* s_k is introduced we write the wave function as

$$\psi_{nlm_lm_s}(r_k, \theta_k, \varphi_k, s_k) = R_{nl}(r_k) Y_l^{m_l}(\theta_k, \varphi_k) \cdot \sigma_{m_s}(s_k), \tag{A.5.4}$$

so that the state of the electron is characterized by four quantum numbers: the *principal quantum number* $n = 1, 2, \ldots$; the *orbital quantum number* $l = 0, 1, 2, \ldots, (n-1)$;† the *orbital magnetic quantum number* $m_l = 0, \pm 1, \pm 2, \ldots, \pm 1$ and the *spin magnetic quantum number* $m_s = \pm\frac{1}{2}$.

The eigenfunctions of the problem with the Hamiltonian (A.5.1) which includes the spin are combinations of the products of the single-electron functions (A.5.4), which as a whole are antisymmetric with respect to the interchange of any two electrons. This requirement leads to the *Pauli prin-*

† For the orbital quantum number $l = 0, 1, 2, 3, 4, 5, 6, 7, 8, \ldots$, the notation $s, p, d, f, g, h, i, j, k, l, \ldots$, is used in spectroscopy.

ciple, according to which in an atom all the electrons differ from each other by at least one of the four quantum numbers n, l, m_l, m_s.

The eigenvalues of the Hamiltonian (A.5.1) depend only upon n and l:

$$W = W_{nl}. \tag{A.5.5}$$

Therefore in accordance with Pauli's principle there may be as many electrons in one energy level of the atom (n, l) as there are different values taken by the numbers m_l and m_s for given n and l.

Generally speaking the values of W_{nl} increase with n and l. If the atom is not excited its electrons are in the lowest energy levels, in so far as this is possible in accordance with Pauli's principle. As the number of electrons in the atoms of the elements of the periodic system successively increases the electron shells are filled in order of increasing energy. At the beginning of the periodic system this increase proceeds in order of successive increase in the quantum numbers n and l (within the possible limits for each n). However, starting with potassium this order is often broken—in the cases when a state with greater l but smaller n has a lower energy than a state with smaller l but greater n. As a result whole groups of elements are formed with incomplete inner shells, e.g. the *iron group* (3*d*) or the *rare earth group* (4*f*).

In the approximation under discussion, which has the Hamiltonian (A.5.1), spin–orbit interaction is not taken into consideration. Therefore the total spin moment of the atom $\boldsymbol{S}$ and the total orbital moment $\boldsymbol{L}$ are independent and can be defined as the sums of the electron moments:

$$\boldsymbol{S} = \sum_k \boldsymbol{s}_k \quad \text{and} \quad \boldsymbol{L} = \sum_k \boldsymbol{l}_k. \tag{A.5.6}$$

The total moments of completely filled shells are obviously zero. But atoms with incomplete inner shells (the atoms of the "transition" elements) have non-zero orbital and spin moments. This results in the paramagnetic properties of these elements.

As a result of summing the electron moments (A.5.6) the total moments of an atom (or ion) can take up a number of values corresponding to different mutual orientations of the electron moments; for example, for a two-electron atom the possible values are

$$L = l_1 + l_2, \quad l_1 + l_2 - 1, \quad \ldots, \quad |l_1 - l_2|$$

and

$$S = 1, 0.$$

These values of the spin and orbital moments correspond to different energy states of the atom. The lowest ("ground") state of an atom (ion) is given by *Hund's rule.*

In the ground state the total spin S of an atom has its maximum possible value for a given electron configuration, the total orbital moment L having the maximum value permitted by Pauli's principle.

For example the triply charged chromium ion Cr^{3+} with the electron configuration

$$Cr^{3+}: \quad (1s^2)(2s^2)(2p^6)(3s^2)(3p^6)\,3d^3 \equiv (A)\,3d^3$$

has in its ground state a spin $S = \frac{3}{2}$ and a total orbital moment $L = 2 + 1 + 0 = 3$. In spectroscopy this state is denoted by ^4F (the figure 4 denotes the number of spin multiplets of the state $2S + 1$).

A.6. Crystal Field Theory

Paramagnetic centres may be introduced into the crystal lattice of some oxides, salts and other compounds in the form of metallic ions. A paramagnetic ion is subject to the action of the electric field of the surrounding ions in the crystal lattice and as a result its energy levels change by being partly or completely split and displaced. The level splitting is largely determined by the interaction with the adjacent ions of the crystal complex; the action of the more distant ions in the lattice can with a high degree of accuracy be neglected.†

For a rigorous solution of the problem it would be necessary to know the nature of the distribution of the electron densities around the paramagnetic ion and the complex surrounding it. Up to the present, however, the precise solution of the Schrödinger equation for a system MX_n (M is a metal ion, X_n is a complex of n surrounding ions) is still unknown, i.e. the true nature of the electron states in such a system is not established. Therefore the problem is solved by perturbation theory, changing from the simplest model of the crystal field to a more detailed one.

The simplest model is one called the *crystal field approximation* (Bethe, 1929), according to which the ions of the complex are looked upon as point charges or dipoles at rest at a certain distance from the paramagnetic centre. The true nature of the distribution of the electron densities ("orbits") of the surrounding ions, the mutual penetration (overlapping) of these "orbits" with the "orbits" of the paramagnetic centre, i.e. the sharing of the electrons in the MX_n complex, are neglected. This approximation corresponds to a model of a purely ionic coupling which, despite the experimentally proven inaccuracy, nevertheless provides qualitatively correct and quantitatively satisfactory results.

The crystal field approximation contains one further assumption, namely that the perturbing action of the complex is considered to be identical for both unexcited and excited states of the paramagnetic ion.‡ This means

† The concentration of paramagnetic ions in the host lattice of a crystal is usually 10^{-3}–10^{-1}%, so their interaction with each other can also be neglected.

‡ It is sufficient to extend this requirement to only the lower excited states.

that the action of the paramagnetic centre on the surrounding crystal complex is neglected.

When using the crystal field approximation we start with states of the paramagnetic ion described by atomic wave functions.

It has been proved experimentally that even in the cases when there is every reason to assume that the coupling of the paramagnetic centre with the crystal complex is ionic in nature the electrons of the centre are nevertheless to a certain degree shared by the whole MX_n complex, i.e. there is also partial covalent bonding.†

The necessity of taking the covalent nature of the coupling into consideration has also been indicated by Van Vleck (1935); the first to do this were Tanabe and Sugano (1954). The *molecular orbital method* they used is as follows (Sugano, 1962). The states of the crystal complex MX_n can be described by overlapping electrons in corresponding molecular orbitals (i.e. molecular wave functions). In this case all possible interactions between electrons in these states are taken into consideration. The most important are the Coulomb and exchange interactions between the shared electrons and also their spin–orbit interaction.

A precise solution of the problem by the molecular orbital method involves considerable difficulties. The results obtained, despite the radical difference of the physical model from that used in the crystal field approximation, contains only small quantitative corrections in the majority of cases of interest to us (Low, 1961).

Therefore in many cases we can use a semi-empirical method for studying the spectra of paramagnetic ions in crystals. In this method the calculations are made in the crystal field approximation and into the results obtained are introduced corrections to allow for the covalent nature of the bonds; the results are thus brought into agreement with experiments.

Great assistance is obtained in the study of the spectra of crystals by examining the symmetry properties of the crystal field in which the paramagnetic ion is located. It has been proved experimentally that the ions of the iron group are usually situated in the lattice of a diamagnetic crystal in an octahedral complex, i.e. the paramagnetic ion is in the centre of an octahedron at whose apexes are the diamagnetic ions of the lattice. Therefore the six closest ions of the lattice are at points with coordinates $(\pm a; 0; 0)$, $(0; \pm b; 0)$, $(0; 0; \pm c)$ if the paramagnetic ion is at the origin (the so-called sixfold coordination complex MX_6) (Fig. A.III.1 a).

If $a \neq b$, the Oz $|001|$ axis of the octahedron is a twofold axis of symmetry and the octahedron has rhombic symmetry. If $a = b$, the $|001|$ axis becomes a fourfold axis of symmetry and the octahedron takes on tetragonal sym-

† The covalent nature of the bond of the paramagnetic ions with the crystal appears in the effect of the nuclei of the surrounding ions on the hyperfine structure of the electron paramagnetic resonance spectra.

metry. If $a = b = c$, the symmetry of the octahedron becomes cubic. The |111| axis is then a threefold axis. When the octahedron is deformed along this axis it loses its cubic symmetry and retains only the trigonal.

In any case the octahedron still has axial symmetry, i.e. symmetry with respect to rotation through an angle 2π around any of the axes. Because of the symmetrical arrangement of the surrounding ions around the paramagnetic ion (at distances $\pm a$, $\pm b$, $\pm c$) the latter is the centre of symmetry of the octahedron (one further element of its symmetry).

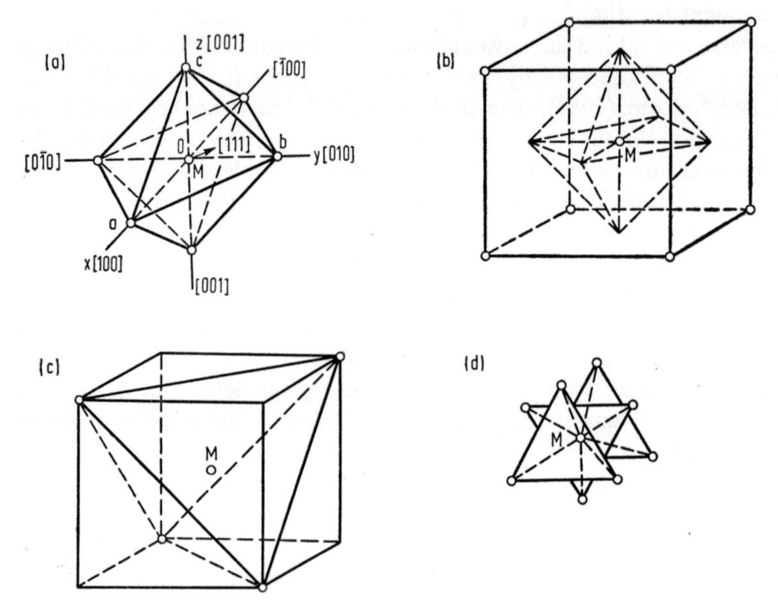

FIG. A.III.1. Paramagnetic ion in elementary cell of a crystal: (a) octahedral complex; (b) cubic complex; (c) tetrahedral complex; (d) trigonal complex. The X ions are denoted by circles ○.

Other cases of arrangement of the complex around the paramagnetic centre are also possible, e.g. the eightfold cubic coordination (MX_8) when the ion M is in the centre of a cube and the eight X ions are at its apexes, or fourfold tetrahedral coordination (MX_4) when the M ion is in the centre of a tetrahedron formed by $4X$ ions (in this case the complex has no centre of symmetry). The connection between these cases and the octahedral is shown in Fig. A.III.1 b, c.

The ions of rare-earth elements generally crystallize in a ninefold coordination (MX_9) having trigonal symmetry (see Fig. A.III.1 d). Here the M ion lies in the centre of an equilateral triangle X_3 above and below which are another two X_3 triangles rotated 60° relative to the equatorial one.

The ions in the complex which are located around the paramagnetic centre create at the centre an electrostatic field whose potential has the same symmetry as the complex itself. This field splits the degenerate levels of the paramagnetic ions.

If a certain energy level of a free ion has $(2L + 1)$-fold degeneracy, then from the point of view of group theory (Landau and Lifshitz, 1963) this means that $(2L+1)$ wave functions belong to the D^L irreducible representation of the group of the operator $\hat{H}_0$, namely of the rotation group K, since the free ion is invariant with respect to any rotations in space.

When the crystal field is introduced the states of the paramagnetic ion are described by the Hamiltonian $\hat{H} = \hat{H}_0 + \hat{H}_{cr}$ which has the same symmetry as the field; this means that the eigenfunctions of the new Hamiltonian transform like irreducible representations of the crystal point symmetry group. The D^L representation corresponding to the given $(2L+1)$ functions, and which was irreducible for the K group, may prove to be reducible for the crystal point group. This corresponds to splitting of the corresponding energy level into several sublevels. The number of sublevels is determined by the number of irreducible representations of the point group into which the reducible representation can be decomposed and the degree of their degeneracy is determined by the dimensions of the irreducible representations obtained.

The expansion of the representations of the rotation group in irreducible representations of point groups as applied to the problem of interest to us was given by Bethe (1929). The notation used there by him for the irreducible representations (Γ_n) corresponding to Mulliken's (1933) notations used by Landau and Lifshitz (1963) are given in Table A.III.1. Below we show the

TABLE A.III.1

Bethe notation	Γ_1	Γ_2	Γ_3	Γ_4	Γ_5	Γ_6	Γ_7	Γ_8
Mulliken notation	A_1	A_2	E	$T_1 \equiv F_1$	$T_2 \equiv F_2$	$E_{1/2}$	$E_{5/2}$	G
Degeneracy	1	1	2	3	3	2	2	4

expansions of the representations D^L of the rotation group in the field of cubic symmetry (Table A.III.2). This case is of the greatest practical interest. Similar tables of the expansions in fields of lower types of symmetry can be found, e.g. in Low's book (1960 b).

It follows from Table A.III.2, for example, that the fourfold degenerate F-state of a free ion ($L = 3, 2L + 1 = 7$) in a crystal field of cubic symmetry splits into one singlet (A_2) and two triplet (T_1 and T_2) states.

TABLE A.III.2

L	Expansion of representations of D^L rotation group into irreducible representations of the cubic symmetry group			
	Term	Bethe notation	Mulliken notation	No. of levels
0	S	Γ_1	A_1	1
1	P	Γ_4	$T_1 \equiv F_1$	1
2	D	$\Gamma_3+\Gamma_5$	$E+T_2 \equiv E+F_2$	2
3	F	$\Gamma_2+\Gamma_4+\Gamma_5$	$A_2+T_1+T_2 \equiv A_2+F_1+F_2$	3
4	G	$\Gamma_1+\Gamma_3+\Gamma_4+\Gamma_5$	$A_1+E+T_1+T_2 \equiv A_1+E+F_1+F_2$	4
5	H	$\Gamma_3+2\Gamma_4+\Gamma_5$	$E+2T_1+T_2 \equiv E+2F_1+F_2$	4
6	I	$\Gamma_1+\Gamma_2+\Gamma_3+\Gamma_4+2\Gamma_5$	$A_1+A_2+E+T_1+2T_2$ $\equiv A_1+A_2+E+F_1+F_2$	6

A.7. The Crystal Field Potential

We shall now find the form of the crystal field Hamiltonian $\hat{H}_{cr}$. If we assume that each outer electron of the paramagnetic ion is subjected to the action of a crystal field with a potential $V(r_k)$ (r_k is the distance from the nucleus of the paramagnetic ion to its kth-electron), then

$$\hat{H}_{cr} = -e \sum_{k=1}^{N} V(r_k), \tag{A.7.1}$$

where summation is carried out with respect to all N outer electrons of this ion.

In the crystal field approximation it is assumed that the potential $V(r)_k$ is created by the static distribution of the point charges or dipoles forming the crystal complex. Therefore the potential $V(r_k)$ satisfies the Laplace equation: $\nabla^2 V(r_k) = 0$. This allows us to expand it into a series in spherical functions:

$$V(r_k) = \sum_{n=0}^{\infty} \sum_{m=-n}^{n} A_n^m r_k^n Y_n^m (\theta_k, \varphi_k), \tag{A.7.2}$$

where $Y_n^m(\theta_k, \varphi_k)$ are the same as the spherical harmonics described in § A.5 and A_n^m are coefficients which depend on the distance between the paramagnetic centre and the nearest ions.

Therefore the crystal field Hamiltonian can be written in the form

$$\hat{H}_{cr} = -e \sum_{n=0}^{\infty} \sum_{m=-n}^{n} \sum_{k=1}^{N} A_n^m r_k^n Y_n^m (\theta_k, \varphi_k). \tag{A.7.3}$$

If we assume, as was shown in § A.6, that the crystal field is identical for all the electrons of the paramagnetic ion, no matter what their state, and introduce the notation

$$V_n^m = A_n^m r^n Y_n^m(\theta, \varphi), \tag{A.7.2a}$$

then (A.7.3) can be written in the form

$$\hat{H}_{cr} = -eN \sum_{n=0}^{\infty} \sum_{m=-n}^{n} V_n^m. \tag{A.7.3a}$$

The number of terms of the expansion (A.7.3a) which are needed to calculate the energy spectrum is comparatively small.

In the first place in the case of ions of the iron group we can eliminate from the discussion all the terms of (A.7.3a) with $n > 4$ and for ions of the rare-earth group with $n > 6$. This can be explained by the fact that when calculating the matrix elements of the crystal field

$$\begin{aligned}\langle lm_l \mid H_{cr} \mid lm_l' \rangle &= -eN \sum_n \sum_m \int \psi_{lm_l}^* V_n^m \psi_{lm_l'} \, dv \\ &= -eN \sum_n \sum_m \int \psi_{lm_l}^* \psi_{lm_l'} V_n^m \, dv\end{aligned} \tag{A.7.4}$$

by means of the wave functions of the d- or f-electrons the products $\psi_{lm_l}^* \psi_{lm_l'}$ can also be expanded in spherical functions, but not higher than the fourth (for d-electrons) or sixth (for f-electrons) order. Therefore if $\hat{H}_{cr}$ contains spherical functions with $n > 4$ (or $n > 6$), then the corresponding terms of the sum (A.7.4) nevertheless vanish because of the orthogonality relations for the spherical harmonics.

Secondly, we need not take into consideration a term with $n = 0$ since it is an additive constant which is not relevant to the spectroscopic calculations (the distances between the levels and not their absolute magnitudes are needed).

Thirdly, the terms of (A.7.4) with odd n give vanishing matrix elements since the product $\psi_{lm_l}^* \psi_{lm_l'}$ is invariant with respect to the inversion of the system of coordinates (the functions $\psi_{lm_l}^*$ and $\psi_{lm_l'}$ have the same parity if they belong to one and the same electron configuration), whilst the odd terms $V_{n=2p+1}^m$ change their sign under this transformation:

$$V_{n=2p+1}^m(x, y, z) \to -V_{n=2p+1}^m(-x, -y, -z).$$

Therefore the expansion (A.7.3a) can be limited to terms with $n = 2$ ($m = 0, \pm 1, \pm 2$) and $n = 4$ ($m = 0, \pm 1, \pm 2, \pm 3, \pm 4$) in the case of d-electrons, and in the case of f-electrons we must also include terms with $n = 6$ ($m = 0, \pm 1, \pm 2, \pm 3, \pm 4, \pm 5, \pm 6$).

Expression (A.7.3a) can be further simplified by allowing for the actual symmetry properties of the crystal field. The spherical harmonics $Y_n^m(\theta, \varphi)$

depend on the angle φ as $e^{im\varphi}$, and so they have axial symmetry when $m = 0$, rhombic when $m = \pm 2$, trigonal when $m = \pm 3$ and tetragonal when $m = \pm 4$ (here the polar axis is respectively a first, second, third and fourth order axis of symmetry).

If the crystal field has rhombic symmetry, only terms with $m = 2$ and multiple values are contained in the expansion (A.7.3); in the case of a trigonal field only terms with $m = 3$ and multiples, and for a tetragonal field only with $m = 4$. In addition, each of these expansions also contains terms with $m = 0$ corresponding to the presence of axial symmetry in all the fields listed.

If we introduce the notation

$$U_n^0 = A_n^0 r^n Y_n^0(\theta, \varphi),$$
$$U_n^{|m|} = r^n[A_n^m Y_n^m(\theta, \varphi) + A_n^{-m} Y_n^{-m}(\theta, \varphi)], \tag{A.7.5}$$

the crystal fields of the different types of symmetry will be described by the following potentials (for *d*-electrons):

$$V_{\text{rhomb}} = U_2^0 + U_2^2 + U_4^0 + U_4^2 + U_4^4,$$
$$V_{\text{trig}} = U_2^0 + U_4^0 + U_4^3, \tag{A.7.6}$$
$$V_{\text{tetrag}} = U_2^0 + U_4^0 + U_4^4.$$

The potential of a field of cubic symmetry, if the polar axis is chosen in the |001| direction, has (for a sixfold coordination) the following form:

$$V_{\text{cub}} = A_4^0 r^4 \{Y_4^0(\theta, \varphi) + \sqrt{\tfrac{5}{14}}\, [Y_4^4(\theta, \varphi) + Y_4^{-4}(\theta, \varphi)]\}, \tag{A.7.7}$$

whilst if the polar axis is chosen in the |111| direction, then

$$V_{\text{cub}} = D_4 r^4 \{Y_4^0(\theta, \varphi) - \sqrt{\tfrac{10}{7}}\, [Y_4^3(\theta, \varphi) + Y_4^{-3}(\theta, \varphi)]\}, \tag{A.7.8}$$

the expression (A.7.8) being exactly equal to (A.7.7) although different in form; the coefficient D_4 is replaced by A_4^0.

Expressions (A.7.6), (A.7.7) and (A.7.8) show that the potentials of fields of lower symmetry can be obtained by adding lower order terms to the cubic field potential in the form (A.7.7) or (A.7.8). This circumstance can be conveniently used in what follows since for crystals containing ions of the iron group the crystal field has basically cubic symmetry with a small addition of a lower type of symmetry (Bleaney and Stevens, 1953).

In many cases it is convenient to use the expressions for the crystal field potential written in Cartesian coordinates. Then the potentials U_n^m can be written (Al'tshuler and Kozyrev, 1961) in the form

$$U_n^m = B_n^m V_n^m, \tag{A.7.9}$$

where

$$V_2^0 = 3z^2 - r^2, \qquad B_2^0 = \frac{1}{4}\sqrt{\frac{5}{\pi}}\,A_2^0;$$

$$V_2^1 = xz, \qquad B_2^1 = \sqrt{\frac{15}{2\pi}}\,|A_2^1|;$$

$$V_2^2 = x^2 - y^2, \qquad B_2^2 = \frac{1}{2}\sqrt{\frac{15}{2\pi}}\,|A_2^2|;$$

$$V_4^0 = 35z^4 - 30r^2z^2 + 3r^4, \qquad B_4^0 = \frac{3}{16\sqrt{\pi}}\,A_4^0;$$

$$V_4^1 = (7z^2 - 3r^2)\,xz, \qquad B_4^1 = \frac{3}{4}\sqrt{\frac{5}{\pi}}\,|A_4^1|;$$

$$V_4^2 = (7z^2 - r^2)(x^2 - y^2), \qquad B_4^2 = \frac{3}{4}\sqrt{\frac{5}{2\pi}}\,|A_4^2|;$$

$$V_4^3 = (x^2 - 3y^2)\,xz, \qquad B_4^3 = \frac{3}{4}\sqrt{\frac{35}{\pi}}\,|A_4^3|;$$

$$V_4^4 = x^4 - 6x^2y^2 + y^4, \qquad B_4^4 = \frac{3}{8}\sqrt{\frac{35}{2\pi}}\,|A_4^4|.$$

The values of the constants A_n^m depend on whether the charges of the complex are considered to be point charges or dipoles; for the first case the values of A_n^m are given by Low (1960 b) and Al'tshuler and Kozyrev (1961). In addition, the values of A_n^m [and also the coefficient D_4 in (A.7.8)] depend on the magnitude of these charges e_i and on the distance R between them and the paramagnetic ion, e.g.

$$A_4^0 = \frac{7\sqrt{\pi}}{3R^5}\,e_i. \tag{A.7.10}$$

The expression for the potential of a cubic field in Cartesian coordinates has the simple form

$$V_{\text{cub}} = D(x^4 + y^4 + z^4 - \tfrac{3}{5}r^4), \tag{A.7.11}$$

where

$$D = \frac{35e_i}{4R^5}.$$

A.8. Crystal Field Matrix Elements

Now that the form of the crystal field potential has been established we can allow for the perturbing action of this field on the energy levels of a free ion. Here, depending on the strength of the crystal field, the resulting perturbation must be taken into account at a quite definite stage in the series of successively decreasing terms of the complete Hamiltonian (A.4.1).

In the case of a weak field (rare-earth ions) $\hat{H}_{cr}$ is allowed for after considering the spin–orbital interaction $\hat{H}_{LS}$. The states of the "unperturbed" system (with the Hamiltonian $\hat{H}_0 + \hat{H}_{LS}$) are characterized by definite values of the total moment J of the ion. The calculation of the crystal field matrix elements contained in the secular equation is therefore carried out in the representation of the total angular momentum operator $\hat{\boldsymbol{J}}$, and the matrix elements themselves take the form

$$(H_{cr})^{(J)}_{J_z J_z'} \equiv \langle J, J_z \,|\hat{H}_{cr}|\, J, J_z' \rangle = \int \Psi^*_{JJ_z} \hat{H}_{cr} \Psi_{JJ_z'}\, dv. \qquad \text{(A.8.1)}$$

In the case of an intermediate field (ions of elements in the iron group) $\hat{H}_{cr}$ is allowed for before the spin–orbit interaction is considered. The states of the unperturbed system (with the Hamiltonian $\hat{H}_0$) are characterized by definite values of the orbital moment L and the spin moment S of the ion. The matrix elements of the crystal field are calculated in the representation of the orbital momentum operator $\hat{\boldsymbol{L}}$ (in the approximation in question the crystal field has no direct effect on the spin variables):

$$(H_{cr})^{(L)}_{M_L M_L'} \equiv \langle L, M_L \,|\hat{H}_{cr}|\, L, M_L' \rangle = \int \Psi^*_{LM_L} \hat{H}_{cr} \Psi_{LM_L'}\, dv. \qquad \text{(A.8.2)}$$

Finally in a strong field it is possible that the effect of the crystal field exceeds the energy of the Coulomb and exchange interactions of the electrons in the ion. In this case we first allow for the action of $\hat{H}_{cr}$ on the system with an unperturbed Hamiltonian $\hat{H} = \sum_k^N [(\hat{p}_k^2/2m) - (Ze^2/r_k)]$ and then in turn we take into consideration the perturbation introduced by the Coulomb and exchange interactions of the electrons $\left(\text{with } \frac{1}{2}\sum_k^N \sum_{j\neq k} e^2/r_{kj} \text{ as}\right.$ the Hamiltonian), the spin–orbit interaction ($\hat{H}_{LS}$), and other interactions.

Calculation of the matrix elements of the secular equation for weak and intermediate fields, generally speaking, is a difficult problem.

Fortunately, many of the matrix elements are equal to zero. Because the wave functions Ψ_{LM_L} (or Ψ_{JJ_z}) and the potentials U_n^m have, as shown above, the same angular dependence of φ it turns out that

$$\int \Psi^*_{LM_L} U_{n=L} \Psi_{LM_L'}\, dv \neq 0 \quad \text{only when} \quad M_L = M_L' + m. \qquad \text{(A.8.3)}$$

The non-zero matrix elements can be calculated by the *method of operator equivalents* developed by Stevens, Elliott and Judd (Stevens, 1952; Elliott and Stevens, 1953a; Judd, 1955). In this method, which is based on the use of group theory, use is made of the fact that the matrix elements of a crystal field with a potential of the form (A.7.9) are the same, except for a constant coefficient, as the matrix elements of an operator obtained by replacing the variables x, y, z in the potential expression by the operators $\hat{L}_x, \hat{L}_y, \hat{L}_z$ (or $\hat{J}_x, \hat{J}_y, \hat{J}_z$) belonging to the set of states with a given L (or J). To replace the products of the variables x, y, z here we must use the symmetrized combinations of the operators $\hat{L}_x, \hat{L}_y, \hat{L}_z$ [e.g. instead of xy we write $\frac{1}{2}(\hat{L}_x\hat{L}_y + \hat{L}_y\hat{L}_x)$, etc.] since the variables x, y, z commute and $\hat{L}_x, \hat{L}_y, \hat{L}_z$ do not. As an example we give the expression for the operator equivalent for the cubic field potential, which in Cartesian coordinates is of the form

$$V_{\text{cub}} = D(x^4 + y^4 + z^4 - \tfrac{3}{5}r^4). \tag{A.7.11}$$

The operator equivalent to it is (in the $\hat{L}$-representation)

$$\hat{V}_{\text{cub}} = \beta \frac{\overline{r^4}}{20} [35\hat{L}_z^4 - 30L(L+1)\hat{L}_z^2 + 25\hat{L}_z^2 - 6L(L+1) + 3L^2(L+1)^2] + \frac{\beta\overline{r^4}}{8}[\hat{L}_+^4 + \hat{L}_-^4], \tag{A.8.4}$$

where

$$\overline{r^4} = \int_0^\infty r^4|R(r)|^2 r^2\, dr; \quad \hat{L}_\pm = \hat{L}_x \pm i\hat{L}_y,$$

and β is the proportionality coefficient mentioned above; its value needs to be calculated only once for ions of a given type.

The calculation of the matrix elements of operators of the (A.8.4) type is comparatively simple if we use the following relations (Landau and Lifshitz, 1963):

$$\hat{L}_\pm Y_L^{M_L} = \sqrt{L(L+1) - M_L(M_L \pm 1)}\, Y_L^{M_L \pm 1},$$
$$\hat{L}_z Y_L^{M_L} = M_L Y_L^{M_L} \tag{A.8.5}$$

(where the quantities L, M_L are written in $\hbar$ units). The expressions for the operator equivalents for the potentials of different crystal fields, the values of the matrix elements of these operators and the proportionality coefficients are given in the literature (Low, 1960b; Bleaney and Stevens, 1953; Al'tshuler and Kozyrev, 1961; Stevens, 1952; Elliott and Stevens, 1953a; Judd 1955).

A.9. The Splitting of the Energy Levels of a Single-electron Ion in an Intermediate Field of Cubic Symmetry

Let us now examine the results obtained from the solution of the secular equation for a medium field. We shall consider the crystal field to be cubic, and start our discussion with the single-electron case. As an example we may take the Ti^{3+} ion which has the electron configuration Ti^{3+}:(A) $3d^1$.

In accordance with Hund's rule the ground state of this ion is the term ^2D which has fivefold orbital degeneracy (values of the projection of the orbital moment $m_1 = 0, \pm 1, \pm 2$). Let us examine the splitting of this level in the crystal field.

The non-zero matrix elements of the cubic field (A.7.7) will, in accordance with (A.8.3), be only those that connect the wave functions of states that differ from each other by $\Delta m_l = 0, \pm 4$ (since the expression (A.7.7) for the cubic field contains terms only with $m = 0, \pm 4$). For the D term ($l = 2$) the only non-zero matrix elements of $(H_{cr})^{(1)}_{m_l m_l'}$ are

$$(H_{cr})^{(2)}_{ii}, \quad i = 0, \pm 1, \pm 2, \quad (m_l = m_l');$$

$$(H_{cr})^{(2)}_{-2,2} \qquad (m_l = m_l' - 4);$$

$$(H_{cr})^{(2)}_{2,-2} \qquad (m_l = m_l' + 4).$$

Therefore we see that each of the wave functions $\psi_{2,0}$; $\psi_{2,1}$, $\psi_{2,-1}$ forms matrix elements only with itself, so these functions are eigenfunctions of the cubic field Hamiltonian. The functions $\psi_{2,2}$ and $\psi_{2,-2}$, however, are not separately eigenfunctions of this Hamiltonian since each of them is coupled to a second one, but their linear combinations are eigenfunctions. The secular equation is

$$\begin{vmatrix} W - H^{(2)}_{0,0} & 0 & 0 & 0 & 0 \\ 0 & W - H^{(2)}_{1,1} & 0 & 0 & 0 \\ 0 & 0 & W - H^{(2)}_{-1,-1} & 0 & 0 \\ 0 & 0 & 0 & W - H^{(2)}_{2,2} & H^{(2)}_{-2,2} \\ 0 & 0 & 0 & H^{(2)}_{2,-2} & W - H^{(2)}_{-2,-2} \end{vmatrix} = 0. \tag{A.9.1}$$

Calculation of the non-zero matrix elements shows (Bleaney and Stevens, 1953) that $(H_{cr})^{(2)}_{1,1} = (H_{cr})^{(2)}_{-1,-1}$; $(H_{cr})^{(2)}_{2,2} = (H_{cr})^{(2)}_{-2,-2}$; $(H_{cr})^{(2)}_{2,-2} = (H_{cr})^{(2)}_{-2,2}$ and in addition: $(H_{cr})^{(2)}_{2,2} = \frac{1}{2}[(H_{cr})^{(2)}_{0,0} + (H_{cr})^{(2)}_{1,1}]$ and $(H_{cr})^{(2)}_{2,-2} = \frac{1}{2}[(H_{cr})^{(2)}_{0,0} - (H_{cr})^{(2)}_{1,1}]$. Using these relations we find that the secular equation has two multiple roots: W_1 and W_2, which means splitting of the fivefold degenerate ^2D term into two sublevels: the doublet ^2E and the triplet $^2F_2(^2T_2)$ (see Table A.II.2).

The doublet state ^2E corresponds to the wave functions with the conventional notation

$$e \equiv d_\gamma = \begin{cases} e^a = \Psi_{2,0}, \\ e^b = \dfrac{1}{\sqrt{2}}(\Psi_{2,2} + \Psi_{2,-2}), \end{cases} \tag{A.9.2}$$

and the triplet $^2F_2$ corresponds to

$$t_2 \equiv d_\varepsilon = \begin{cases} t_2^0 = \dfrac{1}{\sqrt{2}}(\Psi_{2,2} - \Psi_{2,-2}), \\ t_2^+ = \Psi_{2,1}, \\ t_2^- = \Psi_{2,-1}. \end{cases} \tag{A.9.3}$$

The corresponding corrections to the energy level $W\,(n = 3,\, l = 2)$ are

$$W_1(^2E) = \int \Phi^*(e^a, e^b)\, \hat{H}_{\mathrm{cr}} \Phi(e^a, e^b)\, dv = 6Dq \tag{A.9.4}$$

and

$$W_2(^2F_2) = \int \Phi^*(t_2^0, t_2^\pm)\, \hat{H}_{\mathrm{cr}} \Phi(t_2^0, t_2^\pm)\, dv = -4Dq, \tag{A.9.5}$$

where $\Phi(e^a, e^b)$ and $\Phi(t_2^0, t_2^\pm)$ are the linear combinations of the degenerate wave functions (A.9.2) and (A.9.3) and the cubic potential (A.7.11) is used in the expression for $\hat{H}_{\mathrm{cr}}$. Therefore the triplet sublevel $^2F_2$ is the lower one and the doublet ^2E the upper one.

The numerical coefficients (+6) and (−4) can be obtained when calculating the angular integrals in (A.9.4) and (A.9.5) by the operator equivalent method.

The coefficient $D = 35e_i/4R^5$ comes into the equation (A.7.11) for the cubic potential (e_i denotes the charge of an ion of the crystal lattice, and R the distance to it from the paramagnetic ion). This coefficient cannot be obtained by dierct calculation since the value of R is not known precisely.

The coefficient q is the radial part of the integrals (A.9.4) and (A.9.5) and is equal to

$$q = \frac{2e}{105} \int_0^\infty R_{3d}^2(r) \cdot r^4 \cdot r^2 \cdot dr = \frac{2\overline{r^4}e}{105}, \tag{A.9.6}$$

where $R_{3d}(r)$ is the radial part of the $3d$-function of a free single-electron ion. Because there is no precise information on the nature of the function $R_{3d}(r)$, i.e. on the true distribution of the electron density around the paramagnetic ion in the crystal, the coefficient q cannot be obtained by direct calculation either.

In view of what has been said the coefficients D and q have to be calculated empirically. Although they cannot be determined separately from an experiment their product D_q which depends upon the degree of coupling of the paramagnetic ion with the crystal can be determined from an analysis of the optical spectrum of the crystal by comparing the total magnitude of the splitting

$$W_1 - W_2 = 10Dq \equiv \Delta$$

with the frequency of the transition between the ground $^2F_2$ and the first excited ^2E state.

For example, the absorption spectrum of the complex $[Ti(H_2O)_6]^{3+}$ consisting of one band with a maximum at 4900 Å (20,400 cm^{-1}) is ascribed to the transition $^2F_2 \rightarrow {}^2E$, from which $Dq = 2040$ cm^{-1} (Moffitt and Ballhausen, 1956).

The quantity Dq is called the *crystal field parameter* and is an important measure of the degree of splitting of the levels in the crystal, indicating the "strength" of the crystal field (to be more precise the strength of its coupling with a given paramagnetic ion).

The splitting of the lower excited states of the free ion in a cubic crystal field can be treated in the same way as the splitting of the ground state.

A.10. The Splitting of the Energy Levels of a Many-electron Ion in an Intermediate Field of Cubic Symmetry

A calculation of the splitting of the levels of a many-electron ion in an intermediate field can be carried out by the same method as for a single-electron ion† (Bleaney and Stevens, 1953). In this case instead of the single-electron wave $3d$ functions characterized by the values of the quantum numbers l and m_l of a single electron we must use many-electron wave functions characterized by the values of the quantum numbers L and M_L of a many-electron ion.

Here we give the final results of calculating the splitting of the ground states of ions of the iron group in an intermediate cubic field (see Table A.III.3). A qualitative picture of the splitting of all the levels can be easily obtained from group theory (see Table A.III.2).

The values given in Table A.III.3 for the splittings are given in Dq units, i.e. expressed in terms of the crystal field parameter of a single-electron ion.

Ions of the iron group contain from one to ten electrons in the outer $3d$-

† Here we have in mind only the simplest method of calculating the splitting of the levels of many-electron ions in the clearly defined case when $e^2/r_{\kappa j} \gg H_{cr} \gg H_{LS}$. In actual crystal fields the orders of magnitude of $e^2/r_{\kappa j}$ and H_{cr} approach each other, so situations arise which lie between the strong and intermediate field cases defined in § A.4. Because of this a more general method has been developed (Tanabe and Sugano, 1954; Low, 1961) for calculating the splitting, in which the electrostatic interaction of the electron $e^2/r_{\kappa j}$, and H_{cr} compete with each other. This method is outlined by McClure (1959).

shell. According to Hund's rule the ground states of free ions with $3d$-electrons can be only S-, D- and F-terms.

The S-level is orbitally non-degenerate and should not split in the crystal field (the weak splitting of the spin components which nevertheless occurs will be explained in § A.12).

The fivefold degenerate D-level is split in a cubic field into a doublet and a triplet, and the sevenfold degenerate F-level into a singlet and two triplets

TABLE A.III.3. THE SPLITTING OF THE GROUND STATES OF IONS OF THE IRON GROUP IN AN INTERMEDIATE CUBIC FIELD (MOFFITT AND BALLHAUSEN, 1956)

Ground state of free ion	Ions	States formed in a cubic field[a]	Distance from unperturbed level in Dq units
$3d^1:{}^2D$	Ti^{3+}	${}^2E$	+6
		${}^2F_2$	−4
$3d^2:{}^3F$	V^{3+}	${}^3A_2$	+12
		${}^3F_2$	+2
		${}^3F_1$	−6
$3d^3:{}^4F$	V^{2+}, Cr^{3+}, Mn^{4+}	${}^4F_1$	+6
		${}^4F_2$	−2
		${}^4A_2$	−12
$3d^4:{}^5D$	Mn^{3+}, Cr^{2+}	${}^5F_2$	+4
		${}^5E$	−6
$3d^5:{}^6S$	Mn^{2+}, Fe^{3+}	${}^6A_1$	0
$3d^6:{}^5D$	Fe^{2+}, Co^{3+}	${}^5E$	+6
		${}^5F_2$	−4
$3d^7:{}^4F$	Co^{2+}	${}^4A_2$	+12
		${}^4F_2$	+2
		${}^4F_1$	−6
$3d^8:{}^3F$	Ni^{2+}	${}^3F_1$	+6
		${}^3F_2$	−2
		${}^3A_2$	−12
$3d^9:{}^2D$	Cu^{2+}	${}^2F_2$	+4
		${}^2E$	−6

[a] Do not confuse the notation of the F-levels of the free ions with the notation of the F_1 and F_2 sublevels in the crystal field taken from group theory.

(see Table A.III.2). The location of these sublevels in relation to each other depends on the number of d-electrons in a given ion.

It was shown in § A.9 that the D-level of an ion with a single d-electron (${}^2D$) splits in a cubic field so that the triplet is the lower level and the doublet the upper one. The same splitting occurs in ions with a d^6-configuration (the

^5D term), which differs from the d^1-configuration by the addition of the closed d^5-shell (d^5; ^6S) on which the crystal field has no effect.

The d^9-configuration (the ^2D term) can be looked upon for the sake of argument as the combination of a complete d^{10}-shell with one positive d-electron. Therefore the sign of the parameter q, and thus also the splitting of the ^2D-level of an ion with d^9-electrons, will be opposite to the case of $^2D(d^1)$, i.e. the doublet will be the lower level and the triplet the upper one. By adding to d^9 a closed shell with d^5 positive electrons we obtain the configuration d^4, whose ground state ^5D splits in the same way as ^2D (d^9).

By using similar arguments we can understand why the splitting of the ^4F (d^3) and ^3F (d^8) levels are the same and their lower level is a singlet, whilst

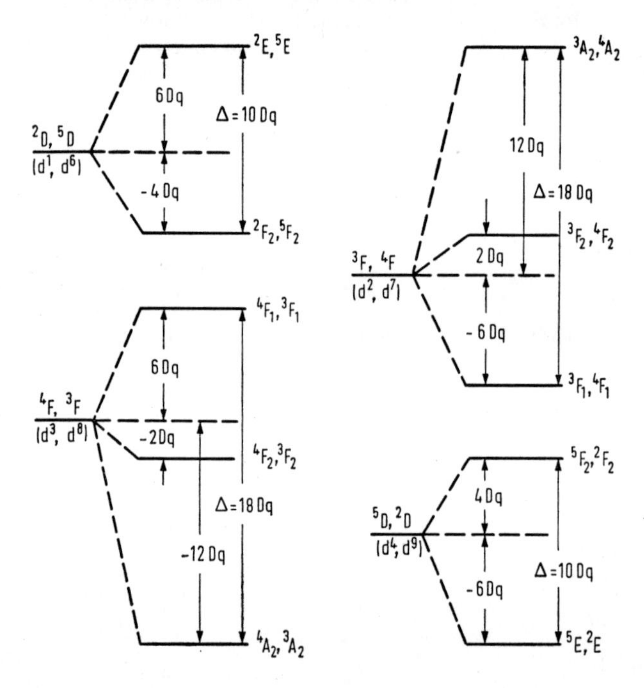

FIG. A.III.2. The splitting of the ground state of a d^n configuration in a cubic field.

the opposite is true (the triplet is the ground sublevel) for the ^4F (d^7) and ^3F (d^2) levels.

Figure A.III.2 shows diagrams of these splittings; the total magnitude of the splitting is always denoted by Δ. In all cases the order of magnitude of Δ is $\sim 10^4$ cm^{-1}.

The lower energy levels obtained in a cubic field correspond to the following states (the quantization axis is taken in the direction of the principal axis of symmetry of the cubic field, i.e. along the [001] axis):

$$3d^1(^2F_2),\ 3d^6(^5F_2): \begin{cases} \Psi_{2,1}, \\ \Psi_{2,-1}, \\ \dfrac{1}{\sqrt{2}}(\Psi_{2,2} - \Psi_{2,-2}); \end{cases}$$

$$3d^4(^5E),\ 3d^9(^2E): \begin{cases} \Psi_{2,0}, \\ \dfrac{1}{\sqrt{2}}(\Psi_{2,2} + \Phi_{2,-2}); \end{cases} \tag{A.10.1}$$

$$3d^2(^3F_1),\ 3d^7(^4F_1): \begin{cases} \sqrt{\tfrac{3}{8}}\,\Psi_{3,-1} + \sqrt{\tfrac{5}{8}}\,\Psi_{3,3}, \\ \sqrt{\tfrac{3}{8}}\,\Psi_{3,1} + \sqrt{\tfrac{5}{8}}\,\Psi_{3,-3}, \\ \Psi_{3,0}; \end{cases}$$

$$3d^3(^4A_2),\ 3d^8(^3A_2): \frac{1}{\sqrt{2}}(\Psi_{3,2} - \Psi_{3,-2}).$$

It is interesting to note that the mean values of the z-component of the orbital moment of ions for the ground states obtained in a crystal field are less than for the ground states of free ions. For example, for a $3d^1$ (or $3d^6$) configuration in three states corresponding to the ground state F_2 the following mean values are obtained for the quantities L_z:

$$\begin{aligned} \langle L_z\rangle_1 &= \int \Psi^*_{2,1}\hat{L}_z\Psi_{2,1}\,dv = 1, \\ \langle L_z\rangle_2 &= \int \Psi^*_{2,-1}\hat{L}_z\Psi_{2,-1}\,dv = -1, \\ \langle L_z\rangle_3 &= \int \left\{\frac{1}{\sqrt{2}}(\Psi^*_{2,2} - \Psi^*_{2,-2})\right\}\hat{L}_z\left\{\frac{1}{\sqrt{2}}(\Psi_{2,2} - \Psi_{2,-2})\right\}dv = 0, \end{aligned} \tag{A.10.2}$$

i.e. an ion in the crystal field in an F_2 ground state may be characterized by the value $L = 1$ although this level originated from a D-term with $L = 2$.

In the case of a $3d^3$ (or $3d^8$) configuration we obtain

$$\langle L_z\rangle = \int \left\{\frac{1}{\sqrt{2}}(\Psi^*_{3,2} - \Psi^*_{3,-2})\right\}\hat{L}_z\left\{\frac{1}{\sqrt{2}}(\Psi_{3,2} - \Psi_{3,-2})\right\}dv = 0 \tag{A.10.3}$$

and the ground A_2 level is orbitally non-degenerate, as though $L = 0$ and not $L = 3$ (as should be the case for the F-term of a free ion).

This effect, which is called "quenching" of the orbital moment in the crystal field, can be explained physically by the fact that because of the effect of the crystal field the orbital moment of a free ion cannot orient itself freely in space and is quantized along the principal axis of symmetry of this field, the number of permitted orientations thus being reduced.

A.11. The Optical Spectra of Paramagnetic Crystals

We have found how splitting of the lower ground energy levels of paramagnetic ions in an intermediate cubic field occurs. Similar arguments are also applicable to the excited levels of the basic electron configuration of a paramagnetic ion ($3d^n$).

The system of levels of the basic electron configuration occupied the region in the energy scale from 0 to $\sim 10^5$ cm^{-1} where all transitions in the microwave, optical and ultraviolet bands, the cases of interest to us, take place. A complete picture of the splitting of these levels in the crystal field can be obtained from the energy spectrum of the paramagnetic in an intermediate cubic field after allowing for the influence of the non-cubic nature of the field, the spin–orbit and spin–spin interactions and the influence of the external magnetic field.

For paramagnetic crystals containing rare-earth ions we have the case of a weak field so the whole order of the calculation is different: the action of the crystal (trigonal) field on the system of the free ion levels caused by the spin–orbit coupling is taken into consideration first; then the corrections due to the spin–spin interaction and the presence of the external field are introduced. In this case, however, it is almost always necessary to deal only with the levels that arise from the basic electron configuration of the free ion ($4f^n$).

Setting aside for the time being the detailed calculation of these corrections we shall now discuss the relevant data on the energy levels and transitions in the optical region for paramagnetic crystals used in lasers.

A.11.1. Pink Ruby

This crystal is formed by substitution in the crystal lattice of α-corundum (α-Al_2O_3) of a small number of Al^{3+} ions by paramagnetic Cr^{3+} ions (in pink ruby there is about 0·05–0·07% by weight of Cr_2O_3 in relation to the Al_2O_3, i.e. about 2×10^{19} ions of Cr^{3+} per cm^3). The Cr^{3+} ions, which have an ionic radius of $\sim$0·6 Å, isomorphically substitute Al^{3+} ions with a radius of $\sim$0·5 Å in the corundum lattice (Olt, 1961).

The slight difference in the ion radii leads to the Cr^{3+} ion located in the Al_2O_3 lattice in an octahedron of six O^{2-} ions being not in the centre of the octahedron but slightly displaced from it along the [111] trigonal axis. Therefore the Cr^{3+} ion is subject to the considerable action of the trigonal C_{3v} field as well as the action of the cubic field (Zaripov and Shamonin,

1956). All the places occupied by Cr^{3+} ions the corundum lattice are equivalment.

In a purely cubic field the energy levels of the ground and first excited states of the Cr^{3+} ion, without taking the spin–orbit interaction into consideration, would split as shown in Fig. A.III.3 (Tanabe and Sugano, 1954), where the strength of the crystal field (i.e. the parameter Dq) is plotted along the abscissa and the corresponding values of the energy along the ordinate.† The diagram of the levels is plotted so that the position of the $^4A_2$ ground state

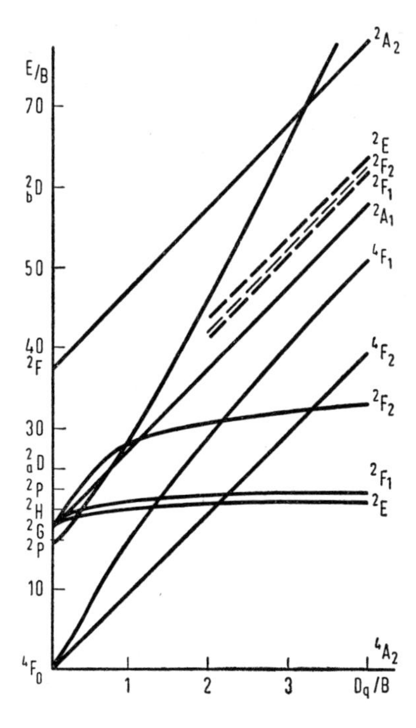

FIG. A.III.3. The splitting of the levels of the Cr^{3+} ion in a crystal field of cubic symmetry.

always coincides with the abscissa. The diagram shown for the energy levels of a Cr^{3+} ion is realized in a MgO crystal where the crystal field parameter is $Dq \approx 1800\ cm^{-1}$ (Low, 1961) and the field is cubic.

We notice that the position of the ^2E, $^2F_1$ and $^2F_2$ levels (Fig. A.III.3) changes only a little when the crystal field increases, whilst the position of the $^4F_1$ and $^4F_2$ levels is strongly dependent on the field. Therefore the inhomogeneities of the crystal field present in a single crystal lead to considerable broadening of the $^4F_1$ and $^4F_2$ levels (turning them into bands) but have little effect on the width of the ^2E, $^2F_1$ and $^2F_2$ levels.

† The scale on the abscissa and the ordinate is in conventional B units which define the magnitude of the Coulomb interaction of the electrons. In the case of Cr^{3+} ions $B \approx 900\ cm^{-1}$.

The cubic component of the field in ruby is approximately the same as in MgO (Low, 1961)† but the presence of the trigonal field around the Cr^{3+} ions and the spin–orbit interaction lead to a further shift and splitting of the levels (small in magnitude). Qualitatively a picture of the splitting in the trigonal field can be obtained from group theory, e.g. by using the tables given by Low (1960b). The spin–orbit coupling for ions of the iron group is comparatively slight, so allowing for it does not introduce any significant changes into the picture of the splitting of the Cr^{3+} levels in ruby.

The effect of the trigonal field and the spin-orbit coupling for Cr^{3+} in ruby has been calculated in detail (Tanabe and Kamimura, 1958; Sugano and Tanabe, 1958; Low, 1958 and 1960a). The diagram obtained on the basis of these papers for the lower energy levels of pink ruby (those responsible for the optical transitions) is shown in basic outline in Fig. A.III.4.

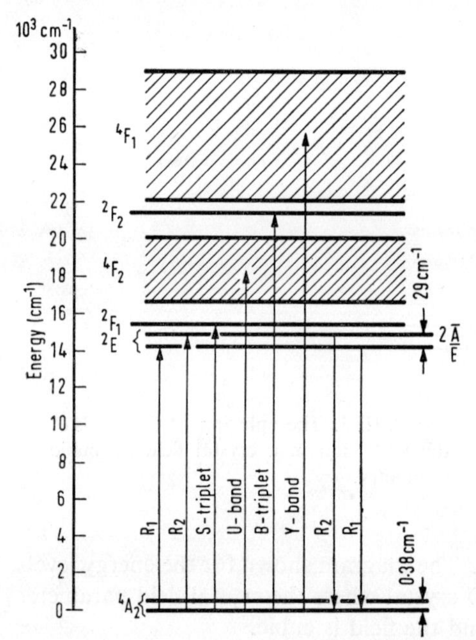

FIG. A.III.4. Diagram of the energy levels of the Cr^{3+} ion in ruby.

The lower $^4A_2$ level (which has fourfold spin degeneracy in a cubic field) is split in ruby by the action of the trigonal field and the spin–orbit interaction into two doublets,‡ which are 0·38 cm^{-1} apart. The other levels shown in Fig. A.III.3 also split, but Fig. A.III.4 shows only the splitting of the ^2E quartet into the doublets $E_{3/2} \equiv 2\bar{A}$ and $E_{1/2} \equiv \bar{E}$ which are $\sim$29 cm^{-1} apart. Since the Cr^{3+} ion contains an odd number of d-electrons all the levels

† The parameter Dq for Cr^{3+} in ruby is slightly greater than 1800 cm^{-1}.
‡ This splitting will be calculated in detail in § A.13.

that form in ruby have even spin degeneracy which can be completely eliminated only in an external magnetic field† (it cannot be eliminated by any deformations of the crystal or by placing it in electrostatic fields of any strength).

When we discuss the scheme of the energy levels of Cr^{3+} in ruby we must bear in mind the following. Under the influence of the crystal field the electron states corresponding to the energy levels shown in the diagram (Fig. A.III.4) are "mixtures" (combinations) of states with wave functions of the same symmetry belonging to different levels of the free ion. For example, there may be mixtures of states with $^2F_2$ symmetry originating from the ^2D, ^2F, ^2G and ^2H levels of the free Cr^{3+} ion; of states with $^2F_1$ symmetry originating from the ^2P, ^2F, ^2G and ^2H levels, etc. According to Low's (1960a) data the ^2E, $^2F_1$ and $^2F_2$ levels shown in Fig. A.III.4 correspond to the following states:

$$^2E \quad 100\%\,^2G,$$

$$^2F_1 \quad 47\%\,^2H + 41\%\,^2G + \cdots + (?),$$

$$^2F_2 \quad 49\%\,^2D + 32\%\,^2H + \cdots + (?).$$

Mixing of the states characterized by different values of the quantum numbers M_L and M_S within the limits set by the values of L and S also occurs as the result of the spin–orbit interaction. The matrix elements of the operator of the spin–orbit interaction are defined by the relations

$$\lambda\langle LSM_LM_S|(\hat{L}\cdot\hat{S})|LSM_LM_S\rangle = \lambda M_LM_S,$$

$$\lambda\langle LSM_LM_S|(\hat{L}\cdot\hat{S})|LS(M_L\pm 1)(M_S\mp 1)\rangle = \frac{\lambda}{4}[L(L+1)-M_L(M_L\pm 1)]^{1/2} \times [S(S+1)-M_S(M_S\mp 1)]^{1/2}, \tag{A.11.1}$$

TABLE A.III.4. THE ABSORPTION OF PINK RUBY AT 300°K IN THE VISIBLE REGION

Transition	λ (in Å)	ν (in cm^{-1})	Notation
$^4A_2 \to \bar{E}(^2E)$	6943	14,401	R_1-line
$^4A_2 \to 2\bar{A}(^2E)$	6929	14,430	R_2-line
$^4A_2 \to {}^2F_1$	6752, 6689, 6588	14,795, 14,950, 15,178	S_1, S_2, S_3-triplet
$^4A_2 \to {}^4F_2$	~5000–6000	~16,700–20,000	U-band
$^4A_2 \to {}^2F_2$	4762, 4745, 4683	20,993, 21,068, 21,352	B_1, B_2, B_3-triplet
$^4A_2 \to {}^2F_1$	~3500–4500	~22,000–28,850	Y-band

† The so-called "Kramers" degeneracy; see § A.12.

from which it can be seen that states having different M_L and M_S are coupled, i.e. mixing occurs.

It is important to take mixing of states into consideration when determining the probabilities of transitions between levels. Transitions which could be completely forbidden for certain pairs of "pure" states prove to be partly permitted for mixed ones.

The theoretical treatment of the energy spectrum of ruby shows that the most intense lines in its optical absorption and fluorescence spectra† can be ascribed to the following transitions:

TABLE A.III.5. THE FLUORESCENCE OF PINK RUBY AT 300°K IN THE VISIBLE REGION

Transition	λ (in Å)	ν (in cm^{-1})	Notation
$\bar{E}(^2E)\rightarrow{}^4A_2$	6943	14,401	R_1-line
$2\bar{A}(^2E)\rightarrow{}^4A_2$	6929	14,430	R_2-line

The general appearance of the absorption and luminescence spectra of pink ruby is shown in Fig. A.III.5 (McClure, 1959).

Ruby is an anisotropic uniaxial crystal and its absorption and fluorescence display considerable dichroism (difference of properties for two directions of polarization). Therefore Fig. A.III.5 shows two forms of the spectra: in polarized light with the direction of the vector $\boldsymbol{E}$ parallel to the principal axis C_3 of the crystal and in light with perpendicular polarization ($\boldsymbol{E} \perp C_3$).

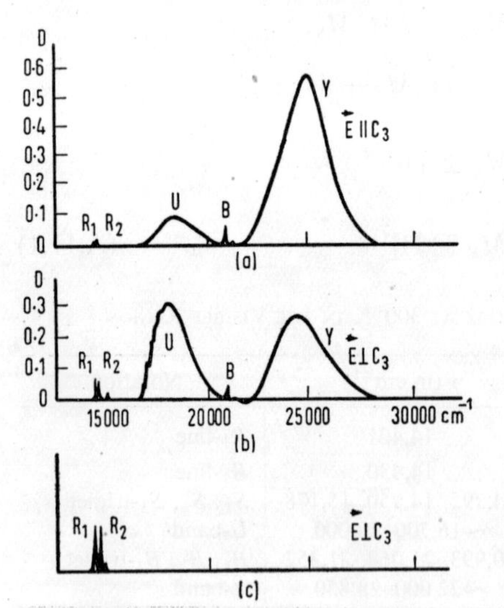

FIG. A.III.5. Optical spectra of pink ruby. (a), (b) absorption at 77°K; (c) fluorescence.

† Detailed information about these spectra can be found in papers by Low (1960a), Deutschbein (1932) and Low (1961).

A.11.2. Crystals doped with Ions of Rare-earth Elements

As has already been indicated, for rare-earth ions in crystals we have the case of a weak field, i.e. $H_{LS} \gg H_{cr}$. This is caused on the one hand by good screening of the 4*f*-electrons by the outer filled $5s^2$ and $5p^6$ shells and by the fact that the wave functions of the 4*f*-electrons do not in general stretch as far as those of the 3*d*-electrons; these effects lead to weakening of the action of the crystal field on the ion ($H_{cr} \sim 10^2$ cm^{-1}). On the other hand rare-earth ions have far larger spin–orbit coupling than ions of the iron group ($\sim 10^3$ cm^{-1} instead of $\sim 10^2$ cm^{-1}).

In the weak field case we must examine the initial splitting of the levels of a free ion caused by spin–orbit coupling and characterized by definite values of the total moment of the ion, i.e. solve the splitting problem in a representation diagonal in $\hat{J}^2$ and $\hat{J}_z$.

In the majority of crystals the rare-earth ions are in a field of trigonal symmetry (Al'tshuler and Kozyrev, 1961) (e.g. C_{3v}), although other cases are also possible. In a trigonal field $(2J + 1)$-fold degenerate levels of ions containing an odd number of electrons split into $(J + \frac{1}{2})$ doublets, and the levels of ions containing an even number of electrons into singlets and doublets (Table A.III.6).

TABLE A.III.6. SPLITTING OF LEVELS WITH $(2J + 1)$-FOLD DEGENERACY IN A TRIGONAL FIELD[a]

Odd number of electrons			Even number of electrons		
J	$(2J+1)$	splitting	J	$(2J+1)$	splitting
1/2	2	1(2)	0	1	1(1)
3/2	4	2(2)	1	3	2 = 1(1) + 1(2)
5/2	6	3(2)	2	5	3 = 1(1) + 2(2)
7/2	8	4(2)	3	7	5 = 3(1) + 2(2)
9/2	10	5(2)	4	9	6 = 3(1) + 3(2)
11/2	12	6(2)	5	11	7 = 3(1) + 4(2)
13/2	14	7(2)	6	13	9 = 5(1) + 2(2)
15/2	16	8(2)	7	15	10 = 5(1) + 5(2)
			8	17	11 = 5(1) + 6(2)

[a] The "splitting" column shows: number of levels = number of singlets (1) + number of doublets (2).

The splitting in the crystal field ($\sim 10^2$ cm^{-1}) does not basically disturb the general arrangement of the energy levels of the free ions of the rare-earth elements: the forms of the splitting in different crystals are generally very similar. Therefore the scheme of the energy levels of trivalent ions in a $LaCl_3$ crystal (Diecke, 1961) given in Fig. A.III.6 can also be used for many other crystals with slight changes. When using this scheme we must remember that although the terms indicated in it have ascribed to them definite values of the quantum numbers S, L and J (e.g. $S = \frac{3}{2}$, $L = 6$ and $J = \frac{9}{2}$ for the $^4I_{9/2}$ level of Nd^{3+}), in fact because of the strong spin–orbit interaction these terms even in free ions correspond to a mixture of states with different S and L, but with the same quantum number J. In addition, mixing of states with different J may also occur in crystals.

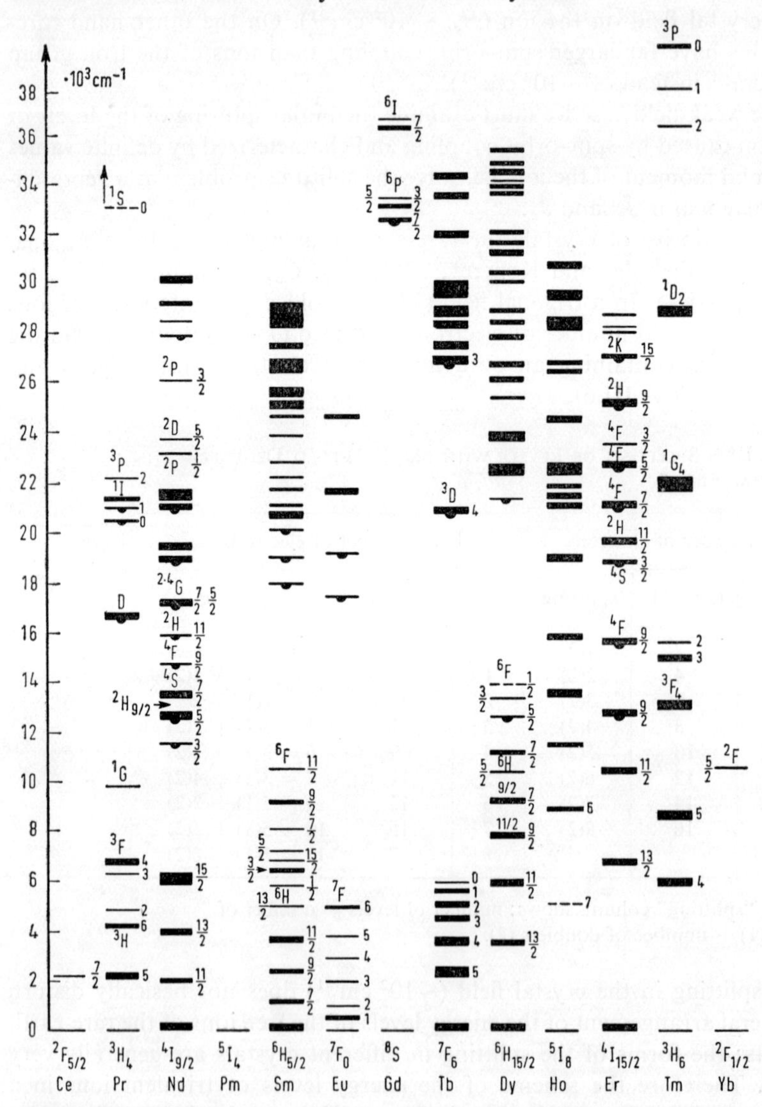

FIG. A.III.6. Diagram of energy levels of trivalent rare-earth ions (in a $LaCl_3$ crystal).

A.12. Crystal Paramagnetic Resonance Spectra. The Spin Hamiltonian

Because of the effect of the departure of the crystal field from cubic symmetry, the spin–orbit and spin–spin interactions, and also the effect of an external magnetic field, the energy levels of a paramagnetic crystal in an intermediate cubic field (the case of ions of the iron group) undergo further shifts and splittings, although they are slight. Some of the levels are rather close together and paramagnetic resonance can be observed in the microwave band as a result of transitions between them.

The fine structure of the energy spectrum can be calculated by using the Hamiltonian

$$\hat{H}' = \hat{H}_{NC} + \hat{H}_{LS} + \hat{H}_{SS} + \hat{H}_{H} \qquad \text{(A.12.1)}$$

(here $\hat{H}_{NC}$ is the contribution to the Hamiltonian due to departure from a cubic field).

As has been shown above, the levels of states determined by the crystalline cubic field are so far apart (up to 10^4 cm^{-1}) that at room temperature and below only the lowest of them is populated and its sublevels may be of interest for investigating transitions at microwave frequencies.† We shall therefore examine the perturbation of the Hamiltonian (A.12.1), which can be rewritten in the form

$$\hat{H}' = \hat{H}_{NC} + \lambda(\hat{\boldsymbol{L}}\cdot\hat{\boldsymbol{S}}) + \varrho(\hat{\boldsymbol{L}}\cdot\hat{\boldsymbol{S}})^2 + \beta((\hat{\boldsymbol{L}} + 2\hat{\boldsymbol{S}})\cdot\boldsymbol{H}_0), \qquad \text{(A.12.2)}$$

on only the lowest of the levels of the ion due to the cubic field.

Direct calculations of the levels of this Hamiltonian would be extremely laborious and all the calculations would have to be repeated for the different paramagnetic ions in a multiplicity of crystals.

Because of this Pryce and Abragam (Pryce, 1950; Abragam and Pryce, 1951) have developed a formal but far simpler—and most important a single, semi-empirical—method of calculating the intervals between the energy levels of the Hamiltonian (A.12.2). The "spin Hamiltonian" method they first developed for ions of the iron group was then extended by Elliott and Stevens (1951 and 1953a) to the case of ions of rare-earth elements.

In the spin Hamiltonian method expression (A.12.2) is transformed to a form containing only the spin operators and external field variables, and also certain parameters which are essentially determined by the symmetry of the crystal field. It is this transformed expression that is called the "spin Hamiltonian" and its eigenvalues are the final corrections to the energy levels

† Sometimes resonance transitions can also be observed in higher levels by populating by intense optical pumping (Al'tshuler and Kozyrev, 1961).

allowing for all the interactions we have considered in the crystal and external fields.

The spin Hamiltonian for ions of the ion group is derived separately for three different cases:

1. When the ground state of the paramagnetic ion in a cubic field is orbitally non-degenerate (but is not an S-level).
2. When there is orbital degeneracy of the ground state in a cubic field.
3. For ions in an S-state.

The spin Hamiltonian of rare-earth ions is treated separately.

A.12.1. Orbitally Non-degenerate Ground State†

We shall examine the perturbing action of the Hamiltonian (A.12.2) on the lowest level, neglecting its spin degeneracy for the time being, i.e. we shall consider that only the orbital variables of the ion are perturbed. This formal approach leads to the corrections obtained from perturbation theory (for non-degenerate levels) remaining operator expressions which, however, contain only spin (but not orbital) variables. These operator expressions form the spin Hamiltonian.

It turns out here that the first approximation correction to the ground state does not contain the contribution from the terms of (A.12.2) that depend upon $\hat{L}$ (including also the contribution from $\hat{H}_{NC}$ since the crystal field acts on the ion only via the orbital variables). In fact when calculating the diagonal matrix elements of these terms we find the expression‡ (see (A.10.1))

$$\langle 0 \,|\, \hat{\boldsymbol{L}} \,|\, 0 \rangle = \int \left\{ \frac{1}{\sqrt{2}} (\Psi^*_{3,2} - \Psi^*_{3,-2}) \right\} \hat{\boldsymbol{L}} \left\{ \frac{1}{\sqrt{2}} (\Psi_{3,2} - \Psi_{3,-2}) \right\} dv,$$

which is equal to zero [(this is shown above for $\langle 0 \,|\hat{L}_z|\, 0 \rangle$; see (A.10.3)].

Therefore the first approximation correction depends only upon the Zeeman term in (A.12.2) which gives

$$\langle 0 \,|\beta \cdot 2(\hat{\boldsymbol{S}} \cdot \boldsymbol{H}_0)|\, 0 \rangle = 2\beta(\boldsymbol{H}_0 \cdot \langle 0|0 \rangle \, \hat{\boldsymbol{S}}) = 2\beta(\boldsymbol{H}_0 \cdot \hat{\boldsymbol{S}}). \qquad \text{(A.12.3)}$$

The second approximation correction is given by the expression

$$-\sum_{n \neq 0} \frac{|\langle 0 \,|\hat{H}'|\, n \rangle|^2}{W(n) - W(0)} . \qquad \text{(A.12.4)}$$

The terms of (A.12.4) that do not contain orbital variables vanish since $\langle 0|n \rangle = 0$ when $n \neq 0$. The non-zero terms contain the matrix elements

† Holds for the $3d^3$ and $3d^8$ configurations in the iron group; see (A.10.1).

‡ Here we denote the states by $n = 0, 1, 2, \ldots$, in order of rising energy of the corresponding orbital levels; $W(n+1) > W(n)$.

$\langle 0 | \hat{\boldsymbol{L}} | n \rangle$ and can be found for each special case. In a general treatment it is sufficient to note that the final result when calculating the sum of the corrections (A.12.3) and (A.12.4) is a quadratic formula† from $\boldsymbol{H}_0(H_x, H_y, H_z)$ and $\hat{\boldsymbol{S}}(\hat{S}_x, \hat{S}_y, \hat{S}_z)$, i.e. an expression of the form

$$\beta^2 \Lambda_{ij} H_i H_j + \beta g_{ij} H_i \hat{S}_j + D_{ij} \hat{S}_i \hat{S}_j, \tag{A.12.5}$$

where $i, j, = x, y, z$ and Λ_{ij}, g_{ij}, D_{ij} are coefficients forming the tensors Λ, G and D.

Limiting ourselves to the second approximation [since $W(n) - W(0) \approx 10^4 \text{ cm}^{-1}$] and omitting from (A.12.5) the term $\beta^2 \Lambda_{ji} H_i H_j$ as it is an irrelevant constant we obtain the *spin Hamiltonian*

$$\hat{H}_{\text{sp}} = D_{ij} \hat{S}_i \hat{S}_j + \beta g_{ij} H_i \hat{S}_j, \tag{A.12.6}$$

which is generally written in tensor form:

$$\hat{H}_{\text{sp}} = \hat{\boldsymbol{S}} D \hat{\boldsymbol{S}} + \beta \boldsymbol{H}_0 G \hat{\boldsymbol{S}}. \tag{A.12.7}$$

Expressions (A.12.6) and (A.12.7) now contain only the spin variables and the components of the external magnetic field, and the behaviour of the spin system in the magnetic field can be described.

The term $\hat{\boldsymbol{S}} D \hat{\boldsymbol{S}}$ gives the initial (zero-field) splitting of the spin levels in the absence of an external magnetic field, which is caused by the effect of the spin–orbit interaction in the non-cubic crystal field.

The term $\beta \boldsymbol{H}_0 G \hat{\boldsymbol{S}}$ gives the Zeeman splitting of the spin levels, and differs from the case of free ions in that the spectroscopic splitting factor G is no longer equal to the constant $g = 2{\cdot}0023$ but is a tensor. If the principal values of this tensor (i.e. the extreme values of the G-factor along any three axes x, y, z at right angles to each other) are G_x, G_y and G_z and the magnetic field $\boldsymbol{H}_0$ has in relation to these axes the direction cosines l, m, n, the splitting factor $G_{(lmn)}$ is defined by the relation

$$G_{(lmn)} = \sqrt{l^2 G_x^2 + m^2 G_y^2 + n^2 G_z^2}. \tag{A.12.8}$$

The anisotropy of the splitting factor G is caused by the crystal field affecting the spin moment via the spin–orbit coupling and forcing it to be quantized along the main crystal axis, i.e. a higher-order axis of symmetry. Therefore when the magnetic field is superimposed the spin is already coupled with the crystal field and not arbitrarily orientated in space as it would be for a free ion. Because of this the effect of the Zeeman splitting of the levels depends on the orientation of the crystal in the magnetic field.

† We are neglecting the term containing ϱ^2 since it is very small; this term need be included only for ions in an S-state.

In crystal fields of trigonal and tetragonal symmetry the tensors D and G are diagonal and have the form

$$D = \begin{pmatrix} -\frac{1}{3}D & 0 & 0 \\ 0 & -\frac{1}{3}D & 0 \\ 0 & 0 & \frac{2}{3}D \end{pmatrix} \text{ and } G = \begin{pmatrix} g_{\perp} & 0 & 0 \\ 0 & g_{\perp} & 0 \\ 0 & 0 & g_{\parallel} \end{pmatrix}. \quad \text{(A.12.9)}$$

Here the spin Hamiltonian can be transformed to the simple form†

$$\hat{H}_{sp} = D(\hat{S}_z^2 - \tfrac{1}{3}\hat{S}^2) + \beta[g_{\parallel} H_z \hat{S}_z + g_{\perp}(H_x \hat{S}_x + H_y \hat{S}_y)]. \quad \text{(A.12.10)}$$

In fields with a symmetry lower than trigonal or tetragonal the term $E(\hat{S}_x^2 - \hat{S}_y^2)$ must be added to (A.12.10) and $g_{\parallel}$ and $g_{\perp}$ must be replaced by G_x, G_y, G_z.

A.12.2. Orbitally Degenerate Ground State. Effective Spin

In this case we must examine the action of the Hamiltonian (A.12.2) on the ground state using the perturbation theory method for a degenerate system.

The final result of this kind of treatment can, however, be qualitatively predicted from two general theorems.

The Kramers theorem (Landau and Lifshitz, 1963): A system containing an odd number of electrons when placed in an electric field has energy levels of at least twofold degeneracy.

The Jahn–Teller theorem (as applied to paramagnetic ions in crystals) (Jahn and Teller, 1937; Low, 1960b): If the ground state of a paramagnetic ion with a given distribution of charges in the complex around it is degenerate, then the whole complex *spontaneously* deforms so that the degeneracy is removed as far as possible.

Of course, if the system has only twofold Kramers degeneracy, this cannot be removed by any deformation of the complex of electric charges.

From the physical point of view the Jahn–Teller effect can be explained by the fact that when the complex is deformed its symmetry is reduced and, as a result, so is the splitting of the levels of the paramagnetic ion. In this case levels appear which are lower than the original degenerate one (since the centre of gravity of the split levels remains at its previous position) and the system can then occupy an energetically more favourable state.

Using these theorems we can say straight away that if the ground state of a paramagnetic ion in a cubic field is orbitally degenerate splitting always occurs as the result of departures of the field from cubic symmetry (which

† We recall that nowhere have we allowed for nuclear interactions which, generally speaking, also make a contribution to (A.12.7) or (A.12.10).

may be inherent in the crystal itself or appear because of the Jahn–Teller effect) and the spin–orbit coupling and all the levels formed are either singlets or have not more than twofold degeneracy.

It has been proved that the intervals between these sublevels are ~ 10–10^2 cm^{-1}. Since this gap cannot be significantly reduced by the action of attainable external fields it becomes clear that in ions with an even number of electrons (and therefore with singlet levels) we cannot observe transitions at microwave frequency (though they can sometimes be observed in the millimetre band).

In ions with an odd number of electrons all the levels, including the lower one, are Kramers doublets and when a magnetic field of suitable strength is superimposed transitions can be observed in the microwave band. Since the nearest excited level is a considerable distance from the lower level ($\sim 10^2$ cm^{-1}) the latter can be looked upon as an isolated doublet by describing it by a certain *effective spin* $S' = \frac{1}{2}$ (although the real spin S of the ion may generally not be $\frac{1}{2}$).

The behaviour of this kind of system with an effective spin S' in a magnetic field can be described by the spin Hamiltonian

$$\hat{H}_{\text{sp}} = \beta \boldsymbol{H}_0 \cdot G \cdot \hat{\boldsymbol{S}}', \quad (M_S = \pm\tfrac{1}{2}), \tag{A.12.11}$$

where the quantity G, as in (69.7), is a tensor, the values of G being far greater than 2 for $S > S' = \frac{1}{2}$.

A.12.3. The Spin Hamiltonian for Ions in an S-State

If the outer electron shell is half-full, then according to Hund's rule the ground state of a free ion is an orbitally non-degenerate S-state ($L = 0$). Examples of this case are Mn^{2+} and Fe^{3+} (^6S-state).

When acted upon by the crystal field an S-level should, it would appear, remain unsplit since the field of the electric charges affects only the orbital properties of the ion. It has been proved experimentally, however, that a small splitting nevertheless occurs.

This effect can be explained if we remember the term proportional to ϱ^2 which we omitted when deriving the spin Hamiltonian (A.12.5) [see the footnote to the derivation of the expression (A.12.5)]. This term is obtained at the expense of the spin–spin interaction described by the Hamiltonian $\varrho(\hat{\boldsymbol{L}} \cdot \hat{\boldsymbol{S}})^2$, from whose form it can be understood that the spin–spin interaction via the spin–orbit coupling is also subject to the effect of the crystal field, although only slightly.

It can be shown that the behaviour of the energy levels of crystals containing paramagnetic ions in an S-state in an external magnetic field can be described by the spin Hamiltonian (Al'tshuler and Kozyrev, 1961)

$$\hat{H}_{sp} = g\beta(\boldsymbol{H}_0 \cdot \hat{\boldsymbol{S}}) + \frac{a}{6}\left[\hat{S}_x^4 + \hat{S}_y^4 + \hat{S}_z^4 - \frac{1}{5}S(S+1)(3S^2+3S-1)\right] \tag{A.12.12}$$

for a crystal field with cubic symmetry and by the Hamiltonian

$$\hat{H}_{sp} = \beta[g_{\parallel} H_z \cdot \hat{S}_z + g_{\perp}(H_x\hat{S}_x + H_y\hat{S}_y)] + D\left[\hat{S}_z^2 - \frac{1}{3}S(S+1)\right] + E(\hat{S}_x^2 - \hat{S}_y^2) + \frac{a}{6}\left[\hat{S}_x^4 + \hat{S}_y^4 + \hat{S}_z^4 - \frac{1}{5}S(S+1)(3S^2+3S-1)\right] \tag{A.12.13}$$

for trigonal and tetragonal fields.

A.12.4. The Spin Hamiltonian for Rare-earth Ions

As has been shown above (§ A.11.2) in the case of rare earths the structure of the energy levels is essentially determined by the dominance of the spin–orbit coupling over the crystal field effect, and also by the symmetry of this field, which is most often trigonal. Therefore the order of calculating the energy levels is generally changed: after we have calculated the action of the crystal field on the levels of a free ion caused by the spin–orbit coupling it only remains to calculate the corrections introduced by the Hamiltonian.

$$\hat{H}' = \hat{H}_{SS} + \hat{H}_H. \tag{A.12.14}$$

As a result of these interactions and the Jahn–Teller effect the levels of the rare-earth ions in the crystals turns out to consist of doublets and singlets in the case of an even number of 4*f*-electrons† and only of doublets in the case of an odd number of 4*f*-electrons, the distances between these levels being at least several cm^{-1}. At low temperatures only the lowest levels will be sufficiently populated and, if they are doublets, transitions in the microwave region can be observed between the split components when an external magnetic field is applied to the crystal.

If each doublet has corresponding to it states precisely defined by the values of the projection of the total moment J_z, then magnetic dipole tran-

† See Table A.III.6.

sitions could be observed only in the case when $J_z = \pm\frac{1}{2}$, since these transitions are subject to the selection rule $\Delta J_z = 0, \pm 1$. As a result of spin–orbit interactions in crystal fields which depart from a simple symmetry, for each energy level there may be corresponding "mixed" states in which, as well as the state with a certain value of J_z, there is also a mixture of states with other values of the projections of the moment, e.g. $J_z \pm 4$ or $J_z \pm 6$. As a result transitions can also be observed for doublets with the value $J_z \neq \pm\frac{1}{2}$.

Elliott and Stevens have shown (1952, 1953a and 1953b) that all the experimental results for rare-earth ions can be described by the spin Hamiltonian

$$\hat{H}_{\text{sp}} = \beta[g_{\parallel} H_z \hat{S}_z + g_{\perp}(H_x \hat{S}_x + H_y \hat{S}_y)] + D[\hat{S}_z^2 - \tfrac{1}{3}S(S+1)], \qquad \text{(A.12.15)}$$

if we use the effective spin value of $S' = \frac{1}{2}$.

It is now clear that for transitions within the limits of doublets with $J_z \neq \frac{1}{2}$ the g factor must be far greater than 2. In general, unlike the case of ions of the iron group, in which $g \approx 2$ ($g_{\parallel} \cong g_{\perp}$), with rare-earth ions the splitting factor is noticeably anisotropic and varies within wide limits (Low, 1960b).

A.13. Calculating Spin Hamiltonian Levels

The levels of the spin Hamiltonian are found by solving the secular equation containing the matrix elements

$$(\hat{H}_{\text{sp}})_{M_S, M_S'} \equiv \langle S, M_S | \hat{H}_{\text{sp}} | S, M_S' \rangle = \int \Psi^*_{S, M_S} \hat{H}_{\text{sp}} \Psi_{S, M_S'} \, dv, \qquad \text{(A.13.1)}$$

where

$$M_S = \pm\tfrac{1}{2}, \ldots, \pm S \qquad \text{(A.13.1 A)}$$

(the calculation is carried out in the representation diagonal in $\hat{S}^2$ and $\hat{S}_z$).

The spin Hamiltonian can be conveniently written in a slightly different form from that in § A.12. First we rewrite it in terms of the spin variables

$$\hat{S}_z \text{ and } \hat{S}_{\pm} = \hat{S}_x \pm i\hat{S}_y, \qquad \text{(A.13.2)}$$

since when calculating the matrix elements we can use the relations

$$\hat{S}_z \Psi_{S, M_S} = M_S \Psi_{S, M_S},$$

$$\hat{S}_{\pm} \Psi_{S, M_S} = \sqrt{S(S+1) - M_S(M_S \pm 1)}\, \Psi_{S, M_S \pm 1}. \qquad \text{(A.13.3)}$$

In addition polar coordinates are introduced, selecting as the polar axis the principal axis of symmetry of the crystal field, which coincides with the z quantization axis of the spin Hamiltonian. Then the components of the external constant magnetic field $\boldsymbol{H}_0$ can be written in the form:

$$\begin{aligned} H_x &= H_0 \sin\theta \cos\varphi, \\ H_y &= H_0 \sin\theta \sin\varphi, \\ H_z &= H_0 \cos\theta, \end{aligned} \tag{A.13.4}$$

where $H_0 = |\boldsymbol{H}_0|$ and the angles θ and φ define the direction of the vector $\boldsymbol{H}_0$ in the x, y, z system of axes.

When calculating the energy levels the angle φ can be arbitrarily chosen since the z-axis is the axis of symmetry of the magnetic field; it is generally set equal to zero. Then instead of (A.13.4) we have

$$H_x = H_0 \sin\theta, \quad H_y = 0, \quad H_z = H_0 \cos\theta. \tag{A.13.5}$$

Finally, for convenience the following notation is introduced:

$$h\nu_0 \equiv g\beta H_0, \quad h\nu_D \equiv D. \tag{A.13.6}$$

This makes it possible to obtain the distances between the energy levels in frequency units, which is very convenient.

We shall illustrate the direct calculation of the spin Hamiltonian levels with some examples which are of practical interest.

1. *Ruby.* The paramagnetic resonance spectrum of ruby was first investigated by Manenkov and Prokhorov (1955) and then by Geusic (1956) and Zaripov and Shamonin (1956). Detailed calculations of the levels of the spin Hamiltonian of Cr^{3+} in ruby are given by Schultz-Dubois (1959) and Weber (1959).

The spin Hamiltonian of the Cr^{3+} ions in ruby is of the form† (Manenkov and Prokhorov, 1955)

$$\hat{H}_{sp} = D[\hat{S}_z^2 - \tfrac{1}{3}S(S+1)] + g_{\parallel}\beta H_z \hat{S}_z + g_{\perp}\beta(H_x\hat{S}_x + H_y\hat{S}_y). \tag{A.13.7}$$

The spin of the Cr^{3+} ion is $S = \frac{3}{2}$. For calculation the following values of the parameters are taken as the most reliable (Schultz-Dubois, 1961):

$$\begin{aligned} 2|D| &= (0{\cdot}3831 \pm 0{\cdot}0002)\ \text{cm}^{-1} = (11{\cdot}493 \pm 0{\cdot}006)\ \text{GHz}, \\ g_{\parallel} &= 1{\cdot}9840 \pm 0{\cdot}0006, \\ g_{\perp} &= 1{\cdot}9867 \pm 0{\cdot}0006. \end{aligned} \tag{A.13.8}$$

It has been proved experimentally by a number of authors (Geusic, 1956; Schultz-Dubois, 1959) that $D < 0$.

† The constant E is so small that the corresponding term can be neglected (Geusic, 1956).

Allowing for the slight anisotropy of the splitting factor ($g_{||} \cong g_{\perp} = g$) and using the notation of (A.13.2), (A.13.5) and (A.13.6) we obtain

$$\hat{H}_{sp} = h\nu_D(\hat{S}_z^2 - \tfrac{5}{4}) + h\nu_0\hat{S}_z \cos\theta + \tfrac{1}{2}h\nu_0(\hat{S}_+ + \hat{S}_-)\sin\theta. \qquad \text{(A.13.9)}$$

By using (A.13.3) it is easy to show that the secular equation is of the form

$$\begin{vmatrix} W - h\left(\frac{3}{2}\nu_0\cos\theta + \nu_D\right) & -\frac{\sqrt{3}}{2}h\nu_0\sin\theta & 0 & 0 \\ -\frac{\sqrt{3}}{2}h\nu_0\sin\theta & W - h\left(\frac{1}{2}\nu_0\cos\theta - \nu_D\right) & -h\nu_0\sin\theta & 0 \\ 0 & -h\nu_0\sin\theta & W + h\left(\frac{1}{2}\nu_0\cos\theta + \nu_D\right) & -\frac{\sqrt{3}}{2}h\nu_0\sin\theta \\ 0 & 0 & -\frac{\sqrt{3}}{2}h\nu_0\sin\theta & W + h\left(\frac{3}{2}\nu_0\cos\theta - \nu_D\right) \end{vmatrix} = 0. \qquad \text{(A.13.10)}$$

Cancelling out h everywhere and expanding the determinant we obtain an equation in the fourth powers of the frequency intervals $\nu = W/h$ between the levels:

$$\nu^4 - 2(\nu_D^2 + \tfrac{5}{4}\nu_0^2)\nu^2 + 2\nu_0^2\nu_D(1 - 3\cos^2\theta)\nu$$
$$+ \nu_D^4 + \tfrac{9}{16}\nu_0^4 + \tfrac{1}{2}\nu_D^2\nu_0^2(1 - 6\cos^2\theta) = 0. \qquad \text{(A.13.11)}$$

The roots of this equation define the positions of the spin levels as functions of the parameters (D and g) of the substance and the external field (H_0 and θ).

Different cases of (A.13.11) have been calculated elsewhere (Schultz-Dubois, 1961; Weber, 1959). We shall limit ourselves here to the simplest ones.

(a) $H_0 = 0$—*the zero-field splitting* of the levels. Equation (A.3.11) is reduced to

$$\nu^4 - 2\nu_D^2\nu^2 + \nu_D^4 = 0 \quad \text{or} \quad (\nu^2 - \nu_D^2)^2 = 0. \qquad \text{(A.13.12)}$$

It gives two doublet levels $\nu_{1,2} = -|\nu_D|$ and $\nu_{3,4} = |\nu_D|$, which are split by an amount

$$\Delta\nu_{zf} = 2|\nu_D| = (11{\cdot}493 \pm 0{\cdot}006)\ \text{GHz},$$

rom which

$$2|D| = (0{\cdot}3831 \pm 0{\cdot}0002)\ \text{cm}^{-1}.$$

(b) When $H_0 = H_z \neq 0$ and $\theta = 0°$ (A.13.11) reduces to

$$\nu^4 - 2(\nu_D^2 + \tfrac{5}{4}\nu_0^2)\nu^2 - 4\nu_0^2\nu_D\nu + \nu_D^4 + \tfrac{9}{16}\nu_0^4 - \tfrac{5}{2}\nu_D^2\nu_0^2 = 0 \qquad \text{(A.13.13)}$$

and gives four divergent levels (Fig. A.III.7).

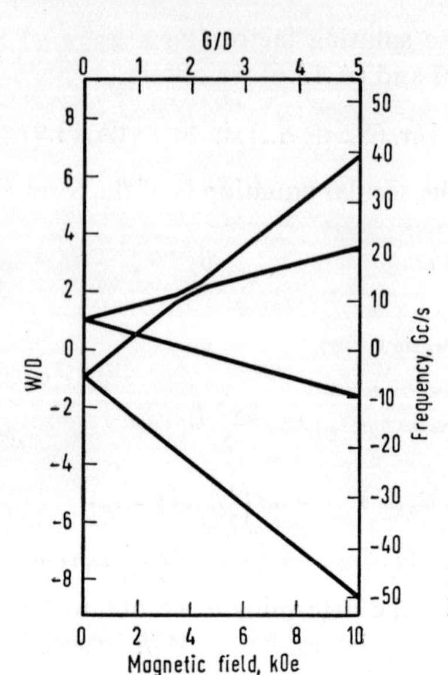

FIG. A.III.7. Diagram of the spin Hamiltonian levels of ruby ($\theta = 0°$).

(c) When $H_0 \neq 0$ and $\theta = 90°$ (A.13.11) reduces to

$$\nu^4 - 2(\nu_D^2 + \tfrac{5}{4}\nu_0^2)\,\nu^2 + 2\nu_0^2\nu_D\nu + \nu_D^4 + \tfrac{9}{16}\nu_0^4 + \tfrac{1}{2}\nu_D^2\nu_0^2 = 0. \tag{A.13.14}$$

A diagram of the levels is given in Fig. A.III.8.

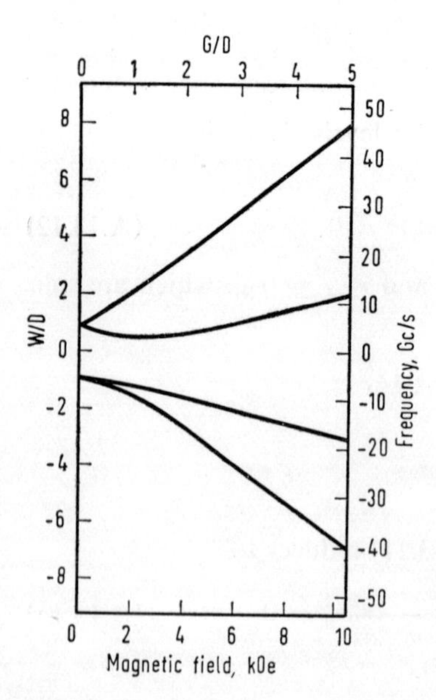

FIG. A.III.8. Diagram of the spin Hamiltonian levels of ruby ($\theta = 90°$).

(d) When $\theta = 54{\cdot}74°$, $\cos^2\theta = \frac{1}{3}$ and from (A.13.11) we obtain

$$\nu^4 - 2(\nu_D^2 + \tfrac{5}{4}\nu_0^2)\,\nu^2 + \nu_D^4 + \tfrac{9}{16}\nu_0^4 - \tfrac{1}{2}\nu_D^2\nu_0^2 = 0. \qquad \text{(A.13.15)}$$

This equation gives four symmetrically divergent levels (Fig. A.III.9) which can be conveniently used in the so-called "symmetrical push-pull" maser operation (Makhov, Kikuchi *et al.*, 1958).

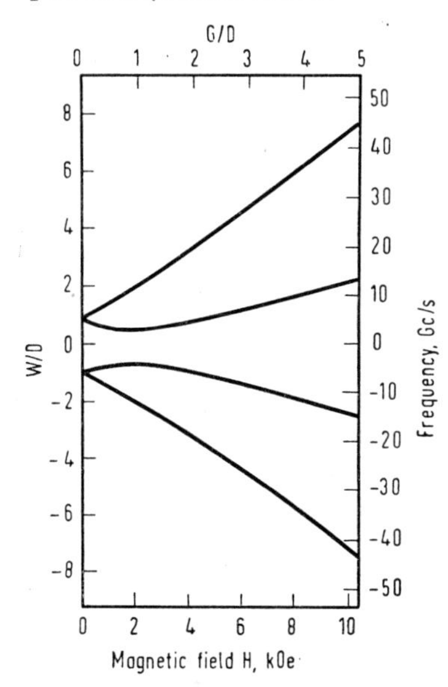

FIG. A.III.9. Diagram of the spin Hamiltonian levels of ruby used in a maser with symmetrical push-pull operation ($\theta = 54{\cdot}74°$).

2. *Rutile*. This name is given to crystals of titanium oxide TiO_2 which have D_{4h} tetragonal symmetry. The Ti^{4+} ion in the rutile lattice is surrounded by six O^{2-} ions which form a rhombic field around it (D_{2h} symmetry).

There are two non-equivalent positions of the Ti^{4+} ions in the rutile lattice. The field symmetry is identical in both of them (D_{2h}) but the axes of symmetry are differently orientated relative to the principal (optical) axis of the crystal [001].

For one of these positions the axes of symmetry have the directions

$$\begin{array}{ll} x_1 & [\bar{1}\ 1\ 0] \\ y_1 & [0\ 0\ 1] \\ z_1 & [1\ 1\ 0] \end{array} \qquad \text{(type I site),}$$

and for the other

$$\begin{array}{ll} x_2 & [\bar{1}\ 1\ 0] \\ y_2 & [0\ 0\ 1] \\ z_2 & [\bar{1}\ 1\ 0] \end{array} \qquad \text{(type II site).}$$

The Ti^{4+} ions can be partially substituted in the rutile lattice by Cr^{3+} or Fe^{3+} ions with the result that the crystal acquires paramagnetic properties. The concentration of the paramagnetic ions is usually $\sim 10^{19}$ cm^{-3} (10^{-2}–10^{-1} % by weight of Cr_2O_3 or Fe_2O_3 relative to the TiO_2).

The behaviour of the lower spin levels of Cr^{3+} in rutile can be described by the spin Hamiltonian (Gerritsen, Harrison *et al.*, 1959)

$$\hat{H}_{sp} = \beta g(\boldsymbol{H}_0 \cdot \hat{S}) + D[\hat{S}_z^2 - \tfrac{1}{3}S(S+1)] + E(\hat{S}_x^2 - \hat{S}_y^2) \quad \text{(A.13.16)}$$

with the following parameter values:

$$D = 0{\cdot}55 \text{ cm}^{-1}, \quad g = 1{\cdot}97 \text{ (isotropic)}, \quad E = 0{\cdot}27 \text{ cm}^{-1}.$$

Since the two non-equivalent positions which the Cr^{3+} ions may occupy in the rutile lattice differ from each other by rotation of the x- and z-axes of symmetry by $\pi/2$ the spectrum of the energy levels is repeated when the crystal is rotated in the magnetic field around the optic axis y every 90°.

The free Fe^{3+} ions, as was shown above, are in the $^6S_{5,2}$ state. In the rutile crystal field with D_{2h} symmetry this state splits into three Kramers doublets.

The zero-field splitting measured experimentally (Okaya *et al.*, 1960) is

$$\Delta_1 = (43{\cdot}3 \pm 0{\cdot}1) \text{ GHz} \cong 1{\cdot}45 \text{ cm}^{-1} \quad (\Delta M_S = \pm\tfrac{3}{2} \to \pm\tfrac{1}{2}),$$

$$\Delta_2 = (81{\cdot}3 \pm 0{\cdot}1) \text{ GHz} \cong 2{\cdot}7 \text{ cm}^{-1} \quad (\Delta M_S = \pm\tfrac{5}{2} \to \pm\tfrac{3}{2}).$$

The position of the Fe^{3+} energy levels in rutile can be described by the spin Hamiltonian (Carter and Okaya, 1960)

$$\hat{H}_{sp} = g\beta(\boldsymbol{H}_0 \cdot \hat{S}) + D\left(\hat{S}_z^2 - \frac{35}{12}\right) + E(\hat{S}_x^2 - \hat{S}_y^2) \quad \text{(A.13.17)}$$

$$+ \frac{a}{6}\left(\hat{S}_x^4 + \hat{S}_y^4 + \hat{S}_z^4 - \frac{707}{16}\right) + \frac{7}{36} F\left(\hat{S}_z^4 - \frac{95}{14}\hat{S}_z^2 + \frac{81}{16}\right),$$

where the parameters are

$$D = (20{\cdot}35 \pm 0{\cdot}1) \text{ GHz}; \quad E = (2{\cdot}21 \pm 0{\cdot}07) \text{ GHz}; \quad a = (1{\cdot}1 \pm 0{\cdot}2) \text{ GHz};$$

$$F = (-0{\cdot}5 \pm 0{\cdot}3) \text{ GHz}; \quad g = (2{\cdot}000 \pm 0{\cdot}005) \text{ (isotropic)}.$$

A diagram of the energy levels for the case $\boldsymbol{H}_0 \parallel [110]$ axis of the crystal calculated for Fe^{3+} ions occupying type I sites in the rutile lattice is shown in Fig. A.III.10 (Schultz-Dubois, 1959). This case is denoted by $\theta = 0°$ ($\boldsymbol{H}_0 \| z$).

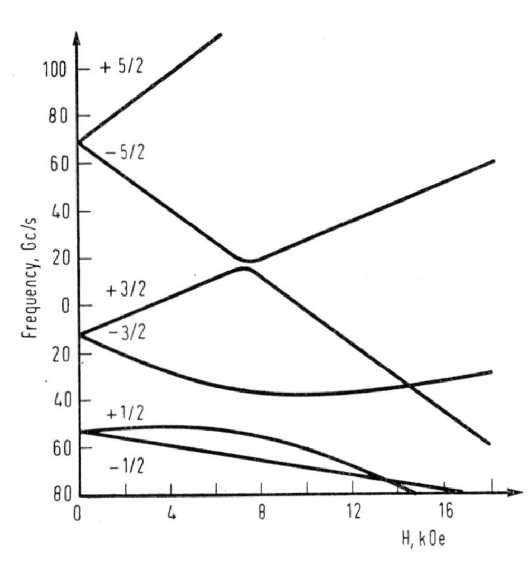

FIG. A.III.10. Diagram of the spin Hamiltonian levels of Fe^{3+} ions in rutile. The diagram shows Fe^{3+} at type I lattice sites for $\theta = 0°$ and at type II lattice sites for $\theta = 90°$.

A diagram of the levels for the same ions in the case of $\theta = 90°$, i.e. $\boldsymbol{H}_z \| [\bar{1}10]$ axis, is shown in Fig. A.III.11 (Carter and Okaya, 1960); for Fe^{3+} ions at type II sites the arrangement shown in Fig. A.III.10 obtains. For all the Fe^{3+} ions in rutile the arrangement of the levels both for $\theta = 0°$ and for $\theta = 90°$ is obtained by superimposing Fig. A.III.10 and Fig. A.III.11 on each other.

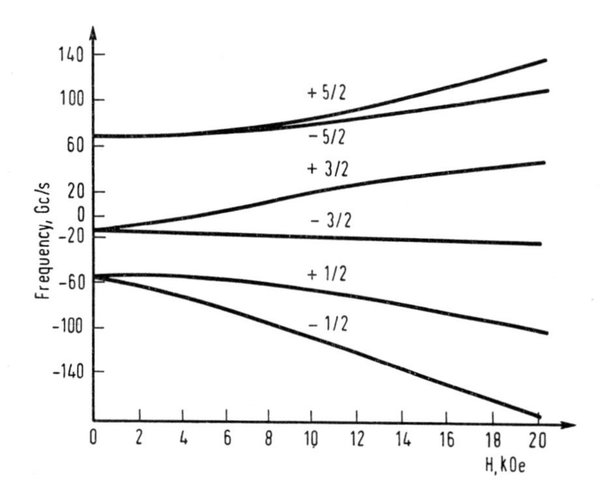

FIG. A.III.11. Diagram of the spin Hamiltonian levels of Fe^{3+} ions in rutile (the Fe^{3+} is at type I lattice sites for $\theta = 90°$ and type II lattice sites for $\theta = 0°$).

With a 73° orientation of the external field $\boldsymbol{H}_0$ in relation to the [110] axis for Fe^{3+} ions located at type I sites a symmetrical arrangement is obtained which can be used for symmetrical push-pull maser operation (Carter, 1961) (Fig. A.III.12).

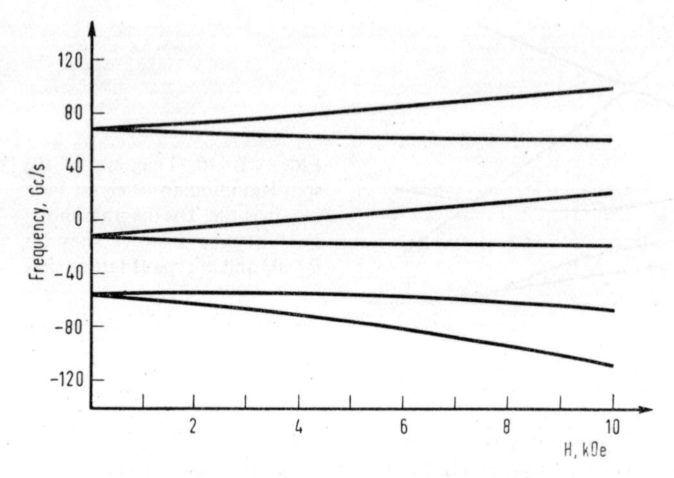

FIG. A.III.12. Diagram of spin Hamiltonian levels of Fe^{3+} ions in rutile used in symmetrical push-pull maser operation (type I lattice sites $\theta = 73°$).

References

ABRAGAM, A. (1961) *Principles of Nuclear Magnetism*, Clarendon Press, Oxford.

ABRAGAM, A. and PRYCE, M.H.L. (1951) *Proc. Roy. Soc.* A **205,** 135.

AISENBERG, S. (1963) *Appl. Phys. Lett.* **2,** 187.

ALEKSANDROV, A.P., KHANIN, YA.I. and YASHCHIN, E.G. (1960) *Zh. eksp. i teoret. fiz.* **38,** 1334; *Soviet Physics JETP*, **11,** 960 (1960).

AL'TSHULER, S.A. and KOZYREV, B.M. (1961) *Electron Paramagnetic Resonance* (in Russian), Fizmatgiz; Translated by Scripta Technica Inc., Academic Press, N.Y. (1964).

ANDREW, E.R. (1955) *Nuclear Magnetic Resonance*, Cambridge University Press.

ANDRONOV, A.A., VITT, A.A. and CHAIKIN, S.E. (1959) *Theory of Oscillations* (in Russian), Fizmatgiz; ANDRONOV, A.A. and CHAIKIN, C.E. (1949) *Theory of Oscillations*, Ed. S. Lefschetz, Princeton University Press).

ARAMS, F.R. (1960) *Proc. IRE*, **48,** 108.

ARTMAN, J.O. BLOEMBERGEN, N. and SHAPIRO, S. (1958) *Phys. Rev.* **109,** 1392.

ASEEV, B.P. (1955) *Oscillating Circuits* (in Russian), Svyaz'izdat.

AUTLER, S.N. and MCAVOY, N. (1958) *Phys. Rev.* **110,** 280.

BARCHUKOV, A.I., PROKHOROV, A.M. and SAVRANSKII, V.V. (1963) *Radiotekhn. i elektron.* **8,** 438; *Radio Eng. and Electronic Phys.* **8,** 385 (1963).

BARNES, F.S. (1960) *Quantum Electronics*, Columbia University Press, N.Y., 57; (1962) *Proc. IRE*, **50,** 1686.

BASOV, N.G. (1956) Doctorate Thesis (in Russian), Fiz. Inst. Akad. Nauk SSSR.

BASOV, N.G. and KROKHIN, O.N. (1960) *Zh. eksp. i teoret. fiz.* **39,** 1777; *Soviet Physics JETP*, **12,** 1240 (1961); (1962) *Appl. Optics*, **1,** 213.

BASOV, N.G., KROKHIN, O.N. and POPOV, YU.M. (1960) *Uspekhi fiz. nauk*, **72,** 161; *Soviet Physics Uspekhi*, **3,** 702 (1961).

BASOV, N.G., NIKITIN, V.V. and ORAEVSKII, A.N. (1961) *Radiotekhn. i elektron.* **6,** 796.

BASOV, N.G. and PROKHOROV, A.M. (1954) *Zh. eksp. i teoret. fiz.* **27,** 431; (1955a) *Zh. eksp. i teoret. fiz.* **28,** 249; *Soviet Physics JETP*, **1,** 184 (1955); (1955b) *Uspekhi fiz. nauk*, **57,** 485; (1956) *Zh. eksp. i teoret. fiz.* **30,** 560; *Soviet Physics JETP*, **3,** 426 (1956).

BASOV, N.G., ZUEV, V.S. and KRYUKOV, P.G. (1962) *Zh. eksp. i teoret. fiz.* **43,** 353; *Soviet Physics JETP*, **16,** 254 (1963).

BENNETT, W.R. JR. (1961) *Advances in Quantum Electronics*, Columbia University Press, N.Y., 28; (1962a) *Appl. Optics Suppl. on Optical Masers*, 24; (1962b) *Phys. Rev.* **126,** 580.

BENNETT, W.R. JR., FAUST, W.L., MCFARLANE, R.A. and PATEL, C.K.N. (1962) *Phys. Rev. Lett.* **8,** 470.

BERGMANN, S.M. (1960) *J. Appl. Phys.* **31,** 275.

BESPALOV, V.I. and GAPONOV, A.V. (1965) *"Izv. VUZov"*, *Radiofizika* **8,** 70; *Soviet Radiophysics* **8,** 49 (1966).

BETHE, H.A. (1929) *Ann. d. Phys.* **3,** 133.

BLEANEY, B. and STEVENS, K.W.H. (1953) *Reports on Progress in Physics*, **16,** 108 (The Physical Society, London).

References

Bloembergen, N. (1956) *Phys. Rev.* **104,** 324; (1961) *Solid State Masers. Progress in Low Temperature Physics*, North-Holland, Amsterdam, **3,** p. 396.

Bloembergen, N., Shapiro, S., Pershan, P.S. and Artman, J.O. (1959) *Phys. Rev.* **114,** 445.

Bloom, S. (1957) *J. Appl. Phys.* **28,** 800.

Bloom, S. and Chang, K.K.N. (1957) *RCA Review*, **18,** 578.

Bogle, G.S. (1961) *Proc. IRE*, **49,** 573.

Bolef, D.I. and Chester, P.F. (1958) *IRE Trans.* MTT **6,** 47.

Bondarenko, N.G., Eremina, I.V. and Talanov, V.I. (1964) *Zh. eksp. i teoret. fiz.* **46,** 1500; *Soviet Physics JETP*, **19,** 1016 (1964).

Bondarenko, N.G. and Talanov, V.I. (1964) "*Izv. VUZov*" *Radiofizika* **7,** 313.

Boom, R.W. and Livingston, R.A. (1962) *Proc. IRE* **50,** 274.

Boyd, G.D., Collins, R.J., Porto, S.P.S., Yariv, A. and Hargreaves, W.A. (1962) *Phys. Rev. Lett.* **8,** 269.

Boyd, G.D. and Gordon, J.P. (1961) *Bell System Techn. J.* **40,** 489.

Boyd, G.D. and Kogelnik, H. (1962) *Bell System Techn. J.* **41,** 1347.

Brangaccio, D. J. (1962) *Rev. Sci. Instrum.* **33,** 921.

Bunkin, F.V. (1959) *Radiotekhn. i elektron.* **4,** 886.

Buatcher, P. (1958) *IEE International Convention on Microwave Valves*, 19–23 **May.**

Byerly, E.H., Goldsmith, J. and McMahan, W.H. (1963) *Proc. IEEE* **51,** 360.

Carter, D.L. (1961) *J. Appl. Phys.* **32,** 2541.

Carter, D. and Okaya, A. (1960) *Phys. Rev.* **118,** 1485.

Chester, P.F., Wagner, P.E. and Castle, J.G. Jr. (1958) *Phys. Rev.* **110,** 281.

Collins, R.J., Nelson, D.F., Schawlow, A.L., Bond, W., Garrett, C. G.B. and Kaiser, W. (1960) *Phys. Rev. Lett.* **5,** 303.

Combrisson, J., Honig, A. and Townes, C.H. (1956) *C.R. Acad. Sci.* **242,** 2451.

Condon, E.U. and Shortley, G.H. (1935) *Theory of Atomic Spectra*, Cambridge University Press.

Cook, J. C. (1961) *Proc. IRE*, **49,** 1570.

Cook, J.J., Cross, L.G., Bair, M.E. and Terhune, R.W. (1961) *Proc. IRE*, **49,** 768.

Cummins, H.Z., Abella, I., Heavens, O.S., Knable, N. and Townes, C.H. (1961) *Advances in Quantum Electronics*, Columbia University Press, N.Y., 12.

DeGrasse, R.W., Kostelnick, J.J. and Scovil, H.E.D. (1961) *Bell System Techn. J.* **40,** 1117.

DeGrasse, R.W., Schulz-DuBois, E.O. and Scovil, H.E.D. (1959) *Bell System Techn. J.* **38,** 305.

DeMaria, A.J. and Gagosz, R. (1962) *Proc. IRE*, **50,** 1522.

Deutschbein, O. (1932) *Ann. d. Phys.* **14,** 712.

Devlin, G.E., McKenna, J., May, A.D. and Schawlow, A.L. (1962) *Appl. Optics*, **1,** 11.

Devor, D.P., D'Haenens, I.J. and Asawa, C.K. (1962) *Phys. Rev. Lett.* **8,** 432.

Dicke, R.H. (1958) U.S. Pat. No. 2851652, 9 Sept. 1958.

Diecke, G.H. (1961) *Advances in Quantum Electronics*, Columbia University Press, N.Y., **164.**

Ditchfield, C.R. and Forrester, P.A. (1958) *Phys. Rev. Lett.* **1,** 448.

Elliott, B.J., Schaug-Pettersen, T. and Shaw, H.J. (1960) *J. Appl. Phys.* **31,** 400 s.

Elliott, R.J. and Stevens, K.W.H. (1951) *Proc. Phys. Soc.* A **64,** 205; (1952) *Proc. Roy. Soc.* A **215,** 437; (1953a) *Proc. Roy. Soc.* A **218,** 553; (1953b) *Proc. Roy. Soc.* A **219,** 387.

Etzel, H.W., Gandy, H.W. and Ginther, R.J. (1962) *Appl. Optics*, **1,** 534.

Evtuhov, V. and Neeland, J.K. (1962) *Appl. Optics*, **1,** 517.

FABRIKANT, V.A., VUDYNSKII, M.M. and BUTAEVA, F.A. (1951) Patent Cert. No. 148441 (576749/26 of 18 July 1951).

FAIN, V.M. (1957) *Zh. exp. i teoret. fiz.* **33,** 945; *Soviet Physics JETP,* **6,** 726 (1958); (1958a) *Zh. eksp. i teoret. fiz.* **34,** 1032; *Soviet Physics JETP,* **7,** 714 (1958); (1958b) "*Izv. VUZov*", *Radiofizika,* **1,** 75; (1966) *Zh. eksp. i teoret. fiz.* **50,** 1327; *Soviet Physics JETP,* **23,** 882 (1966).

FAIN, V.M. and KHANIN, YA.I. (1961) *Zh. eksp. i teoret. fiz.* **41,** 1498; *Soviet Physics JETP,* **14,** 1069 (1962).

FAIN, V.M., KHANIN, YA.I. and YASHCHIN, E.G. (1961) *Zh. eksp. i teoret. fiz.* **41,** 986; *Soviet Physics JETP,* **14,** 700 (1962); (1962) "*Izv. VUZov*", *Radiofizika,* **5,** 697.

FEHER, G., GORDON, J.P., BUEHLER, E., GERE, E.A. and THURMOND, C.D. (1958) *Phys. Rev.* **109,** 221.

FIRTH, I.M. (1963) *Physica,* **29,** 857.

FIRTH, I.M. and BIJL, D. (1961) *Nature,* **192,** 860.

FOK, V.A. and VAINSHTEIN, L.A. (1963) *Radiotekhn. i elektron.* **8,** 363; *Radio Eng. and Electronic Phys.* **8,** 317 (1963).

FONER, S., MOMO, L.R. and MAYER, A. (1959) *Phys. Rev. Lett.* **3,** 36.

FONER, S., MOMO, L. R., MAYER, A. and MYERS, R.A. (1960) *Quantum Electronics,* Columbia University Press, N.Y., 487.

FONER, S., MOMO, L.R., THAXTER, J.B., HELER, G.S. and WHITE, R.M. (1961) *Advances in Quantum Electronics,* Columbia University Press, N.Y., 553.

FOX, A.G. and LI, T. (1960) *Proc. IRE,* **48,** 1904; (1961) *Bell System Techn. J.* **40,** 453; (1963) *Proc. IEE,* **51,** 80.

GALANIN, M.D., LEONTOVICH, A.M. and CHIZHIKOVA, Z.A. (1962) *Zh. eksp. i teoret. fiz.* **43,** 347; *Soviet Physics JETP,* **16,** 249 (1963).

GALKIN, L.N. and FEOFILOV, P.P. (1957) *Dokl. Akad. Nauk,* **114,** 745; *Soviet Physics Doklady,* **2,** 255 (1957); (1959) *Optika i spektroskopiya,* **7,** 840; *Optics and Spectroscopy,* **7,** 492 (1959).

GANDY, H.W. and GINTHER, R.J. (1962) *Appl. Phys. Lett.* **1,** 25.

GARRETT, C.G.B., KAISER, W. and BOND, W.L. (1961) *Phys. Rev.* **124,** 1807.

GENKIN, V.N. and KHANIN, YA.I. (1962) "*Izv. VUZov*", *Radiofizika,* **5,** 423.

GEORGE, N. (1963) *Proc. IEEE,* **51,** 1152.

GERRITSEN, H.J., HARRISON, S.E., LEWIS, H.R. and WITTKE, J.R. (1959) *Phys. Rev. Lett.* **2,** 153.

GERRITSEN, H.J. and LEWIS, H.R. (1960) *Quantum Electronics,* Columbia University Press, N.Y., 385.

GEUSIC, J.E. (1956) *Phys. Rev.* **102,** 1252.

GIANINO, P.D. and DOMINICK, F.J. (1960) *Proc. IRE,* **48,** 260.

GINZBURG, V.L. (1947) *Uspekhi fiz. nauk,* **31,** 320.

GORDON, J.P. and WHITE, L.D. (1958) *Proc. IRE,* **46,** 1588.

GORDON, J.P., ZEIGER, H.J. and TOWNES, C.H. (1954) *Phys. Rev.* **95,** 282; (1955) *Phys. Rev.* **99,** 1264.

GORDY, W., SMITH, W.V. and TRAMBARULO, R.F. (1953) *Microwave Spectroscopy,* Wiley, New York.

GORELIK, G.S. (1947) *Dokl. Akad. Nauk SSSR* **58,** 45; (1948) *Uspekhi fiz. nauk,* **34,** 321.

GOUBAU, G. and SCHWERING, F. (1961) *IRE Trans.* AP **9,** 248.

GUREVICH, A.G. (1952) *Cavity Resonators and Waveguides* (in Russian), Izd. "Sov. radio"; (1960) *Ferrites at Ultra-High Frequencies* (in Russian), Fizmatgiz.

GURTOVNIK, A.S. (1958) "*Izv. VUZov*", *Radiofizika,* **1,** 83.

GVOZDODER, S.D. and MAGAZANIK, A.A. (1950) *Zh. eksp. i teoret. fiz.* **20,** 705.

HADDAD, G.I. and ROWE, J.E. (1962) *IRE Trans.,* MTT **10,** 3.

References

HERRIOTT, D.R. (1961) *Advances in Quantum Electronics*, Columbia University Press, N.Y., 44.
HIGA, W.H. and CLAUSS, R.C. (1963) *Proc. IEEE*, **51**, 948.
HIGA, W.H. and WIEBE, E. (1963) *Proc. IEEE*, **51**, 851.
HOSKINS, R.H. (1959a) *Phys. Rev. Lett.* **3**, 174; (1959b) *J. Appl. Phys.* **30**, 797.
HOSKINS, R.H. and BIRNBAUM, G. (1960) *Quantum Electronics*, Columbia University Press, N.Y., 499.
HSU, H. and TITTEL, F.K. (1963) *Proc. IEEE*, **51**, 185.
HUGHES, W.E. and KREMENEK, C.R. (1963) *Proc. IEEE*, **51**, 856.
HUGHES, T.P. and YOUNG, K.M. (1962) *Nature*, **196**, 332.
IVANOV, A.P., BERKOVSKII, B.M. and KATSEV, I.L. (1962) *Inzh.-fiz. zhurnal*, No. 10, 58.
JACOBSOHN, B.A. and WANGSNESS, R.K. (1948) *Phys. Rev.* **73**, 942.
JAHN, H.A. and TELLER, E. (1937) *Proc. Roy. Soc.* A **161**, 220.
JAVAN, A. (1959) *Phys. Rev. Lett.* **3**, 87; (1961) *Advances in Quantum Electronics*, Columbia University Press, N.Y., 18.
JAVAN, A., BALLIK, E.A. and BOND, W.L. (1962) *J. Opt. Soc. Amer.* **52**, 96.
JAVAN, A., BENNETT, W.R. JR. and HERRIOTT, D.R. (1961) *Phys. Rev. Lett.* **6**, 106.
JELLEY, J.V. (1962) *Microwave J.* **5**, 149.
JELLEY, J.V. and COOPER, B.F.C. (1961) *Advances in Quantum Electronics*, Columbia University Press, N.Y., 619.
JOHNSON, L.F. (1962a) *J. Appl. Phys.* **33**, 756; (1962b) *Proc. IRE*, **50**, 1691.
JOHNSON, L.F., BOYD, G.D. and NASSAU, K. (1962a) *Proc. IRE*, **50**, 86; (1962b) *Proc. IRE*, **50**, 87.
JOHNSON, L.F., BOYD, G.D., NASSAU, K. and SODEN, R.R. (1962) *Proc. IRE*, **50**, 213.
JOHNSON, L.F. and NASSAU, K. (1961) *Proc. IRE*, **49**, 1704.
JOHNSON, L.F. and SODEN, R.R. (1962) *J. Appl. Phys.* **33**, 757.
JUDD, B.R. (1955) *Proc. Roy. Soc.* A **227**, 552.
KAISER, W., GARRETT, C.G.B. and WOOD, D.L. (1961) *Phys. Rev.* **123**, 766.
KAPLAN, D.E. and BROWNE, M.E. (1959) *Phys. Rev. Lett.* **2**, 454.
KARASIK, V.R. (1962) *Pribory i Tekh. Eksper.* No. 6, 5; *Instrum. exper. Tech.* No. 6, 1075 (1962).
KARLOV, N.V. and MANENKOV, A.A. (1964) "*Izv. VUZov*", *Radiofizika*, **7**, 5.
KARLOV, N.V. and PROKHOROV, A.M. (1963) *Radiotekhn. i elektron.* **8**, 453.
KARLOVA, E.K., KARLOV, N.V., PROKHOROV, A.M. and SOLOV'EV, E.G. (1963) *Pribory i Tekh. Eksper.* No. 2, p. 107; *Instrum. exper. Tech.* No. 2, 289 (1963).
KEMP, J.C. (1961) *Phys. Rev. Lett.* **7**, 21.
KHALDRE, KH. YU. and KHOKHLOV, R.V. (1958) "*Izv. VUZov*" *Radiofizika*, **1**, 60.
KHANIN, YA.I. (1966) *Izv. Vuzov. Radiofiz.* **9**, 697; *Soviet Radiophysics*, **9**, No. 4 (1967).
KIKUCHI, C., LAMBE, J., MAKHOV, G., TERHUNE, R.W. (1959) *J. Appl. Phys.* **30**, 1061.
KINGSTON, R.H. (1958) *Proc. IRE*, **46**, 916.
KISLIUK, P. and WALSH, D.J. (1962) *Appl. Optics*, **1**, 45.
KISS, Z.J. and DUNCAN, R.C. (1962a) *Proc. IRE*, **50**, 1531; (1962b) *Proc. IRE*, **50**, 1531; (1962c) *Proc. IRE*, **50**, 1532.
KLEPPNER, D., GOLDENBERG, H.M. and RAMSEY, N.F. (1962) *Phys. Rev.* **126**, 603.
KONTOROVICH, V.M. (1960) "*Izv. VUZov*", *Radiofizika*, **3**, 656.
KOROBKIN, V.V. and LEONTOVICH, A.M. (1963) *Zh. eksp. i teoret. fiz.* **44**, 1847; *Soviet Physics JETP*, **17**, 1242 (1963).
KOSTER, G.F. and STATZ, H. (1961) *J. Appl. Phys.* **32**, 2054.
KOTIK, J. and NEWSTEIN, M.C. (1961) *J. Appl. Phys.* **32**, 178.
KRUPNOV, A.F. (1959) "*Izv. VUZov*", *Radiofizika*, **2**, 658.
KRUPNOV, A.F. and SKVORTSOV, V.A. (1963a) *Zh. eksp. i teoret. fiz.* **45**, 101; *Soviet Physics JETP*, **18**, 74 (1964); (1963b) "*Izv. VUZov*", *Radiofizika*, **6**, 513; (1965) *Pribory i Tekh. Eksper.* No. 1, 128.

KUBAREV, A.M. and PISKAREV, V.I. (1964) *Zh. eksp. i teoret. fiz.* **46,** 508; *Soviet Physics JETP,* **19,** 345 (1964).

KUNZLER, J.E., BUEHLER, E., HSU, F.S.L., MATTHIAS, B.T. and WAHL, C. (1961) *J. Appl. Phys.* **32,** 325.

KUNZLER, J.E., BUEHLER, E., HSU, F.S.L. and WERNICK, J.H. (1961) *Phys. Rev. Lett.* **6,** 89.

LANDAU, H.J. and POLLAK, H.O. (1961) *Bell System Techn. J.* **40,** 65; (1962) *Bell System Techn. J.* **41,** 1295.

LANDAU, L.D. and LIFSHITZ, E.M. (1957) *Electrodynamics of Continuous Media* (in Russian), Gostekhizdat; (1960) Translated by J.B. Sykes and J.S. Bell, Pergamon, Oxford. (1957) *Mechanics* (in Russian), Fizmatgiz; (1960) Translated by J.B. Sykes and J.S. Bell, Pergamon, Oxford. (1963) *Quantum Mechanics* (in Russian), Fizmatgiz; (1965) Translated by J.B. Sykes and J.S. Bell, Pergamon, Oxford.

LANDSBERG, G.S. (1957) *Optics* (in Russian), Gosetkhizdat.

LEONTOVICH, A.M. and VEDUTA, A.P. (1964) *Zh. eksp. i teoret. fiz.* **46,** 71; *Soviet Physics JETP,* **19,** 51 (1964).

LI, T. and SIMS, S.D. (1962) *Proc. IRE,* **50,** 464.

LIPSETT, M.S. and STRANDBERG, M.W.P. (1962) *Applied Optics* **1,** 343.

LOW, W. (1958) *Phys. Rev.* **109,** 256; (1960a) *J. Chem. Phys.* **33,** 1162; (1960b) *Solid State Physics,* Supplement 2, Paramagnetic Resonance in Solids, Academic Press, New York; (1961) *Advances in Quantum Electronics,* Columbia University Press, N.Y., 138.

MCCLUNG, F.J. and HELLWARTH, R.W. (1962) *J. Appl. Phys.* **33,** 828; (1963) *Proc. IEEE,* **51,** 46.

MCCLUNG, F.J., SCHWARZ, S.E. and MEYERS, F.J. (1962) *J. Appl. Phys.* **33,** 3139.

MCCLURE, D.S. (1959) *Solid State Physics,* **9,** 399, Academic Press, New York.

MCMURTRY, B.J. and SIEGMAN, A.E. (1962) *Appl. Optics,* **1,** 51.

MCWHORTER, A.L. and MEYER, J.W. (1958) *Phys. Rev.* **109,** 312.

MAIMAN, T.H. (1960a) *Quantum Electronics,* Columbia University Press, N.Y., 324; (1960b) *Nature,* **187,** 493.

MAIMAN, T.H., HOSKINS, R.H., D'HAENENS, I.J., ASAWA, C.K. and EVTUHOV, V. (1961) *Phys. Rev.* **123,** 1151.

MAKHOV, G., CROSS, L.G., TERHUNE, R.W. and LAMBE, J. (1960) *J. Appl. Phys.* **31,** 936.

MAKHOV, G., KIKUCHI, C., LAMBE, J. and TERHUNE, R.W. (1958) *Phys. Rev.* **109,** 1399.

MALYUZHINETS, G.D. (1959) *Uspekhi fiz. nauk,* **69,** 321; *Soviet Physics Uspekhi,* **2,** 749 (1959).

MALYUZHINETS, G.D. and VAINSHTEIN, L.A. (1961a) *Radiotekhn. i elektron.* **6,** 1489; (1961b) *Radiotekhn. i elektron.* **6,** 1247.

MANENKOV, A.A., MARTIROSYAN, R.M., PIMENOV, YU.P., PROKHOROV, A.M. and SYCHUGOV, V.A. (1964) *Zh. eksp. i teoret. fiz.* **47,** 6 (12), 2055; *Soviet Physics JETP,* **20,** 1381.

MANENKOV, A.A. and PROKHOROV, A.M. (1955) *Zh. eksp. i teoret. fiz.* **28,** 762; *Soviet Physics JETP,* **1,** 611 (1955).

MARCUSE, D. (1961) *Proc. IRE,* **49,** 1706.

MARSHALL, F.R. and ROBERTS, D.L. (1962) *Proc. IRE,* **50,** 2108.

MARSHAK, I.S. (1962) *Uspekhi fiz. nauk,* **77,** 229; *Soviet Physics Uspekhi,* **5,** 478 (1962).

MASTERS, J.I. (1962) *Proc. IRE,* **50,** 220.

MASTERS, J.I. and PARRENT, G.B. JR. (1962) *Proc. IRE,* **50,** 230.

MATHIAS, L.E.S. and PARKER, J.T. (1963a) *Appl. Phys. Lett.* **3,** 16; (1963b) *Phys. Lett.* **7,** 194.

MISEZHNIKOV, G.S. and SHTEINSHLEIGER, V.B. (1961) *Radiotekhn. i elektron.* **6,** 1545.

MOFFITT, W. and BALLHAUSEN, C.J. (1956) *Annual Rev. of Phys. Chem.* **7,** 107 (Annual Reviews, Inc., Palo Alto, California).

MORRIS, R.J., KYHL, R.L. and STRANDBERG, M.W.P. (1959) *Proc. IRE,* **47,** 81.

MULLIKEN, R.S. (1933) *Phys. Rev.* **43,** 279.

References

NAGY, A.W. and FRIEDMAN, G.E. (1963) *Proc. IRE*, **50**, 2504.

NEDDERMAN, H.C., KIANG, Y.C. and UNTERLEITNER, F.C. (1962) *Proc. IRE*, **50**, 1687.

NEIMARK, YU.I. (1949) *Izd. LKVVIA*.

NELSON, D.F. and BOYLE, W.S. (1962) *Appl. Optics*, **1**, 181.

NELSON, D.F. and COLLINS, R.J. (1961) *J. Appl. Phys.* **32**, 739.

OKAYA, A., CARTER, D. and NASH, F. (1960) *Quantum Electronics*, Columbia University Press, N.Y., 389.

OKWIT, S. and SMITH, J.G. (1961) *Proc. IRE*, **49**, 1210.

OKWIT, S., SMITH, J.G. and ARAMS, F.R. (1961) *Proc. IRE*, **49**, 1078.

OLT, R.D. (1961) *Electronics*, **34**, 88.

ORAEVSKII, A.N. (1959) *Radiotekhn. i elektron.* **4**, 719; (1963a) *Trudy Fiz. in-ta Akad. Nauk*, **21**, 3; (1963b) "*Izv VUZov*", *Radiofizika*, **6**, 5.

OSTROVSKII, L.A. and YAKUBOVICH, E.I. (1964) *Zh. eksp. i teoret. fiz.* **46**, 963; *Soviet Physics JETP*, **19**, 656 (1964); (1965) "*Izv. VUZov*", *Radiofizika* **8**, 91; *Soviet Radiophysics* **8**, 63 (1966).

PAANANEN, R.A., TANG, C.L., HORRIGAN, F.A. and STATZ, H. (1963) *J. Appl. Phys.* **34**, 3148.

PATEL, C.K.N., BENNETT, W.R. JR., FAUST, W.L. and MCFARLANE, R.A. (1962) *Phys. Rev. Lett.* **9**, 102.

PATEL, C.K.N., FAUST, W.L. and MCFARLANE, R.A. (1962) *Appl. Phys. Lett.* **1**, 84.

PORTIS, A.M. (1953) *Phys. Rev.* **91**, 1071.

PORTO, S.P.S. and YARIV, A. (1962) *Proc. IRE*, **50**, 1542.

POWERS, J.K. and HARNED, B.W. (1963) *Proc. IEEE*, **51**, 605.

PROKHOROV, A.M. (1958) *Zh. eksp. i teoret. fiz.* **34**, 1658; *Soviet Physics JETP*, **7**, 1140 (1958).

PRYCE, M.H.L. (1950) *Proc. Phys. Soc.* A **63**, 25.

PURCELL, E.M. and POUND, R.V. (1951) *Phys. Rev.* **81**, 279.

RABINOWITZ, P., JACOBS, S. and GOULD, G. (1962) *Appl. Optics*, **1**, 513.

RACAH, G. (1942) *Phys. Rev.* **61**, 537.

RAUTIAN, S.G. and SOBEL'MAN, I.I. (1961) *Zh. eksp. i teoret. fiz.* **41**, 2018; *Soviet Physics JETP*, **14**, 1433 (1962).

READY, J.F. and HARDWICK, D.L. (1962) *Proc. IRE*, **50**, 2483.

REDFIELD, A.G. (1955) *Phys. Rev.* **98**, 1787.

RIGROD, W.W., KOGELNIK, H., BRANGACCIO, D.J. and HERRIOTT, D.R. (1962) *J. Appl. Phys.* **33**, 743.

ROBERTS, R.W., BURGESS, J.H. and TENNEY, H.D. (1961) *Phys. Rev.* **121**, 997.

RODAK, M.I. (1959) *Radiotekhn. i elektron.* **4**, 891.

RÖSS, D. (1963) *Proc. IEEE*, **51**, 468.

SANDERS, J.H. (1959) *Phys. Rev. Lett.* **3**, 86.

SCHAWLOW, A.L. (1961) *Advances in Quantum Electronics*, Columbia University Press, N.Y., 50.

SCHAWLOW, A.D. and DEVLIN, G.E. (1961) *Phys. Rev. Lett.* **6**, 96.

SCHAWLOW, A.L. and TOWNES, C.H. (1958) *Phys. Rev.* **112**, 1940.

SCHAWLOW, A.L., WOOD, D.L. and CLOGSTON, A.M. (1959) *Phys. Rev. Lett.* **3**, 271.

SCHULTZ-DUBOIS, E.O. (1959) *Bell System Techn. J.* **38**, 271.

SCOVIL, H.E.D., FEHER, G. and SEIDEL, H. (1957) *Phys. Rev.* **105**, 762; (1958) Phys. Rev. **109**, 312.

SHAMFAROV, YA.L. and SMIRNOVA, T.A. (1963) *Radiotekhnika i elektronika*, **8**, 1567.

SHIMODA, K., WANG, T.C. and TOWNES, C.H. (1956) *Phys. Rev.* **102**, 1308.

SHTEINSHLEIGER, V.B. (1955) *Wave Interaction Phenomena in Electromagnetic Resonators* (in Russian), Oborongiz; (1959) *Radiotekhn. i elektron.* **4**, 1947; (1962) *Radiotekhn. i elektron.* **7**, 1253.

SHTEINSHLEIGER, V.B., AFANAS'EV, O.A., MISEZHNIKOV, G.S. and ROZENBERG, YA.I. (1964) *Pribory i Tekh. Eksper.* No. 5, 136; *Instrum. exper. Tech.* No. 5, 1062 (1965).

SHTEINSHLEIGER, V.B., MISEZHNIKOV, G.S. and AFANAS'EV, O.A. (1962) *Radiotekhn. i elektron.* **7,** 874; *Radio Eng. and Electronic Phys.* **7,** 828.
SIEGMAN, A.E. and MORRIS, R.J. (1959) *Phys. Rev. Lett.* **2,** 302.
SINGER, J.R. (1959) *Masers*, Wiley, New York.
SLATER, J.C. (1950) *Microwave Electronics*, Van Nostrand, Princeton, N.J.
SLEPIAN, D. and POLLAK, H.O. (1961) *Bell System Techn. J.* **40,** 43.
SNITZER, E. (1961) *Phys. Rev. Lett.* **7,** 444.
SOROKIN, P.P. and STEVENSON, M.J. (1960) *Phys. Rev. Lett.* **5,** 557; (1961) *Advances in Quantum Electronics,* Columbia University Press, N.Y., 65.
STATZ, H. and DEMARS, G. (1960) *Quantum Electronics*, Columbia University Press, N.Y., 530.
STEPANOV, B.I. and GRIBKOVSKII, V.P. (1964) *Uspekhi fiz. nauk*, **82,** 201; *Soviet Physics Uspekhi*, **7,** 68 (1964).
STEPANOV, B.I., IVANOV, A.P., BERKOVSKII, B.M. and KATSEV, I.L. (1962) *Optika i spektroskopiya*, **12,** 533; *Optics and Spectroscopy*, **12,** 298 (1962).
STEVENS, K.W.H. (1952) *Proc. Phys. Soc.* A **65,** 209.
STICKLEY, C.M. (1963) *Appl. Optics*, **2,** 855.
STIGLITZ, M.R. and MORGENTHALER, F.R. (1960) *J. Appl. Phys.* **31,** 37S.
STOICHEFF, B.P. and SZABO, A. (1963) *Appl. Optics*, **2,** 811.
STRATTON, J.A. *et al.* (1956) *Spheroidal Wave Functions*, The Technology Press of MIT and John Wiley, N.Y.
SUGANO, S. (1962) *Appl. Phys.* Suppl. to vol. 33, 303.
SUGANO, S. and TANABE, Y. (1958) *J. Phys. Soc. Jap.* **13,** 880.
TABOR, W.J. (1963) *Proc. IEEE*, **51,** 1143.
TANABE, Y. and KAMIMURA, H. (1958) *J. Phys. Soc. Jap.* **13,** 394.
TANABE, Y. and SUGANO, S. (1954) *J. Phys. Soc. Jap.* **9,** 753.
TANG, C.L. (1962) *Appl. Optics*, **1,** 768.
TANG, C.L., STATZ, H. and DEMARS, G. (1963) *Appl. Phys. Lett.* **2,** 222.
THORP, J.S. (1961) *Advances in Quantum Electronics*, Columbia University Press, N.Y., 602.
THORP, J.S., PACE, J.H. and SAMPSON, D.F.J. (1961) *J. Electronics and Control*, **10,** 13.
TITCHMARSH, E. (1937) *Introduction to the Theory of Fourier Integrals*, Clarendon Press, Oxford.
TROITSKII, V.S. (1958a) *Radiotekhn. i elektron.* **3,** 1298; (1958b) *Zh. eksp. i teoret. fiz.* **34,** 390; *Soviet Physics JETP*, **7,** 271 (1958); (1959) "*Izv. VUZov*", *Radiofizika*, **2,** 377.
TROUP, G. (1959) *Masers*, Methuen, London.
VAINSHTEIN, L.A. (1953) *Diffraction of Electromagnetic and Sound Waves at the Open End of a Waveguide* (in Russian), Izd. "Sov. Radio"; (1957) *Electromagnetic Waves* (in Russian), Izd. "Sov. Radio": (1963a) *Zh. eksp. i teoret. fiz.* **44,** 1050; *Soviet Physics JETP*, **17,** 709 (1963); (1963b) *Zh. eksp. i teoret. fiz.* **45,** 684; *Soviet Physics JETP*, **18,** 471 (1964).
VAN VLECK, J.H. (1935) *J. Chem. Phys.* **3,** 807.
VUYLSTEKE, A.A. (1960) *Elements of Maser Theory*, Van Nostrand, Princeton, N.J.
WAGNER, P.E., CASTLE, J.G. and CHESTER, P.F. (1960) *Quantum Electronics*, Columbia University Press, N.Y., 509.
WEBER, J. (1957) *Phys. Rev.* **108,** 537; (1959) *Rev. Mod. Phys.* **31,** 681.
WEGER, M. (1960) *Bell System Techn. J.* **39,** 1012.
WESSEL, G.K. (1959) *Proc. IRE*, **47,** 590.
WIEDER, I. and SARLES, L.R. (1961) *Phys. Rev. Lett.* **6,** 95.
WHITE, A.D. and GORDON, E.I. (1963) *Appl. Phys. Lett.* **3,** 197.
WHITE, A.D. and RIGDEN, J.D. (1963a) *Appl. Phys. Lett.* **2,** 211; (1963b) *Proc. IEEE*, **51,** 943.
WITTKE, J.P. (1957) *Proc. IRE*, **45,** 291.

References

WOODBURY, E. J. and NG, W. K. (1962) *Proc. IRE* **50,** 2367.

YARIV, A. (1962) *Proc. IRE,* **50,** 1699.

YARIV, A., PORTO, S. P. S. and NASSAU, K. (1962) *J. Appl. Phys.* **33,** 2519.

YASHCHIN, E. G. (1960) "*Izv. VUZov*", *Radiofizika,* **3,** 989.

ZARIPOV, M. M. and SHAMONIN, YU. YA. (1956) *Zh. eksp. i teoret. fiz.* **30,** 291; *Soviet Physics JETP,* **3,** 171 (1956).

ZHABOTINSKII, M. E. and ZOLIN, V. F. (1959) *Radiotekhn. i elektron.* **4,** 1943.

ZVEREV, G. M., KARLOV, N. V., KORNIENKO, L. S., MANENKOV, A. A. and PROKHOROV, A. M. (1962) *Uspekhi fiz. nauk,* **77,** 61; *Soviet Physics Uspekhi,* **5,** 401 (1962).

Index

Index

RANDALL LIBRARY-UNCW
3 0490 0048681 0